California's Social Problems

SECOND EDITION

The second edition of this text is dedicated to:

The late Richard Pelletier
CFH, *San Diego State University*
and

Bob, C.J., and Chris Glynn
JAG, *Madera College*

California's Social Problems

SECOND EDITION

Editors
Charles F. Hohm
San Diego State University

James A. Glynn
Madera College

PINE FORGE PRESS
Excellence and Innovation for Teaching

Copyright © 2002 by Pine Forge Press.

All rights reserved. No part of this book may be reproduced or utilized in any form or by any means, electronic or mechanical, including photocopying, recording, or by any information storage and retrieval system, without permission in writing from the publisher.

For information:

Pine Forge Press
A Sage Publications Company
2455 Teller Road
Thousand Oaks, California 91320
E-mail: sales@pfp.sagepub.com

Sage Publications Ltd.
6 Bonhill Street
London EC2A 4PU
United Kingdom

Sage Publications India Pvt. Ltd.
M-32 Market
Greater Kailash I
New Delhi 110 048 India

Printed in the United States of America

Library of Congress Cataloging-in-Publication Data

Main entry under title:

California's social problems / edited by Charles F. Hohm and James A. Glynn.— 2nd ed.
   p. cm.
Originally published: New York: Longman in collaboration with the California Sociological Association, c1997.
Includes bibliographical references and index.
  ISBN 0-7619-8713-4
  1. Social problems—California. 2. California—Social conditions.
3. California—Ethnic relations. I. Hohm, Charles F., 1947- . II. Glynn, James A.
  HN79.C2 C36 2001
  361.1'09794—dc21   2001003376

01 02 03 04 05 06 07 9 8 7 6 5 4 3 2 1

Acquiring Editor:	Steven Rutter
Editorial Assistant:	Kirsten Stoller
Production Editor:	Denise Santoyo
Typesetter/Designer:	Janelle LeMaster
Indexer:	Molly Hall
Cover Designer:	Jane Quaney

CONTENTS

Foreword *Earl Babbie*	x
Acknowledgments	xii
Prologue	1

Part I
CHALLENGES TO CALIFORNIA COMMUNITIES   7

1. HOMELESSNESS *Ronald W. Fagan*	15
Homelessness in the United States and California	16
Alternative Explanations for Homelessness in California	20
Future Prospects for Homelessness in California	22
Solutions to the Problem of Homelessness	23
Conclusion	30
2. JUVENILE DELINQUENCY AND URBAN GANGS *Anne Hendershott*	35
California's Historical Roots of Juvenile Crime and Justice	42
California's Urban Youth Gangs: Social Problem or Moral Panic?	44
Theories on the Causes of California Gang Delinquency	47
Ethnicity and Juvenile Crime in California	51

Increased Minority Confinement Rates in California and the Nation	52
Solutions to the Problem of Juvenile Delinquency and Gangs	56

3. RACE AND ETHNIC RELATIONS — 63
Phylis Cancilla Martinelli

The History of Race Relations in California	63
Race Relations in California Today	67
Race Relations in California as Compared to Those in the United States	72
California's Problem in Global Context	73
Alternative Explanations for Race Relations in California	75
Future Prospects for Race Relations in California	76
Solutions to the Problem of Race Relations	77

Part II
CHALLENGES TO SEXUALITY AND HEALTH — 85

4. RESISTING HETEROSEXISM AND HOMOPHOBIA: THE GAY AND LESBIAN MOVEMENT FOR EQUAL RIGHTS — 91
Peter M. Nardi

A History of Lesbian and Gay Oppression and Resistance	92
Lesbian and Gay Oppression and Resistance in California Today	99
Lesbian and Gay Oppression and Resistance in California as Compared With the United States	103
Lesbian and Gay Issues in a Global Context	106
Alternative Explanations for Lesbian and Gay Oppression and Resistance in California	108
Future Prospects for Lesbian and Gay Oppression and Resistance in California	111
Solutions to the Problem of Lesbian and Gay Oppression and Resistance	112

5. HIV/AIDS — 117
Kevin D. Kelley

The Basic Science	117
History of HIV/AIDS in California	119
Course of HIV Infection in California	120
The Problem of HIV/AIDS in California Today	122
Comparison of California's Situation to the United States	125
California's Problem in a Global Context	127
Alternative Explanations for the Spread of HIV and AIDS	129
Future Prospects: AIDS Projections in California	132
Possible Solutions to AIDS in California	133

Part III
CHALLENGES TO CALIFORNIA INSTITUTIONS — 139

6. PROBLEMS OF FAMILIES — 149
Robin Wolf, Robin Franck, and James A. Glynn

- The History of Family Trends in California — 150
- Family Trends in California Today — 154
- Family Trends in California as Compared to the United States — 158
- Alternative Explanations for Family Trends — 165
- The Family in Global Context — 168
- Future Prospects — 171
- Suggestions for the Future — 171

7. INEQUALITY AND POVERTY — 177
Robert Enoch Buck

- How Unequal Is It? — 179
- Social Classes and Inequality — 183
- The Causes of Poverty — 192
- Conclusions — 198

8. POVERTY AND WELFARE REFORM — 203
Robert Enoch Buck

- The Welfare State — 204
- The Outcomes of Welfare Reform — 207
- Outcomes for Program Participants — 215
- Continuing Problems and Policy Issues — 216
- Conclusions — 227

9. THE CONTINUING CRISIS IN HIGHER EDUCATION: THREATS TO CALIFORNIA'S MASTER PLAN — 235
James L. Wood and Lorena T. Valenzuela

- The Problem in California Today — 238
- Distance Learning as a Strategic Plan — 239
- Related Strategic Plans — 241
- Origin of the Plans: California and the United States — 244
- California's Problem in Global Context — 253
- Alternative Explanations for the Problem — 254
- Future Prospects: What To Do? — 255
- Solutions and Suggestions: Strategies for Distressed Campuses — 255
- Summary and Conclusion — 260

Part IV
CHALLENGES TO GROWTH AND THE ENVIRONMENT 267

10. POPULATION GROWTH AND ENVIRONMENTAL DEGRADATION 273
James A. Glynn and Charles F. Hohm

History of Population Growth and Environmental Degradation 273
The Problem in California Today 278
California's Population Compared to the United States 283
California's Problem in Global Context 286
Alternate Explanations for the Problem 288
Future Prospects 291
Solutions to the Problems 296

11. PROPOSITIONS 187 AND 227: A NATIVIST RESPONSE TO MEXICANS 303
Adalberto Aguirre, Jr.

History of the Problem in California 303
Nativism in California and the United States 305
Mexican Immigrants and Nativism 307
California's Immigrant Population 308
Immigrants and Work 309
Segmentation of Mexican Immigrant Workers 311
The Nativistic Context 313
Concluding Remarks 317

12. DISASTERS 325
Harvey E. Rich and Loretta I. Winters

Defining Disaster 326
History of Disaster Research 329
The California Experience 330
California Disasters Compared to the United States 331
California, the Developed World, and the Developing World 332
Sociological Perspectives for Examining Disaster 334
Natural Disasters in California Today 337
Planning 339
Communication 339
The Media 342
Emergent Roles and Groups 343
Disasters and the Future 345

Coping With Disasters	346
Conclusion	349
Epilogue: Deregulating Energy in California	353
James A. Glynn	
The History of the Problem	354
The Problem in California Today	355
California's Problem Compared With the United States	356
California's Problem in Global Context	357
Alternative Explanations for the Problem	359
Future Prospects	360
Solutions/Suggestions	361
Glossary and Index	365
About the Editors	383
About the Contributors	385

FOREWORD
Earl Babbie

Bigger than life, California is often seen as a leading indicator for the nation, whether concerning surfers or unmarried mothers. Thus, while this book focuses specifically on social problems in the Golden State, most of the analyses place the California experience in a national and sometimes international context. While it will, naturally, be of special interest to sociologists teaching social problems courses in California, our colleagues elsewhere in the nation may find it useful as well.

Some of California's social problems—such as poverty and population growth—are currently worse than in other states. In other arenas—such as educational entitlement—California is doing better than other states. However, there is probably nothing in California's experience, good and bad, that is irrelevant to the rest of the nation. Thus, these insightful analyses of social problems in California will be useful to sociologists anywhere in the country.

The book covers the full range of social problems. Many are age-old concerns such as poverty, crime, and pollution. Others—child abuse, for example—seem to be newer problems since they have most recently been identified as social problems in public consciousness, even though the conditions involved have existed for a long time and may have been even worse in the past. Each topic is presented with historical attention to how circumstances emerge from common acceptability to be defined as social problems, how preferred explanations and remedies fluctuate, and how good times can turn sour.

In all these respects, the book maintains a level of currency that always brings the reader up to date on present conditions and our prospects for the future. An epilogue deals with the energy crisis that was manifesting in rolling blackouts as the book was going to press. While social problems in California and elsewhere will, of course, have changed further as the book is read by students and scholars, the essays in this book offer the tools for understanding and dealing with the future.

ACKNOWLEDGMENTS

The authors would like to thank Steve Rutter, publisher of Pine Forge Press, for his faith in this project. He had the foresight to recognize California as the bellwether state where fads and fashions, as well as social problems, originate for the nation. We'd also like to recognize the organizational skills of his assistant, Kirsten Stoller, who held the project together. Thanks also to Sage production staff members Kathryn Journey, editorial assistant, Denise Santoyo, production editor, Jaqueline Tasch, copy editor, and Nick Alexander, the graphic artist who created the graphs and figures for this edition. We also thank Linda Carlson Hohm for her careful readings of the entire manuscript. Finally, the book would simply not exist without the support of Blaise Simqu, Vice President of Sage Higher Education Division. In addition, we would like to acknowledge the contributions of prepublication reviewers: Terry R. Kandal, California State University, Los Angeles, and Gordon Clanton, San Diego State University.

PROLOGUE

The state of California is only one of fifty states in the United States, but the Golden State is hardly average. California has the largest and most diverse population of any state in the union. If California were a sovereign country, its population would make it the 31st most populous nation. Its economy would rank fifth in the world (Liedtke, 2001). California is also seen by many as a bellwether state. Fads, trends, and problems originate in California and later surface in other parts of the United States. From ocean surfing in the 1960s; to the tax revolt embodied in Proposition 13 in the late 1970s; to rapid advances in industrial technology, both in research and development; to anti-immigrant measures; to breakthroughs in biotechnology; California generally takes the lead, for better and—unfortunately—for the worse, as well.

As many of the chapters in this book will show, some problems appear to be worse in California than in other states. For example, the chapter on poverty and inequality demonstrates that the poverty rate in California is increasing much faster than it is in the United States in general. The chapter on the poor and welfare reform shows that California's success in helping the needy through the welfare-to-work program is not as successful as the efforts of other states. Another chapter shows that California's teen pregnancy rate is one of the highest in the nation and that poverty is highly associated with teen pregnancies. Yet another example of how California is different from the rest of the nation is the racial and ethnic diversity of the state's population and the relatively high fertility rate associated with many of California's ethnic groups. Because of the above, California will continue to add many more people to its ranks through natural growth than will other states.

Historically, most textbooks used in social problems courses in American colleges and universities have focused on the United States. However, some have followed a trend toward taking a global perspective, such as Glynn, Hohm, and Stewart's *Global Social Problems*. Although the global approach to the study of social problems has its benefits, it also has a downside: that is, the tendency to overlook social problems that are right in our backyard.

This book is designed to fill that gap for California college students by addressing numerous pressing social problems in the Golden State. When used in conjunction with another textbook with a national or international theme, *California's Social Problems* should provide students with a well-rounded orientation to the area of social problems research for California. Also, we feel that when California students study social problems in their home state, the subject will resonate and be more meaningful than if they studied national and international problems. Of course, this book can also be used in introductory courses in sociology, social welfare, human services, race and ethnic relations, and similar subjects.

The authors of the chapters, each of whom specifically wrote for this text, were chosen because of their dedication to undergraduate education. Each is also an expert in the topic of her or his chapter. And, collectively, the authors represent a cross-section of California's system of higher education, including the University of California, the California State University system, California's community colleges, and private colleges and universities.

The second edition of the text has been reorganized into four sections: Challenges to California Communities, Challenges to Sexuality and Health, Challenges to California Institutions, and Challenges to Growth and the Environment. Each section also has an introduction in which the content of each chapter is summarized, and there is a combined glossary and index at the end of the book. The chapters that make up each section follow the same general outline and use the definition of a *social problem* provided by Glynn, Hohm, and Stewart (1996):

> [Social problems] can be defined as situations, policies, or trends that are (1) distressing or threatening to large numbers of people, (2) contrary to the mores (essential practices) or moral beliefs of the society, and (3) partly or wholly correctable through the action of social groups. (p. 3)

When defining social problems, it is important to note that just because something is immoral or illegal, it does not necessarily follow that a *social* problem exists. For example, a social problem does not exist if a local butcher short-weights a customer. Such a practice may break a law and may be punishable, but it can be corrected by simple police action. However, if several million cars emit enough carbon monoxide to pollute the atmosphere, kill vegetation, and harm the health of all animal life, it's a very different situation. Obviously, police cannot arrest every driver, impound every car, or fine every automobile manufacturer. However, the collective will of the people can be demonstrated during an election when voters approve or disapprove an initiative to limit or eliminate the pollutants that are discharged from cars that are registered in the state.

It is also important to note that mores may change over time and space. What have been, are, and will be defined as social problems in California depend, in part, on the views, perceptions, and norms of the population. For example, today, the problem of child maltreatment in California commands much attention, and numerous laws have been passed in the last 15 years to protect the rights of children. Many believe that we are in the midst of an epidemic of child abuse, given the numerous reports of child maltreatment that are filed yearly in the state of California. However, the actual incidence of child maltreatment, when controlling for population size, may be *less* now than it was 25 years ago. It is important to note that although child maltreatment was not defined as a problem in California in decades past, that does not mean that it did not occur. It did occur, but it was "behind closed doors." The way children were treated by their parents was considered to be a private affair. In short, practices that are defined as child abuse today were seen as "discipline" a few decades ago. What is considered a problem, then, depends on the prevailing norms. Even now, there is no universal consensus about what constitutes child maltreatment. A person from Germany or any of the Scandinavian countries would be shocked to see a mother swat her child in a supermarket because the child was misbehaving. In a number of European countries, not only is spanking seen as abusive, it is against the law.

Likewise, smoking cigarettes in restaurants or workplaces was commonplace in the Golden State well into the 1980s. Coming to California nowadays often produces a kind of culture shock to visitors from states that still permit that practice. So what is or what is not a social problem depends on the prevailing norms of a certain place and a certain time.

Part I of this text considers the social problems of homelessness, juvenile delinquency and urban gangs, and race and ethnic relations. Racial and ethnic tensions represent yet another example of how a phenomenon that is defined as problematic can change over time. The state of California is beset by racial and ethnic tensions, and a vast majority of Californians are aware of these tensions. Does this mean that racial and ethnic relations are a problem only today and that this problem did not exist historically? That is hardly the case. The record shows that in the past, Californians of European ancestry treated indigenous natives and immigrants from Asian, African, and Latin American backgrounds with contempt and scorn. Californians of European ancestry viewed the inequitable relations between them and other races and ethnic groups as being inevitable and normal, certainly not problematic. Minority groups in California did not share the same opinion, but their views were not considered at that time.

Part II deals with problems associated with heterosexism, homophobia, and the AIDS epidemic. The first cases of AIDS were discovered in California, and because the disease was associated with gay men, significant research on the ailment was delayed until the disorder began to show up in other sectors of the population. Society's initial lack of concern helped to unite the gay and lesbian communities and lead them to form activist organizations to promote equal treatment.

Part III considers the institutions of the family, the economy (and peripherally the government), and education. Each of these institutions is intricately linked to the nation's problem of a widening gap between the rich and the poor. Adding to the growth of a two-tiered society will be the changing demographics of the family. Whereas families headed by single mothers are projected to increase by 2% nationally between the mid-1990s and 2005, such living arrangements will grow by 24% in California (Dortch, 1996). As the reader will see in Chapter 6, the creation of the single-mother family is practically a formula for poverty.

Also, denial of an entitlement can be considered a social problem. The nation has assumed for most of the past century that every child should attend school up to a certain age (usually defined by each state). Over forty years ago, Californians extended that entitlement to postsecondary education (meaning education beyond high school). In 1960, Californians approved a broad policy, called the Master Plan for Higher Education, often simply referred to as the Master Plan.

The Master Plan was forged by educators, legislators, and the state's governor. It virtually guaranteed higher education to any Californian who desired to continue studying beyond high school. In fact, it produced the "most affordable, accessible, diverse, and highly respected system of higher education in the United States" (California Citizens Commission on Higher Education, 1998, p. ix).

The net result of emphasis on higher education can be seen from a recent survey of the percentage of college graduates who inhabit major metropolitan areas of the nation. Five of the top 12 cities are located in California, with San Jose leading the list. Whereas the national average for major cities is 25.6% college graduates, San Jose boasts 42.4%. Other California locations among the select dozen are San Francisco (39.5%), Oakland (34.3%), San Diego (34%), and Orange County (31.9%). Falling just outside the top 12, with 31.4% college graduates, is Sacramento (Associated Press, 2000). The Master Plan accomplished this success by using tax dollars to create a tripartite system of public higher education, consisting of the 108 campuses in the California Community College system, the 23 campuses in the California State University (CSU) system, and the 9 campuses in the University of California (UC) system (with a 10th campus now being added in Merced).

Initially, there was no fee to attend community colleges in California, and the fees to attend a CSU or UC campus were very small. However, the picture changed somewhat after the elections of 1978, when Proposition 13 passed, giving property tax relief to homeowners. Reduced property tax income severely affected all higher education institutions throughout the state. Facilities could not be built fast enough to keep up with population growth, colleges began relying more on poorly paid part-time instructors, and the number of, and selection of, courses was cut back. Since that time, all segments of higher education have assessed fees of various sorts.

It is projected that, by the year 2006, the state's community college system will add 408,000 more students (a 37% increase), the CSU system will add 106,000 more students (a 31% increase), and the UC will add 36,500 more students (a 24% increase) (California Citizens Commission on Higher Education, 1998). In Chapter 9, it is pointed out that to accommodate this growth of the student population, an increasing number of classes is being offered via distance education, a mode of delivery that may prove problematic.

Although the California system of higher education is the most successful in the world, many potential students are currently denied admission, and an even greater number of future potential students will be denied admission. There are a number of reasons for this, but the primary reason is financial. The California Postsecondary Education Commission has determined that enrollment in the three systems dropped from about 63% of all high school graduates in the mid-1980s to about 55% in the late 1990s (California Citizens Commission on Higher Education, 1998, p. 51). As higher education increasingly becomes the open door to well-paying jobs, this change in the state's colleges will certainly contribute to the growing gap between rich and poor. For example, in the 1970s, 68.7% of the children of the richest quarter of the population were in college, and 26.3% of the children of the poorest quarter were also in attendance. By the 1990s, the respective percentages had changed: 75.4% of the richest and 25.5% of the poorest youngsters were then college students (California Citizens Commission on Higher Education, 1998, p. 53).

Part IV deals with the problems of population growth, including California's disproportionate share of immigrants, and the environment, including the effects of natural disasters. While the Golden State benefits from a "global brain drain" (foreign-trained engineers, computer scientists, and medical personnel coming to the state where the citizenry can capitalize on their talents), it also suffers from an influx of poorly trained and undereducated immigrants. Many of these are undocumented or unauthorized migrants. It has been estimated that 5,000 unauthorized migrants enter the United States each day, but 4,000 of them are apprehended as they cross the border. Still, about 2 million unauthorized migrants (about 40% of all people who are in the country illegally) live in California, nearly triple the number in Texas, the state with the second-highest total (Martin & Midgley, 1999). A recent study by the United States/Mexico Border Counties Coalition (2001) demonstrates that counties along the U.S.-Mexico border incur many expenses dealing with illegal immigrants and that the federal government reimburses less than 12% of these costs. The research team showed that California's border counties (Imperial and San Diego counties) spend $55.7 billion a year on migrants, more than twice the amount spent by Arizona and Texas and more than 11 times as much as New Mexico.

The growth of the immigrant and minority populations within the state has led to nativistic responses from the dominant group. The result has been a pattern of discrimination against certain minorities, repatriation of Mexican nationals (as well as many U.S. citizens of Mexican origin) during the Great Depression, and recent ballot initiatives aimed at curtailing immigration and imposing penalties on illegal immigrants and their employers.

Finally, the book concludes with an epilogue that discusses a social problem that was not evident when the book was first envisioned: the energy crisis. This concluding section shows how governmental mismanagement, corporate greed, and public impassivity have combined to create a situation that will negatively affect the state for years to come. And, as our definition of a social problem clearly indicates, this is a condition that can only be solved by collective social action.

These are just a few examples of the kinds of problems that are unique to California. We believe that three conditions in the state exacerbate these issues. First, the size of California's population and its continued growth put exceptional pressure on natural resources, air quality, and availability of water and energy. Second, geographical and geological features put the state at particular risk for earthquakes, drought, and illegal immigration. The huge number of residents and their concentration in several urban areas increase the effects of natural disasters. And, third, California's population is much more racially, ethnically, and economically diverse than the population of any other state. Although this accounts for much of the strength of California society, it also contributes to a lack of consensus about norms.

Consequently, we firmly believe that there is a need for a book on California's social problems.

—Charles F. Hohm
San Diego State University

—James A. Glynn
Madera College

REFERENCES

Associated Press. (2000, December 20). State rates high in degrees. *Fresno Bee*, p. A21.
California Citizens Commission on Higher Education. (1998). *A state of learning: California higher education in the twenty-first century.* Los Angeles: Center for Governmental Studies.
Dortch, S. (1996). California's next decade. In *America's hotspots: Migration trends in the 1990s.* New York: American Demographics.
Glynn, J. A., Hohm, C. F., & Stewart, E. W. (1996). *Global social problems.* New York: HarperCollins.
Liedtke, M. (2001, June 16). California's economy is world's 5th. *Fresno Bee*, p. A21.
Martin, P., & Midgley, E. (1999). *Immigration to the United States.* Washington, DC: Population Reference Bureau.
United States/Mexico Border Counties Coalition. (2001). *Illegal immigrants in U.S. border counties: Cost of law enforcement, criminal justice, and emergency medical services* [On-line]. Available: www.bordercounties.com

Part I

CHALLENGES TO CALIFORNIA COMMUNITIES

In this section we will be looking at homelessness, juvenile delinquency and urban gangs, and race and ethnic relations.

The first chapter, "Homelessness," by Ronald W. Fagan, tackles the most visible of California's social problems: its homeless men, women, and children. Fagan points out that California is often referred to as the homeless capital of the United States because there are more homeless in the Golden State than in any other state. The author indicates that the public has become increasingly aware of the problems of the poor and homeless in the last two decades. This is largely due to the increased visibility of the homeless caused by a number of things: the decline in available low-cost shelter; the relative decline in social welfare and other entitlement programs; the decriminalization of public drunkenness, vagrancy, and the offenses listed in other "public nuisance" statutes; and the emergence of new types of homeless people, particularly women and children. Although Californians are increasingly aware of homelessness, they have mixed sentiments toward the homeless. On the one hand, they think government should help alleviate the problem. On the other hand, they believe the problem results from the irresponsible behavior of the homeless.

Fagan discusses the difficulty of ascertaining how many people are homeless. How homelessness is defined and how it is measured have a very real effect on the outcome. Although no systematic study has been done to count the number of homeless in California, the state estimates that there are about 170,000 to 360,000 homeless in California on any given day. When discussing the characteristics of the homeless, the author points out that in the 1950s and 1960s, the homeless were primarily single men older than 50 years of age, living alone, supported by Social Security pensions or the wages of temporary labor. Many of them suffered from disabilities, including alcoholism and other substance abuse. The homeless today share some characteristics with the homeless of the 1950s and 1960s. However, beginning in the mid-1970s, the homeless became far more heterogeneous than their skid row predecessors. For example, today, the homeless are younger, they are ethnically diverse, and they are more likely to be women and children. Also, families are much more likely to be part of the homeless population than in the past. Although shelter is the most obvious need of the homeless, they also are likely to be hungry, unemployed, underemployed, or underpaid.

Fagan uses the functionalist, conflict, and interactionist perspectives as alternative ways of explaining the existence of homelessness. The author asserts that the housed poor are different from the homeless poor because the former typically can live with parents or relatives and receive financial help from them, they are able to find at least marginal or part-time employment, and they avoid alcoholism, drug use, mental illness, and crime. Fagan points out the recurrent belief among many people that homelessness is primarily a consequence of personal disabilities or failures. However, most of the people who are unemployed, mentally ill, alcohol or drug abusing, and physically disabled are not homeless; neither are most single mothers with children. These characteristics and situations do make people more vulnerable to homelessness. However, people are homeless primarily because of factors in our society that cause poverty and lack of affordable housing. Fagan describes how affordable housing in California has become less and less available, how much of the low-cost housing in the state has been torn down, and how the federal government's contribution to low-cost housing has decreased markedly.

In the section on "Solutions to the Problem of Homelessness," Fagan presents legal approaches, housing approaches, economic approaches, and personal illness and disability approaches. Legal approaches to solving homelessness center on passing legislation that prohibits aggressive panhandling as well as sleeping, loitering, or storing personal belongings in public places. Many cities in California are passing such laws. Homeless advocacy groups have also used the law in an attempt to gain relief for the homeless.

Housing approaches to solving homelessness have focused on five interrelated areas: preserving existing subsidized housing; providing incentives to the private sector to increase construction of affordable housing; constructing new affordable housing through community-based, nonprofit development groups, making housing affordable through rent control, subsidies, and vouchers; and preventing homelessness and promoting tenants' rights.

Economic approaches to solving the problem center on ways to give help to the homeless without giving them cash. For example, in many California cities, organizations sell vouchers that people can give to the homeless instead of giving them cash that might be used to purchase alcohol or drugs. The homeless can redeem the vouchers for food at local businesses. Economic approaches also deal with helping able-bodied adults develop long-term recovery skills, reconfiguring the welfare system so that the working poor do not face discrimination, and developing affordable child care systems.

Finally, the last approach to solving homelessness focuses on ways of helping homeless people with significant disabilities, including mental and physical illnesses and alcohol and other substance abuse problems. For example, the mentally ill need integrated community and residential treatment and rehabilitation programs; also, psychiatric professionals must show a greater willingness to treat the severely mentally ill. Also needed are better community housing, better state and local funding, and, for some of the mentally ill, long-term hospitalization. However, Fagan points out that it is common for some homeless to have a number of disabilities (such as mental health and substance abuse problems), which can exclude them from some assistance programs. He points out that our society needs to find innovative ways of helping homeless people with such combined disabilities.

The second chapter, "Juvenile Delinqency and Urban Gangs," by Anne Hendershott, commences with a historical overview of juvenile crime in California. Hendershott points out that the California history of juvenile crime and justice reveals patterns and cyclical changes in attitudes toward youth and how they are to be treated. The "lock 'em up and throw away the key" view that is so prevalent today in California has been in vogue at certain times in California's past. Hendershott argues that an analysis of California's juvenile justice system reveals that the emphasis has changed from punishment in the earlier days of the juvenile justice system, to an opposite emphasis on protection for juveniles, back to today's renewed emphasis on punishment. Proposition 21, which was passed overwhelmingly by California voters on March 7, 2000, is presented as an example of the current emphasis on punishment rather than rehabilitation. This proposition makes it easier for the state's prosecutors to send juvenile offenders from juvenile court to adult court for processing and sentencing. The law also enhances the already severe penalties for documented juvenile gang members involved in criminal activity. The irony is that, since 1988, juvenile crime rates in California have been steadily declining, not increasing. It appears that the media campaign promoting Proposition 21 and the "growing gang menace" was a success. California voters were hoodwinked into believing that juvenile crime is increasing in the Golden State. Hendershott points out that treating juveniles as adults represents a movement back to a time—the late 1800s—before juvenile courts were established.

The author discusses some of the progressive innovations to California juvenile justice that came about in the 1960s, such as community-based juvenile correctional facilities. The central idea was that rehabilitation of youth could be done more effectively outside of conventional correctional facilities. Supporters of community-based programs argue that children are better served when they are housed in the community. Today, in California, as in the rest of the nation, community-based programs are increasingly opposed by citizens and lawmakers who view such programs as "too soft on juvenile criminals."

In the section "California's Urban Youth Gangs: Social Problem or a Moral Panic," Hendershott discusses the emergence of Latino youth groups in Los Angeles in the 1940s and how the white community experienced a "moral panic," even though the activities of the youth groups were social and not criminal. The result was the Zoot Suit riots, in which 200 white uniformed servicemen went into the Latino barrios and beat up Latino youths. Hendershott points out that as a result, Latino gangs emerged in an attempt to protect their turf. The 1965 Watts riot in Los Angeles had the same solidifying effect and fostered the emergence of youth gangs in the African American community.

In the section "Theories on the Causes of California Gang Delinquency," Hendershott discusses a number of differing explanations of juvenile involvement in gangs. One explanation asserts that juvenile gangs are increasingly involved in the distribution and sale of illegal drugs and that adolescent cohorts serve as recruiting and training grounds for adult criminal enterprises. Another explanation for gang membership is economic and social dislocation caused by an economy and occupational structure that has undergone tremendous transformation, a transformation that often results in the loss of jobs that pay decent wages. In the anthropological view, gangs appeal not to adolescents' desire for money and fame but to their deep-seated longing for the tribal group processes that sustained and nurtured their ancestors. Hendershott discusses a few other theories in addition to these and lists the theories' strengths and weaknesses.

The author discusses the relationship between ethnicity and gang membership, citing studies that show structural variables such as income and social class but fail to explain the fact that minority youth, specifically African Americans and Latinos, are more likely to join gangs than are white youths. One explanation for this difference is that poor white youths in California lack the ideology or the history of being singled out for their membership in a white ethnic group and, as a result, do not have the same propensity to join gangs. Hendershott points out that a growing proportion of California's population is made up of non-white ethnic groups. Thus, we can expect gang membership to increase as a result of this demographic shift.

Hendershott also points out that, in 1997, a disproportionate number of juvenile minority group members were likely to be in residential placement in the state of California. For nearly a decade, minorities have accounted for 51% of the juvenile population in California but 81% of all confined youth. Hendershott asserts that although overrepresentation does not necessarily imply discrimination, there are concerns in California that minorities are overrepresented at all stages of the juvenile justice system, compared with their proportion in the population.

Hendershott concludes her chapter by presenting three different solutions to the problem of juvenile delinquency and gangs in California: (a) step up enforcement efforts and impose appropriately severe sentences, (b) help strengthen communities in fighting crime, and (c) attack juvenile crime at its root, poverty. The author then discusses the strengths and weaknesses of these proposed solutions.

Chapter 3, "Race and Ethnic Relations," by Phylis Cancilla Martinelli, begins with a history of race relations in California, including a discussion of how California's patterns of racial separation and suspicion of immigrants are also found globally. The state's history of race relations starts with the initial contact between Spaniards and Native Americans, long before California was part of the United States. The first fissure emerged in the late 1700s as the Spanish established their missions and forts, coming in contact with a variety of tribes and several major linguistic groups. Because these various tribes spoke different languages, they were unable to unify against the invaders. The Spanish goal was to convert tribes to Catholicism and get them to serve as serfs on Spanish ranchos. Although the Spanish were patronizing to Native Americans, they at least regarded them as human beings with rights and privileges under the system of Spanish law. This was not true of Anglo Americans, who began a large-scale movement into California in the 1840s; they regarded Native Americans as almost subhuman. For some Anglos, extermination and removal from their land was the only fate acceptable for tribes, and the population of Native Americans in California was decimated during this period.

African Americans faced a mixed reception on their arrival in California. Those who found manufacturing jobs in Oakland and Los Angeles were able to establish middle-class lives. However, blacks often faced discrimination stemming from the ideology developed to justify slavery, and various types of legislation prevented them from competing in the marketplace and from living next to whites.

Japanese, Chinese, Korean, and Filipino immigrants also confronted nativist ideology in which whites were viewed as being superior to other ethnic groups. Legislation banned additional Asians from coming to the United States and California. Finally, Mexicans were also viewed with hostility by whites. They were seen as closer to Indians in the rigid racial classification system.

In the section "Race Relations in California Today," Martinelli describes how California's population is becoming racially diverse and how non-Hispanic whites, who once were in the vast majority, are losing their demographic edge. The author suggests, however, that racial diversity does not necessarily result in contact. Many whites have retreated to exclusive suburbs or rural counties. According to Martinelli, the increasing separation between ethnic groups along with economic downturns (such as the recession in the early 1990s) could result in future racial incidents like the 1992 Los Angeles riot. Aftershocks continue with anti-immigrant legislation in 1994, a tortured debate on affirmative action, and racially polarized reactions to the O. J. Simpson case. Proposition 227 (passed in 1998), which curtailed bilingual programs in California that served large numbers of immigrant children, represents the latest divisive issue in the Golden State. The author indicates that Los Angeles clearly rests on a racial fault line. Because many trends in Los

Angeles foreshadow events in the rest of America, it is important to examine race relations there. Martinelli cites demographic research showing that Los Angeles is extremely segregated and that territorially based ethnic tension is endemic there. Although white-black tensions are very real in Los Angeles, as evidenced by the Rodney King[1] trial, conflict between other ethnic groups also characterizes Los Angeles. In the Watts riots of 1965, blacks were the group most arrested or killed; in the 1992 riot following the Rodney King verdict, Latinos were the group most arrested and killed. Also, the animosity between blacks and Korean shopkeepers was shown by the number of Korean businesses that were burned in the 1992 Los Angeles riot and by the number of Korean shopkeepers who have been killed in Southern California since 1990.

Comparing California to the nation, Martinelli notes that the same racial divisions that characterize the Golden State are found elsewhere, too. Members of the urban underclass have been left behind in the ghettos of large American cities such as Los Angeles, Chicago, and New York. They suffer from the evaporation of lower middle-class jobs with good pay, which occurred as the American economy changed from a base of manufacturing to a base of high technology. The jobs left for inner-city residents are mostly in service industries, pay little, and offer no prospects for promotion. The author also points out that minorities, both in California and the nation, receive less adequate health care and education than whites do.

Why do some ethnic groups fare better economically than others? Martinelli discusses the view that differences in the success of racial groups are linked to hereditary traits. This explanation has been used for more than a century, with one of its most recent manifestations being the controversial book, *The Bell Curve*, by Herrnstein and Murray (1996). Martinelli summarizes the studies of social scientists on this theory and notes that the vast majority find it lacking. Most sociologists agree that race is a recent social category, varying from culture to culture and within one culture over time. Thus, sociology looks at structural patterns and ideologies to explain differences.

Discussing "Future Prospects for Race Relations in California," Martinelli finds reasons for both optimism and skepticism. She cites some recent demographic analyses of California cities showing that younger cities such as San Jose and Anaheim are much less segregated than older cities such as Los Angeles and San Francisco. This bodes well for California, as a very high percentage of future growth will occur in the small to mid-size cities. That being said, the prevalence of the urban underclass and the violence and human carnage that accompany it are cause for alarm.

In Martinelli's section "Solutions to the Problem of Race Relations," one possibility is coalition politics, which seeks to connect concerned individuals from different races. The author discusses instances of success and failure of coalition politics in an attempt to learn what tactics are most successful.

NOTE

1. Rodney King was stopped on a Los Angeles Freeway for allegedly driving erratically, and he was beaten by a group of policemen. This incident was captured on videotape and fueled a tremendous controversy. When Rodney King was found guilty and the police who were involved in the beating did not get punished, a series of riots broke out in Los Angeles.

REFERENCE

Herrnstein, R., & Murray, C. (1996). *The bell curve: Intelligence and class structure in American life.* New York: Free Press.

CHAPTER 1

HOMELESSNESS

Ronald W. Fagan

The inscription on the Statue of Liberty reads, "Give me your tired, your poor, your huddled masses yearning to be free, The wretched refuse of your teeming shore, Send these, the homeless, tempest-tossed to me: I lift my lamp beside the golden door." Yet, for many poor and **homeless** people in America, the promises have not been fulfilled. Along with food, health, and personal safety, there is probably no greater human need than shelter. The homeless tend to be the poorest of the poor. They are America's most visible social problem. We see them clustered in our inner-city areas. We see them huddled against the cold as they seek warmth from fires in trash cans and from steam grates. We see the bag ladies pushing all their possessions in a shopping cart or the strangely dressed man talking to himself as he walks down the street. We encounter them as they beg for money or scavenge for food or recyclables in trash dumpsters. We see them trying to find shelter in doorways, abandoned buildings, parks, subway and bus terminals, and encampments along the road. Increasingly, we are seeing young families and adolescents who appear to be homeless.

California is often referred to as the Golden State. More than 34 million people live in California, representing 12.5% of the entire U.S. population. Median household income in California is $40,600, and median per capita income is $20,498. California has the fifth-largest economy in the world. But there is another side to the Golden State. The U.S. Bureau of the Census (using a definition of poverty based on income by family size) estimated that, in 1999, California had the 10th-highest poverty rate in the country with 14.6% of its residents defined as poor (compared to 11.9% of U.S. residents). California ranks 45th in the nation in the percentage of children in poverty (20%); only five states have worse records. About 6% of Californians are unemployed (State of California, 2000).

With a homeownership rate of 55.7% (compared to 65% for the country as a whole), a median house value of $221,520, and median monthly rent of $561, it's not difficult to see why California can also be described as the homeless capital of the United States. Six of the seven least affordable rental markets in the country for low-income families are in California (Housing California, 2000). There are probably more homeless people in California than in any other state. An estimated 360,000 homeless people (about 35% of them families and 25% children) reside in California, with more than 1 million people experiencing one episode of homelessness each year (California Department of Housing and Community Development, 2000).

Since the late 1970s and early 1980s, the American public has become increasingly aware of the problems of the poor and homeless. A number of factors have caused this heightened concern. The most significant factor was an increase not only in the number of homeless people but also in their visibility throughout major urban areas. This increased visibility was due to a number of factors, including a decline in availability of low-cost shelter and housing and a relative decline in social welfare and other entitlement programs. Decriminalization of public drunkenness, vagrancy, and other "public nuisance" statutes meant that fewer of the homeless were in jail. Finally, new types of homeless people, particularly women and children, emerged. Also significant in the new awareness was the rise of advocates (both individuals and groups), positive mass media coverage, and some changes in beliefs about who is primarily responsible for the problems of the homeless (Barak, 1992; Link et al., 1996).

Public opinions polls consistently show that the public has mixed sentiments toward the homeless. On the one hand, the public tends to view homelessness as a significant social problem and believes the government should do more to address it. On the other hand, the public tends to associate the homeless population with other stigmatized groups and views the primary causes of homelessness as drug and alcohol abuse and what they perceive to be irresponsible behavior (Hoch & Slayton, 1989; Link et al., 1996; Rossi, 1994). These mixed feelings have had a profound impact on public policy toward the homeless.

HOMELESSNESS IN THE UNITED STATES AND CALIFORNIA

THE NUMBER OF HOMELESS

Although it is generally agreed that there was a dramatic and largely unexpected increase in the number of homeless beginning in the 1980s, there is much less consensus regarding the actual number of homeless in the United States and in the individual states, including California. Producing valid estimates of how many people are homeless is difficult for conceptual, methodological, and political reasons (Applebaum, 1990; Blasi, 1994; Breakey & Fischer, 1990; Burt, 1996; Rossi, 1989; Snow, Anderson, & Koegel, 1994; White, 1992).

Conceptually, the estimates depend on how homelessness is defined and whether the definition includes people who cycle in and out of homelessness. *Homelessness* can be narrowly defined as being without suitable, conventional housing (literally homeless) or broadly defined to include being inadequately housed or at high risk for homelessness (Fischer & Breakey, 1991; Rossi, 1994).

Methodologically, counting the homeless is problematic for practical and political reasons. Researchers primarily use two methods to attempt to measure homelessness. *Point-in-time counts* or cross-sectional counts attempt to count all the people who are literally homeless on a given day or during a given week. *Period prevalence counts* examine the number of people who are homeless over a longer period of time. Point-in-time studies tend to overestimate the proportion of people who are chronically homeless and those who cycle in and out of homeless, and they tend to underestimate those who have one-time or a few short homeless periods. Reliance on these types of estimates may result in an overemphasis on services for people with individual disabilities such as mental illness, substance abuse, or significant health problems at the expense of people whose homelessness is primarily the result of poverty and shortage of affordable housing (Wong, 1997).

Researchers have become adept at counting and describing people in shelters and people using traditional health and welfare services. But these sampling methods tend to underestimate the number of homeless because, depending on the location, at least one quarter of all requests for emergency shelter are unmet due to lack of resources, and many of the homeless do not use these types of services on a regular basis. For example, in Los Angeles, although the shelter capacity has tripled in the last 10 years, there are between five and eight homeless people for every available shelter bed (National Coalition for the Homeless, 2000). Researchers also have difficulty counting those who are living on the streets and in automobiles, campgrounds, and rural areas, as well as those living in other nontraditional housing arrangements (living in "doubled-up situations" for example) (Link et al., 1996). For most people, homelessness is neither a short-term crisis nor a long-term pattern. Rather, it is part of an ongoing pattern of residential instability (Wong, 1997).

Finally, the estimates have political ramifications. In the early 1980s, to dramatize the problems of the homeless, homeless advocates estimated that between 1.5 million and 3 million Americans (or about 1 in every 100 Americans) were literally homeless on any given night (Hombs & Snyder, 1982). Wright and Devine (1991), summarizing the various reports, concluded that in the late 1980s, about 500,000 to 750,000 people in the United States were literally homeless on a given day, 1.5 million to 3 million people were homeless at least once in the span of a year, and some additional millions were at high risk of homelessness at a given time.

In a widely cited national telephone survey done in 1990 and updated in 1994, researchers found that 6.5% of respondents (12 million adults) had been literally homeless at some point in their lives, and 3.6% (6.6 million adults) had experienced homelessness between 1989 and 1994 (Link et al., 1996). The National Law Center on Homelessness and Poverty (2000) estimates that more than 700,000 people are homeless on any given night and up to 2 million people experience homelessness during a single year.

No systematic studies have tried to count the number of homeless in California, although some valid studies were done of locations in the state. A well-designed study in Los Angeles County found that on any given night, between 35,500 and 83,900 people were homeless, of which 13,000 were homeless family members and 9,000 were children (Shelter Partnership, 1994). As a rough estimate, many jurisdictions use the figure that .5% to 1% of each state's, county's, or city's population use shelters in a year, 2% use shelters in 3 years, and 4% to 12% have experienced an episode of homelessness within the previous 5 years (Link et al., 1996). Using this standard, it could be estimated that about 170,000 to 360,000 Californians are homeless on any given day, and 1.3 million to 4 million Californians have experienced an episode of homelessness within the past 5 years (California Department of Housing and Community Development, 2000). Although the number of homeless people may be difficult to determine, the more important task is to understand the characteristics of people who are homeless so that we can meaningfully address the needs of this diverse and significant population in our state and in our country.

CHANGING CHARACTERISTICS OF THE HOMELESS

In the 1950s and 1960s, in California and nationally, the homeless were primarily single men older than 50 years of age who lived alone, supported by Social Security pensions or wages from temporary labor. Many suffered from physical and mental disabilities, including alcoholism and other substance abuse. Most of them were not without shelter. They lived primarily in inner-city, skid row areas, where they found inexpensive housing, restaurants, bars, and temporary employment agencies (Bahr, 1973; Bahr & Caplow, 1973; Blumberg, Shipley, & Shandler, 1973; Bogue, 1963; Fagan & Mauss, 1978, 1986; Wallace, 1965; Wiseman, 1970).

The homeless today share some characteristics with the homeless of the 1950s and 1960s. However, beginning in the mid-1970s, the homeless population became far more heterogeneous (Hopper, 1997; Hopper & Baumohl, 1996). One homeless man in Venice, California, described the change this way:

I used to look at [homeless] people, old guys hanging around liquor stores. They'd ask you for change.... So you figure they're just old and alcoholic, what they call a "wino." But now that I'm homeless I see it's not just old people or old alcoholics.... I'd never seen so many young people that were homeless until I became homeless.... Now it's young and old that are homeless because there's no work or maybe something occurred in the family. (quoted in Wolch & Dear, 1993, p. 219)

To many people, the apparent rise in the number of homeless families has been the most dramatic change in the homeless population (Liebow, 1995; U.S. Conference of Mayors, 1999).

U.S. Bureau of the Census statistics do not include homeless people and people not living in shelters because the surveys are done primarily by household. Using the Census Bureau data for 1999, the poor are 9.8% white, 7.7% non-Hispanic white, 23.6% black, 22.8% Hispanic, and 10.7% Asian/Pacific Islander. It is estimated that 25.9% of the poor are Native American/Alaska Native (the Census Bureau says their sample for these groups is not large enough to produce reliable estimates). Almost 1 in 10 families are considered poor, and the majority (53%) are households headed by women with no husband present. About 40% of the poor worked (versus 71% of the general population), but only 12% worked full-time and year-round (versus 47% of the general population). Poverty rates are at least four times higher for people in households headed by women (U.S. Bureau of the Census, 2000). It is estimated that, on average, single men make up about 43% of the national homeless population; families with children, 37%; single women, 13%; children under 18 years of age, 25%; and unaccompanied minors, 7%. Half of the homeless are African American, 31% are white, 13% are Hispanic, 4% are Native American, and 2% are Asian. An average of 20% to 30% of homeless people are considered to be mentally ill, 30% to 40% are dependent on alcohol or other drugs, 21% are employed, 15% to 30% are veterans, and 20% have been incarcerated (National Law Center on Homelessness and Poverty, 2000; U.S. Conference of Mayors, 1999). But it is important to keep in mind that these percentages may vary dramatically depending on where the study is done and how the data are collected. The homeless in California have similar characteristics to the homeless in the nation as a whole, but the percentage of non-whites is probably higher, reflecting their numbers in the state.

Therefore, people who are homeless include people from a diversity of ethnic, racial, and social backgrounds; people of both sexes and a wide age range; families, including families with children; single people; runaway and throwaway youth; people with physical health, mental health, and substance abuse problems; people who are homeless primarily for economic reasons; people whose homelessness is temporary; people whose homelessness is more or less chronic; people with military and criminal records; people who live in a variety of substandard housing situations; people who often lack family and community ties; people who do not work; and people who work full-time at low salaries. All homeless people have distinct characteristics and needs. Despite their diversity, and usually as a result of their lack of an adequate income, virtually all homeless people have tangible material needs for shelter, food, clothing, and medical care. Some homeless people require only limited, temporary assistance, whereas others require extensive assistance (Fosburg & Dennis, 1999).

ALTERNATIVE EXPLANATIONS FOR HOMELESSNESS IN CALIFORNIA

The causes of homelessness are diverse and complex. One can look at homelessness from the three traditional sociological perspectives. From a functionalist perspective, homelessness exists because it serves positive functions for society or at least for some groups in society. For example, some have argued that the shelter movement and its emphasis on short-term assistance makes homelessness functional because it provides employment for the caregivers, and it may encourage people to become dependent on the system (Hoch & Slayton, 1989). On the other hand, if one looks at the problem from a conflict perspective, homelessness exists because some groups are deprived of the opportunity to accumulate the resources necessary for proper food, shelter, and care (Timmer & Eitzen, 1994). Finally, from the interactionist perspective, one has to look at the values, attitudes, and cultural/psychological orientation of the homeless and how these perspectives contribute to their remaining, adapting to, or changing their homeless situation.

Homelessness and poverty are integrally related. But it is important to keep in mind that most poor people are not homeless. In fact, the homeless represent less than 10% of the poor. There are more than 10 million working poor in America. These people walk a tightrope from paycheck to paycheck and can be wiped out by one misstep, such as being laid off from a job, falling ill, or dealing with family disruption (Gibbs, 1995). Many homeless advocates have stated that most people are just a paycheck or two away from homelessness, but the situation is much more complicated. Most families who experience sudden unemployment do not find themselves chronically homeless. People who become homeless have other significant characteristics that contribute to their homelessness, particularly if their homelessness becomes long-term.

Although poverty and homelessness are integrally related, there appears to be no simple direct correlation between homelessness and unemployment. Most people who lose their jobs do not become chronically homeless. The housed poor are different from the homeless poor. The former typically can live with their parents or relatives and receive financial help from them, they are able to find at least marginal or part-time employment, and they avoid alcoholism, drug use, mental illness, and crime (Rossi, 1989).

By definition, homelessness is a housing problem in terms of a lack of affordable housing and the accessibility of this housing to low-income people. But a recurrent belief among many people is that homelessness is primarily a consequence of personal disabilities or failures. But people typically are not homeless because they are unemployed, mentally ill, abusive of alcohol or drugs, or physically disabled. Nor are single mothers with children typically homeless. Most people whom these characteristics describe have places to live. What these characteristics do is make people more vulnerable to homelessness (Timmer & Eitzen, 1994).

The game of musical chairs provides an analogy. Homelessness is a consequence of the imbalance between the number of households that require low-income housing and the number of available units they can afford (Shinn & Gillespie, 1994). The number of affordable housing units (chairs) has been systematically reduced, while the number of players has been increasing. The problems are illustrated by a young homeless woman interviewed in Pasadena, California.

> I lost a job as a cook, and I couldn't pay my rent so I got a live-in job (caring for a convalescing elderly person) just to have a roof over my head. . . . But it never was a permanent home, it was just a few days here and a few weeks there. . . . I've worked all over the Los Angeles area. And depending on where my last job was, that's where my car would be parked. . . . I'd stay either in the back seat of my Toyota or at a homeless shelter or at a friend's house. (quoted in Wolch & Dear, 1993, p. 47)

What major factors are responsible for the shortage of affordable housing? In the early 1980s, with the downturn in the economy, jobs were being lost, housing costs were going up, and housing costs consumed more and more of the typical household budget, thus creating a large population at risk for homelessness. The problem was magnified in California because of the state's expensive rental and housing market. This situation intensified when the California economy experienced a dramatic upturn in the 1990s. During this period, the federal government was also significantly reducing its central role in building and supporting affordable housing. In the name of urban renewal, much of the existing low-income housing was being demolished, never to be rebuilt. Residents in many cities actively resisted attempts to build low-income housing in their communities. Finally, instead of building new low-income housing or rehabilitating existing housing, the government introduced a voucher system that helped people afford existing housing. But vouchers only work where there is enough affordable housing (Shlay & Rossi, 1992). Many cities in

California, for all practical purposes, have stopped accepting applications for subsidized or assisted housing (called Section 8 housing) because the waiting lists have become too long.

In conclusion, Morse (1992) identifies six key factors that contribute to initiating and maintaining homelessness: culture (discrimination and public apathy), institutions (unemployment, shortage or inaccessibility of low-income housing, deinstitutionalization of mental health treatment), communities (urban redevelopment and zoning policies), organizations (social service requirements and inaccessible or inappropriate services), groups (lack of social support), and individuals (impairments, disabilities, and personal choice). Each of these factors is usually insufficient alone to cause homelessness, but the convergence of a number of factors causes homelessness.

FUTURE PROSPECTS FOR HOMELESSNESS IN CALIFORNIA

Some people have argued that although our country excels in productivity and high standard of living, it does less well than many other countries at protecting its citizens and assuring their personal security (Bok, 1997). We have always felt the need to distinguish between the "deserving poor" and the "undeserving poor." The former are often seen as victims of circumstances beyond their control, who deserve compassion and, at least, short-term assistance. The latter are seen as personally responsible for their situation; they merit contempt and control (Handler, 1992; Jencks, 1994).

There is much debate on what are the best strategies to address the problems of the homeless, including whether the primary focus should be on the individual or structural level; whether the goal should be containment, elimination, or prevention of the problem; and whether responsibility for homelessness is primarily a personal or a public or private sector problem. For example, those who believe that homelessness is primarily a personal problem might decide to make it less attractive by reducing the number of shelters and soup kitchens, decreasing social welfare and disability payments, and arresting people who sleep in public areas. But people holding this belief might also focus on rehabilitative services to help those with physical, social, and psychological problems. On the other hand, those who believe that homelessness is caused by structural factors outside the individual might focus on broader policy issues affecting employment, income distribution, and the housing market (Belcher & DiBlassio, 1990; Garrett, 1992; Koegel, Burnam, & Baumohl, 1996). Finally, to prevent homelessness, we need to address the questions of who are the most at risk of becoming homeless, where homeless people come from, and what events precipitate or cause individuals and families to become homeless (Lindblom, 1997).

SOLUTIONS TO THE PROBLEM OF HOMELESSNESS

LEGAL APPROACHES

Historically, one of the most common approaches to dealing with the homeless has been control or punishment through legal means. Beginning in the 16th century in England, it became a crime to be willfully poor, and it was also a crime to give money to "sturdy beggars." Forms of this practice were carried to colonial America (Fischer, 1992). Beginning in the 1980s, many jurisdictions enacted legal statutes that prohibit behaviors often associated with homelessness, such as vagrancy, loitering, panhandling, sleeping in public places, disorderly conduct, or drinking in public. Jurisdictions also use mental health statutes to remove mentally ill individuals from the streets (Isaac & Armat, 1990).

Criminalization of the problem appears to be increasing, particularly as jurisdictions experience frustration with the lack of success of many social service and intervention programs and as businesses and citizens continue to complain about the homeless sleeping, panhandling, and loitering in public places. Jurisdictions are passing new legislation and enforcing legislation already enacted more strictly (American Civil Liberties Union, 1996).

Homeless advocacy groups have also used the law in an attempt to gain relief for the homeless in three primary areas: the adequacy or availability of public assistance, the ability of the homeless to keep their families together (in terms of the adequacy of the relief grants and the ability to keep their children, even if homeless), and right to shelter. Litigation has challenged municipal codes prohibiting encampments; gathering, sleeping, and storing goods in public places; panhandling or begging; loitering; and searching people and seizing their possessions. Other litigation has focused on upholding the rights of homeless people to vote, receive public assistance, and send their children to public school.

One of the most publicized California attempts to control the homeless using laws was Santa Ana's passage in 1990 of an ordinance that made it a crime punishable by up to 6 months in jail to use a sleeping bag or blanket or to store personal effects on public sidewalks, streets, parking lots, and government malls. The California Court of Appeals struck down the ordinance, but in 1995, the California Supreme Court upheld the law, stating there is no "constitutional mandate that sites on public property be made available for camping to facilitate a homeless person's right to travel, just as there is no right to use public property for camping or storing property" (Simon, 1996, p. 156). More than 90 California cities signed briefs supporting Santa Ana's position in the case. Since the ruling, many other cities in California have adopted similar statutes.

Many of the arguments in support of the rights of the homeless have been based on Article I, Section 8, of the U.S. Constitution, which gives the federal government the power to "provide for the general welfare" and on the Equal Protection Clause of the Fourteenth Amendment, which guarantees all people equality and due process under the law (Stoner, 1995). Although the U.S. Supreme Court has not recognized any constitutional guarantee of a "right to shelter," some states and cities have enacted statutory legislation that provides for various degrees of adequate shelter for its residents (Blau, 1992). Fundamental to understanding the federal court's reluctance to rule on homeless rights is the fact that most constitutional protections are based on property rights, and the condition of being homeless means there is no or little property ownership.

HOUSING APPROACHES

In December 1993, a mentally ill woman who had been in and out of homeless assistance programs in Washington, D.C., was found dead one morning in the bus shelter where she had been sleeping. The bus shelter was located outside of the U.S. Department of Housing and Urban Development (HUD), the federal agency charged with the responsibility to administer federal housing programs (Hombs, 1994). The story generated a lot of publicity, and HUD officials responded that more would be done to help the homeless.

Traditionally, the housing needs of the homeless have been approached on three levels: temporary emergency shelters; transitional housing, which allows for longer stays than emergency housing and typically provides for support services; and permanent housing. Historically, a number of different strategies and programs have been used to address the housing needs of the homeless, from large federal, state, county, and city programs to privately funded nonprofit programs (Fosburg & Dennis, 1999).

Homeless housing activists have focused on five interrelated areas: preserving existing subsidized housing; providing incentives to the private sector to increase construction of affordable housing; constructing new affordable housing through community-based, nonprofit development groups; making housing affordable through rent controls, subsidies, and vouchers; and preventing homelessness and promoting tenants' rights (Dreier & Atlas, 1989).

Although there had been many programs for the poor and homeless before 1987, the federal government became formally involved in addressing homelessness that year with the passage of the Stewart B. McKinney Homeless Assistance Act. The McKinney Act originally consisted of 15 programs providing a range of services to homeless people, including emergency shelter, transitional housing, job training, primary health care, education, and some permanent housing. The act has been amended four times. The amendments have tended to focus on expanding the types of services, improving physical and mental health services for the homeless, and increasing access of homeless children to public education.

There have been many criticisms of the McKinney Act, including inadequate funding, but many people feel that the act's greatest weakness is its focus on temporary emergency measures such as providing food, clothing, and shelter, measures that focus on the symptoms of homelessness not its causes (National Coalition for the Homeless, 2000). Some authors feel that an emphasis on short-term relief may, in fact, encourage some people to maintain their homeless status (Hoch & Slayton, 1989). An estimated 9% of shelter users enter nearly five times per year and stay nearly 2 months each time, using 18% of the system's resources. An additional 10% enter the system just over twice a year and spend an average of 280 days per stay (National Alliance to End Homelessness, 2000). Ways must be found to break this dependence on shelters among a minority of the homeless population. This might be done by increasing access to and availability of adequate long-term housing and helping people to get effective treatment for their disabilities so that they can develop adequate employment skills for long-term recovery (Dreier & Atlas, 1989).

It is clear that if ways can be found to reduce unemployment, homelessness will be reduced, but only to the degree to which low-cost affordable housing is available. An estimated 25% to 40% of the homeless are employed. With California's **minimum wage** of $5.75 per hour, the income of a family of four supported by a full-time worker earning minimum wage equals 72% of the 1999 federal poverty threshold. A full-time minimum wage earner currently makes about $11,900 a year.

Even when people do find work, they may not be able to afford adequate housing. In California in 1999, fair market rent for one-bedroom units was $617 and for two-bedroom units, $775. About 45% of renters in California were unable to afford fair market rent for a two-bedroom unit. California workers earning the federal minimum wage ($5.15 per hour) had to work 116 hours per week to afford a two-bedroom unit at the area's fair market rent. The housing wage in California is $14.90 per hour. This is the amount a worker would have to earn to be able to work 40 hours per week and afford a two-bedroom unit at the area's fair market rent. This is 289% of the present federal minimum wage. California ranks 11th in the nation in its lowest fair market rent as a percentage of the minimum wage (Children's Defense Fund, 2000). California ranks 45th in the nation in per capita funding for affordable housing. In January 2001, California raised the minimum wage to $6.25, and it will increase to $6.75 per hour in January 2002. This represents an extra $2,000 a year for thousands of California workers (Creeland, 2000; Housing California, 2000; National Low Income Housing Coalition, 1999).

Clearly, for many Americans, work provides little or no escape from poverty. Despite recent increases, the real value of the minimum wage has decreased almost 20% in the last 20 years. In addition, reductions in the number of unionized workers, losses of relatively well-paying manufacturing jobs coupled with expansion of lower-paying service-sector employment, and increases in temporary and part-time employment have all contributed to a decline in wages for many people.

The benefits of the recent economic growth in our country have not been equally distributed. Wage-based incomes have become increasingly polarized. In California in 1967, for every $4 earned by someone in the top 10% by income, a person in the bottom category made $1. In 2000, the ratio was almost 10 to 1 (Texeira, 2000). Putting the recent gains in the economy in their historical context, it is clear that the living standards of many working families in America have neither fully recovered from the early 1990s recession nor benefitted from the overall growth in productivity (Mishel, Bernstein, & Schmitt, 1998-1999). "A rising tide does not lift all boats, and in the United States today, many boats are struggling to stay afloat" (National Coalition for the Homeless, 2000, p. 3). To extend the analogy, merely continuing to bail out the boat may keep the problem from getting worse, but to address the problem significantly, we need to permanently fix the leaks by finding ways to prevent homelessness (Lindblom, 1997).

Today, fewer than 30% of Americans eligible for low-income housing receive it (U.S. Conference of Mayors, 1999). Only 13% of Californians receive housing assistance, the lowest rate in the nation. Six of the seven least affordable rental markets in the country for low-income families are in California. California stands to lose at least 30,000 units of affordable housing within the next few years unless there is significant intervention and investment by both the private and public sector. California must build at least 220,000 units per year just to meet growth projections. This will be difficult because, for the past 10 years, production averaged less than 110,000 units per year (California Department of Housing and Community Development, 2000; Housing California, 2000). Although the California Department of Housing and Community Development (2000) spent more than $30 million in the past 5 years to expand housing opportunities, there are still critical shortages. Needs include increased government support for public housing and nonprofit housing corporations, increased efforts to preserve and refurbish existing affordable housing, improvements in municipal services to low-income neighborhoods, and increased voucher benefits to make housing more available and affordable to a greater number of people (Jahiel, 1992).

ECONOMIC APPROACHES

There is a wide range of economic approaches to homelessness. On a personal level, everyone has probably been approached for money by a person who appears to be poor and homeless. The decision about whether or not to give money is an individual decision. People may want to help but be concerned about whether the money will be spent on alcohol or drugs rather than on food and shelter. In response to these concerns, many California cities and organizations sell vouchers that people can buy and give to people instead of cash. The vouchers can be redeemed for food and other goods at local businesses. Many people who work with the homeless suggest that concerned people may want to give money to organizations that serve the homeless rather than to people who solicit money directly on the street. Whether or not people choose to give a person money, those who work with the homeless suggest that passers-by should not ignore the homeless or look away as if the person does not exist. Acknowledging a person's existence by looking at them is one important way to reaffirm their humanity at a difficult time in their life (National Coalition for the Homeless, 2000).

Throughout history, a wide variety of private and public programs have attempted to help the poor and the homeless. The English Poor Laws, set up by the government of England in the late 16th and early 17th centuries, divided the poor into two groups. The deserving poor were those deemed unable to work (the disabled, blind, elderly, and children), and they generally were eligible for "outdoor relief," cash or other forms of assistance in their homes. The undeserving poor, individuals who were able to work, received "indoor relief," which amounted to public service employment, usually inside large public facilities called workhouses. The American colonists incorporated much of the framework of the English Poor Laws into American laws. By the early 19th century, counties or municipalities were required to provide for the poor and needy. But reflecting U.S. ambivalence about the poor, by the mid to late 1800s, movements were opposing the practice of unconditional relief unrelated to work (Encarta, 2000; Handler, 1992; Jencks, 1994).

The modern U.S. welfare system dates to the Great Depression of the 1930s. During the Depression, more than two thirds of all households would have been considered poor by today's standards (adjusted for inflation). Being poor was no longer considered primarily a personal failing. Led by President Franklin D. Roosevelt's New Deal, a number of programs were created, many of which are in place in revised form today, including the Social Security Act (SSA), Aid to Families with Dependent Children (AFDC), and Supplemental Security Income (SSI). Today, the U.S. government provides **welfare** in a number of different ways, including direct cash assistance, specific goods (for example, vouchers to cover rent or food), and services (for example, health and child care).

In 1996, in response to increasing criticism that the U.S. welfare system fostered dependence or was failing to raise recipients out of poverty, Congress passed two bills known as the Personal Responsibility and Work Opportunity Reconciliation Act (PRWORA) and the Illegal Immigration Reform and Immigrant Responsibility Act (IIRIRA). Among the provisions of PRWORA was the shifting of primary responsibility from the federal government to the states to determine eligibility requirements and benefit levels through the Temporary Assistance for Needy Families (TANF) program. One of the most significant provisions of TANF is that federal funds cannot be used to provide benefits to families who have been on assistance for

5 years (with some exceptions), and TANF recipients must work after 2 years of assistance. To meet these requirements, states must continue to fund job training and placement, subsidized employment, food stamps, and child care programs (Encarta, 2000).

Although the impact of these sweeping changes has yet to be fully evaluated, since 1996, welfare rolls nationally have dropped about one half due both to the changes in welfare policies and improved labor market conditions. Studies have found that between 50% and 70% of welfare recipients who have left welfare were employed at a point in time. Half of those employed were working at least 30 hours per week on average, but for the most part, they were entering the low end of the labor market (sales, service, or clerical support occupations) with jobs paying between $5.50 and $7 per hour. Although some of these jobs pay more than the minimum wage, the income is usually not enough to raise a family out of poverty (National Governor's Association, 1998). Most people who have left welfare report incomes that are lower than or similar to their combined earnings and benefits before their exit from the welfare rolls. Most families continue to receive some form of public assistance, usually food stamps, child care, and Medicaid, although at rates much lower than the near-universal participation of people on welfare. Some studies have found that most families, despite low-wage jobs and continued use of other forms of income support, believe they are better off and are confident they will not need to return to welfare. At the same time, about a third of families report problems providing enough food and paying utility bills and rent; they say they have to rely on friends and family for additional means of support (Brauner & Loprest, 1999).

The direct impact of these changes in the welfare system on the number and characteristics of the homeless population is difficult to evaluate. For example, research shows that single women with children who were on welfare under the old system tended to have substantial labor market experience (Edin, Harris, & Sandefur, 1998). Therefore, their problem was not their unwillingness or inability to work, but, rather, it was that the jobs they typically were able to hold paid too little to lift them out of poverty. Three main factors determined who did and did not work, who left welfare when they began to work, and who remained on welfare when they worked: human capital (the person's education and work experience), child care constraints, and labor market conditions (unemployment and wage rates) (Bane & Ellwood, 1994). Research shows that education is more important than job experience in helping women and their families stay off welfare once they exit. Education enables them to get better jobs with advancement possibilities. Research shows that women tend to rely on four sources of income: official work (reported to the Internal Revenue Service and other authorities), the food stamp program, unofficial (unreported) work, and cash and other benefits from family, friends, and community resources (Edin & Lein, 1997). When one or more of these sources becomes unavailable, recipients become especially vulnerable to homelessness (Sherman, Amey, Duffield, Ebb, & Weinstein, 1998). For example, a family might avoid homelessness by staying with relatives or friends, but with more people in similar situations, this buffer to homelessness has become a less available option for many of today's poor (Culhane, 1997).

PERSONAL ILLNESS AND DISABILITY APPROACHES

Many homeless people have significant disabilities, including mental and physical illnesses and alcohol and other **substance abuse** problems. But in general, people do not become homeless directly because of these disabilities. For example, only 5% of the estimated 4 million people who have a serious mental illness are homeless at any given point in time (Federal Task Force on Homelessness and Severe Mental Illness, 1992). People with disabilities become homeless because they do not have the financial resources to afford available housing (Wright & Rubin, 1997).

Some of the homeless have a number of disabilities at the same time, which can result in their exclusion from some assistance programs. For example, homeless people may be excluded from getting mental health treatment because of the problems created by their substance abuse, and they may be excluded from substance abuse programs because of the problems created by their **mental illness.** They may also have significant untreated health problems that interfere with their treatment (Beatty & Haggard, 1998; Goldfinger, Susser, Roche, & Berkman, 1998; National Resource Center on Homelessness and Mental Illness, 2000). More than a third of people living in poverty have no health insurance of any kind (National Coalition for the Homeless, 2000).

Provisions in the 1987 Stewart B. McKinney Homeless Assistance Act and its amendments target funding at assistance for homeless people with mental illnesses, including outreach programs to engage disenfranchised people, dual diagnosis treatment programs to help individuals with co-occurring mental illness and substance use disorders, and supported housing arrangements to help people with serious mental illnesses obtain and retain permanent housing. The act also has provisions for health clinics specifically devoted to the needs of the homeless (Lezak & Edgar, 1998). But serious deficiencies remain in both the delivery and number of services available.

The National Association of State Alcohol and Drug Abuse Directors estimates that more than 1 million people are waiting for treatment. Many people feel that it should be easier for substance abusers to receive disability benefits (Rossi, 1989). Others have argued against providing cash grants to known substance abusers, but they recommend providing vouchers or services or entrusting care to a third-party protective payee instead (White, 1992). Recent policy changes have made the situation worse for homeless people with addictive disorders. In 1996, national legislation went into effect that denied SSI and Social Security Disability Insurance (SSDI) benefits and, by extension, access to Medicaid to people whose addictions are considered to be a "contributing factor material to" the determination of their disability status (National Health Care for the Homeless Council, 1997).

One significant problem with assisting the homeless who are mentally ill and some substance abusers is that their illness may prevent them from actively seeking and cooperating with treatment and assistance. Isaac and Armat (1990), for example, propose the creation of a legal status intermediate between confinement to an institution and total freedom in the community as a way to assist these people. The plan would create a form of guardianship, including outpatient commitment, cash management, and legal representation, to get them into treatment but avoid longer term commitment against their will.

Most people with mental illnesses can qualify for SSI and/or housing vouchers, but the disparity between SSI income and vouchers and the cost of housing is so great that many people are virtually shut out of the housing market (Lezak & Edgar, 1998). Some people have suggested creating a shelter allowance, or what has been called the Aid to Families with Dependent Adults policy, whereby people would be paid if they gave housing to someone who is homeless and/or disabled (O'Flaherty, 1996; Rossi, 1989; White, 1992; Wright & Rubin, 1997).

CONCLUSION

"Homelessness is the sum total of our dreams, policies, intentions, errors, omissions, cruelties, kindnesses, all of it recorded, in flesh, in the life on the streets" (Marin, 1987, p. 10). Throughout U.S. history, Americans have discussed and developed a wide variety of policies and programs to try to deal with poverty and homelessness. As a society, it appears that we often do not want to make significant sacrifices, either individually or as a society, to address the problems in meaningful ways. We will begin to deal effectively with homelessness to the degree that we can significantly change attitudes toward the poor and homeless and develop policies that focus on the causes and consequences of their problems. We need to "close the front door" by better preventing the causes of poverty and homelessness, and we need to "open the back door" by concentrating on the needs of people who become poor and homeless. We need to examine the two core components of homelessness: lack of adequate income and lack of affordable housing. We need to improve execution and funding of existing programs and develop new policies and programs that address underserved people. But a focus exclusively on housing or poverty will not meaningfully address all of the problems.

What can readers do to help? While the causes of homelessness are complicated and the people who are homeless have complex needs, there are ways people can make a difference for some of the men, women, and children who are homeless. The National Coalition for the Homeless (2000) lists a number of suggestions including volunteering time at shelters and other service agencies; providing clothing, food, and other goods that the homeless need; giving money to support the work of those who help the homeless; becoming an advocate, working with homeless people to bring about positive changes in policies and programs on the local, state, and federal levels; and educating self, family, and community about the problems of the homeless.

REFERENCES

American Civil Liberties Union. (1996). Homepage [On-line]. Available: aclu.org

Applebaum, R. (1990). Counting the homeless. In J. Momeni (Ed.), *Homelessness in the United States: Data and issues* (pp. 1-16). Westport, CT: Greenwood.

Bahr, H. (1973). *Skid row: An introduction to disaffiliation.* New York: Oxford University Press.

Bahr, H., & Caplow, T. (1973). *Old men drunk and sober.* New York: New York University Press.

Bane, M. J., & Ellwood, D. (1994). *Welfare realities: From rhetoric to reform.* Cambridge, MA: Harvard University Press.

Barak, G. (1992). *A social history of homelessness in America.* New York: Praeger.

Beatty, C., & Haggard, L. (1998). *Legal remedies to address discrimination against people who are homeless and have mental illness.* Rockville, MD: U.S. Department of Health and Human Services.

Belcher, J., & DiBlassio, F. (1990). *Helping the homeless: Where do we go from here?* Lexington, MA: Lexington Books.

Blasi, G. (1994). And we are not seen: Ideological and political barriers to understanding homelessness. *American Behavioral Scientist, 37,* 563-586.

Blau, J. (1992). *The visible poor: Homelessness in the United States.* New York: Oxford University Press.

Blumberg, L., Shipley, T., & Shandler, I. (1973). *Skid-row and its alternatives.* Philadelphia: Temple University Press.

Bogue, D. (1963). *Skid row in American cities.* Chicago: University of Chicago Community and Family Study Center.

Bok, D. (1997). Maximizing the performance of government. In J. Nye, P. Zelikow, & D. King (Eds.), *Why people don't trust government* (pp. 43-55). Cambridge, MA: Harvard University Press.

Brauner, S., & Loprest, P. (1999). *Where are they now?* Washington, DC: The Urban Institute.

Breakey, W., & Fischer, P. (1990). Homelessness: The extent of the problem. *Journal of Social Issues, 46,* 31-47.

Burt, M. (1996). Homelessness: Definitions and counts. In J. Baumohl (Ed.), *Homelessness in America* (pp. 15-23). Phoenix, AZ: Oryx Press.

California Department of Housing and Community Development. (2000). *Raising the roof.* Sacramento: Author.

Children's Defense Fund. (2000). *State of California's children: Children in the states* [On-line]. Available: childrensdefense.org/states/data_ca.html

Creeland, N. (2000, October 22). State expected to OK $1 hike in minimum wage. *Los Angeles Times,* p. A1.

Culhane, D. (1997). Introduction. In D. Culhane & S. Hornburg (Eds.), *Understanding homelessness: New policy and research perspectives* (pp. 5-8). Washington, DC: Fannie May Foundation.

Dreier, P., & Atlas, J. (1989). Grassroots strategies for the housing crisis: A national agenda. *Social Policy, 6,* 25-38.

Edin, K., Harris, K. M., & Sandefur, G. (1998). *Welfare to work: Opportunities and pitfalls.* Washington, DC: American Sociological Association.

Edin, K., & Lein, L. (1997). *Making ends meet: How single mothers survive welfare and low-wage work.* New York: Russell Sage Foundation.

Encarta. (2000). Welfare. In *Encarta Online Encyclopedia 2000.* Available: encarta.msn.com.

Fagan, R., & Mauss, A. (1978). Padding the revolving door: An initial assessment of Uniform Alcoholism and Intoxication Treatment Act in practice. *Social Problems, 46,* 232-237.

Fagan, R., & Mauss, A. (1986). Social margins and social re-entry: An evaluation of a rehabilitation program for skid-row alcoholics. *Journal for the Study of Alcohol, 47,* 413-425.

Federal Task Force on Homelessness and Severe Mental Illness. (1992). *Outcasts on Main Street.* Washington, DC: Interagency Council on Homeless.

Fischer, P. (1992). The criminalization of homelessness. In M. Robertson & M. Greenblatt (Eds.), *Homelessness: A national perspective* (pp. 57-66). New York: Plenum.

Fischer, P., & Breakey, W. (1991). The epidemiology of alcohol, drug and mental disorders among the homeless persons. *American Psychologist, 46,* 1115-1128.

Fosburg, L., & Dennis, D. (1999). *Practical lessons: The 1998 National Symposium on Homelessness Research.* Washington, DC: U.S. Department of Housing and Urban Development and U.S. Department of Health and Human Services.

Garrett, G. (1992). *Responding to the homeless: Policy and practice.* New York: Plenum.

Gibbs, N. (1995, July 3). Getting nowhere. *Time,* pp. 17-20.

Goldfinger, S., Susser, E., Roche, B., & Berkman, A. (1998). *HIV, homelessness, and serious mental illness: Implications for policy and practice.* Delmar, NY: National Resource Center on Homelessness and Mental Illness Policy Research Associates.

Handler, J. (1992). The modern pauper: The homeless in welfare history. In M. Robertson & M. Greenblatt (Eds.), *Homelessness: A national perspective* (pp. 35-46). New York: Plenum.

Hoch, C., & Slayton, R. (1989). *New homeless and old: Community and the skid row hotel.* Philadelphia: Temple University Press.

Hombs, C. (1994). *American homelessness* (2nd ed.). Santa Barbara, CA: ABC-CLIO.

Hombs, M. E., & Snyder, M. (1982). *Homelessness in America: A forced march to nowhere.* Washington, DC: Community for Creative Non-Violence.

Hopper, K. (1997). Homelessness old and new: The matter of definition. In D. Culhane & S. Hornburg (Eds.), *Understanding homelessness: New policy and research perspectives* (pp. 9-67). Washington, DC: Fannie May Foundation.

Hopper, K., & Baumohl, J. (1996). Redefining the cursed word: A historical interpretation of American homelessness. In J. Baumohl (Ed.), *Homelessness in America* (pp. 3-14). Phoenix, AZ: Oryx Press.

Housing California. (2000). *The long wait: The critical shortage of housing in California*. Sacramento: Author.

Isaac, R., & Armat, V. (1990). *Madness in the streets*. New York: Free Press.

Jahiel, R. (1992). *Homelessness: A prevention-oriented approach*. Baltimore: Johns Hopkins University Press.

Jencks, C. (1994). *The homeless*. Cambridge, MA: Harvard University Press.

Koegel, P., Burnam, A., & Baumohl, J. (1996). The causes of homelessness. In J. Baumohl (Ed.), *Homelessness in America* (pp. 24-33). Phoenix, AZ: Oryx Press.

Lezak, A., & Edgar, E. (1998). *Preventing homelessness among people with serious mental illnesses*. Delmar, NY: The National Resource Center on Homelessness and Mental Illness.

Liebow, E. (1995). *Tell them who I am: The lives of homeless women*. New York: Penguin.

Lindblom, E. (1997). Toward a comprehensive homelessness prevention strategy. In D. Culhane & S. Hornburg (Eds.), *Understanding homelessness: New policy and research perspectives* (pp. 265-334). Washington, DC: Fannie May Foundation.

Link, B., Phelan, J., Stueve, A., Moore, R., Bresnahan, M., & Struening, E. (1996). Public attitudes and beliefs about homeless people. In J. Baumohl (Ed.), *Homelessness in America* (pp. 143-148). Phoenix, AZ: Oryx Press.

Marin, P. (1987). Helping and hating the homeless: The struggle at the margins of America. *Harpers, 274*, 26-35.

Mishel, L., Bernstein, J., & Schmitt, J. (1998-1999). *The state of working America 1998-1999*. Washington DC: Economic Policy Institute.

Morse, G. (1992). Causes of homelessness. In M. Robertson & M. Greenblatt (Eds.), *Homelessness: A national perspective* (pp. 3-18). New York: Plenum.

National Alliance to End Homelessness. (2000). Homepage [On-line]. Available: nch.ari.net

National Coalition for the Homeless. (2000). Homepage [On-line]. Available: www.nch.ari.net

National Governor's Association. (1998). *Tracking recipients after they leave welfare*. Washington, DC: Author.

National Health Care for the Homeless Council. (1997). SSI/SSDI study. In *Healing hands* (Vol. 1, pp. 15-23). Nashville, TN: Author.

National Law Center on Homelessness and Poverty. (2000). *Myths and facts about homelessness* [On-line]. Available: nlchp.org/myths.html

National Low Income Housing Coalition. (1999). *Out of reach*. Washington, DC: Author.

National Resource Center on Homelessness and Mental Illness. (2000). *National organizations concerned with health, housing, and homelessness*. New York: Author.

O'Flaherty, B. (1996). *Making room: The economics of homelessness*. Cambridge, MA: Harvard University Press.

Rossi, P. (1989). *Down and out in America: The origins of homelessness*. Chicago: University of Chicago Press.

Rossi, P. (1994). Troubling families: Family homelessness in America. *American Behavioral Scientist, 37*, 342-395.

Shelter Partnership. (1994). *The number of homeless people in Los Angeles city and county July 1992 to June 1993*. Los Angeles, CA: Author.

Sherman, A., Amey, C., Duffield, B., Ebb, N., & Weinstein, D. (1998). *Welfare to what*. Washington, DC: Children's Defense Fund and National Coalition for the Homeless.

Shinn, M., & Gillespie, C. (1994). The roles of housing and poverty in the origins of homelessness. *American Behavioral Scientist, 37*, 505-521.

Shlay, A., & Rossi, P. (1992). Social science research and contemporary studies of homelessness. *Annual Review of Sociology, 18,* 129-160.

Simon, H. (1996). Municipal regulation of the homeless in public spaces. In J. Bauhmol (Ed.), *Homelessness in America* (pp. 149-159). Phoenix, AZ: Oryx Press.

Snow, D., Anderson, L., & Koegel, P. (1994). Distorting tendencies in research on the homeless. *American Behavioral Scientist, 37,* 461-475.

State of California. (2000). *State of the state: California 2000* [On-line]. Available: governor.ca/gov/ca2000/cafunfacts.shtm

Stoner, M. (1995). *The civil rights of homeless people.* New York: Aldine De Gruyter.

Texeira, E. (2000, October 20). Study finds widening gap between rich, poor. *Los Angeles Times,* p. B3.

Timmer, D., & Eitzen, S. (1994). *The homeless in America.* Boulder, CO: Westview.

U.S. Bureau of the Census. (2000). *Poverty in the United States 1999.* Washington, DC: U.S. Department of Commerce.

U.S. Conference of Mayors. (1999). *A status report on hunger and homelessness in America's cities.* Washington, DC: Author.

Wallace, S. (1965). *Skid-row as a way of life.* New Jersey: Bedminister.

White, R., Jr. (1992). *Rude awakenings: What the homeless crisis tells us.* San Francisco: ICS Press.

Wiseman, J. (1970). *Stations of the lost: The treatment of skid row alcoholics.* Englewood Cliffs, NJ: Prentice Hall.

Wolch, J., & Dear, M. (1993). *Malign neglect.* San Francisco: Jossey-Bass.

Wong, Y.-L. (1997). Patterns of homelessness: A review of longitudinal studies. In D. Culhane & S. Hornburg (Eds.), *Understanding homelessness: New policy and research perspectives* (pp. 135-164). Washington, DC: Fannie May Foundation.

Wright, J., & Devine, J. (1991). No fixed address: The nature of homelessness in contemporary America. *National Social Science Journal, 3,* 1-18.

Wright, J., & Rubin, B. (1997). Is homelessness a housing problem? In D. Culhane & S. Hornburg (Eds.), *Understanding homelessness: New policy and research perspectives* (pp. 205-224). Washington, DC: Fannie May Foundation.

CHAPTER 2

JUVENILE DELINQUENCY AND URBAN GANGS

Anne Hendershott

Citizens in California have joined others throughout the nation in demanding harsher treatment for juveniles. Back in 1994, the cover of *Time* magazine demanded that the nation "Lock 'Em Up and Throw Away the Key" (Lacayo, 1994), and since that time, citizen demands for attention to juvenile crime have been taken seriously by California lawmakers, who have adopted longer and harsher sentencing codes. On March 7, 2000, California voters overwhelmingly supported an initiative (Proposition 21) that would make it easier for the state's prosecutors to pass juvenile offenders from juvenile court to adult court for processing and sentencing. The law will also enhance the already severe penalties for documented juvenile gang members involved in criminal activity. The result of the legislation will be a significant increase in the number of juvenile offenders being tried in California's adult courts and incarcerated in local or state correctional facilities. With Proposition 21, California citizens voted in favor of yet another opportunity to strengthen what many juvenile justice researchers have considered the "toughest juvenile sentencing policies in the nation"(Demuro, Demuro, & Lerner, 1988).

This new legislation is aimed at what most Californians believe is a significant social problem: the growing youth crime threat. Strengthening the punitive power of the state to deal with juveniles might be called for during a time of increasing juvenile crime, but the reality is that since 1988 there have been significant declines in criminal offenses committed by juveniles. Although youth gangs can and do pose a significant social problem in some urban neighborhoods throughout California, national and state juvenile crime arrests are now at their lowest level in more than a decade (Snyder & Sickmund, 1999). Yet, California voters may have found Proposition 21 appealing because they have been led to believe that juvenile crime has continued to increase in the state. The media campaign that promoted the new legislation used rhetoric like the "growing gang menace." It played on citizens' fears regarding what the national and state media have been warning of for more than a decade now: the juvenile "super-predator." And the predictions have been ominous. A cover story in *U.S. News & World Report* (Geist & Pope, 1996, p. 29) drew from demographics and crime trends to predict that because of a 23% surge in the teen population in the "violence-prone" age group, the resulting demographic catastrophe would boost the juvenile murder total by 25% by 2005.

By 1999, social scientists knew that none of the predictions about the super-predators were coming true. In fact, the most recent *Juvenile Offenders and Victims: 1999 National Report* asserted that a trend analysis of juvenile homicide arrests leads to the conclusion that "juvenile super-predators are more myth than reality" (Snyder & Sickmund, 1999). The authors caution that "history shows that it is a fool's errand to try and predict future crime trends" by simply using demographic data (Snyder & Sickmund, 1999, p. 134). Still, in the media campaign that accompanied California's Proposition 21, facts were lost in hyperbole. A closer look at the data reveals that juvenile crime, like adult crime, has been declining steadily on both a national and a state level for several years. This is especially true in cities like San Diego, where juvenile felony charges declined by 43% from 4,791 in 1997 to 2,734 in 1999 (cited in Milliken, 2000, p. G-3).

Similar declines in juvenile felony arrests can be seen throughout California. As Table 2.1 demonstrates, California juvenile arrest rates declined 19.7% from 1988 to 1998. Tables 2.2 and 2.3 point to major declines in property offenses and felony drug offenses since 1988. There is also a sharp decline in California juvenile homicide arrests and in juvenile rape arrests. Overall, the arrest rates for the state's juveniles have been declining steadily in nearly every category.

The data on declining crime rates are encouraging, but there remains a disjuncture between that reality and California voters' perceptions of crime. As a result, Californians most likely responded with fear to the graphic media representation of gang-related juvenile crime that continued to be broadcast on the local news. The outcome of this reality-perception disconnect has been legislation that will cost California taxpayers hundreds of millions of dollars while diverting badly needed funds from prevention programs. This is especially unfortunate because many of these juvenile diversion and prevention programs have actually been shown to reduce juvenile crime. In an attempt to dissuade voters from supporting Proposition 21, Judge James Milliken (2000), the presiding judge of the San Diego County Juvenile Court, wrote an editorial for the *San Diego Union-Tribune,* warning that "this legislation could inadvertently increase the number of juveniles who become career criminals" (p. G-3).

TABLE 2.1 California Juvenile Felony Arrests, 1988 and 1998

	1988		1998		
Age Group	Number of Arrests	Rate per 100,000	Number of Arrests	Rate per 100,000	Percentage Change in Rate, 1988-1998
Total felony arrests	550,446	2505.4	508,257	2011.9	−19.7
Juveniles	80,758	2618.1	76,104	2021.4	−22.8
Violent offenses	13,886	450.2	19,646	521.8	15.9
Homicide	389	12.6	308	8.2	−34.9
Forcible rape	543	17.6	412	10.9	−38.1
Robbery	4,850	157.2	6,821	181.2	15.3
Aggravated assault	8,104	262.7	12,105	321.5	22.4
Property offenses	49,061	1590.5	38,308	1017.5	−36.0
Drug offenses	11,646	377.6	7,392	196.3	−48.0
Weapons offenses	2,704	87.7	5,887	156.4	78.3
Other offenses	3,461	112.2	4,871	129.4	15.3

SOURCE: State of California, Office of the Attorney General, 2000.

Despite Judge Milliken's warnings, Proposition 21 passed with more than 65% of California's voters supporting it. This latest law is just the most recent step California has made in reverting to the past by treating juveniles as adults. In many ways, the new legislation is part of the movement across the state and the nation to more punitive responses to juvenile delinquency. As Table 2.4 points out, from 1992 through 1997, legislatures in 47 states and the District of Columbia enacted laws that made their juvenile justice systems more punitive (Office of Juvenile Justice and Delinquency Prevention [OJJDP], 1999).

Areas of change have emerged as states pass laws designed to enhance penalties for juvenile crime. These laws, like the latest ones in California, generally involve expanded eligibility for criminal court processing and adult correctional sanctioning, and reduced confidentiality protections for juvenile offenders (OJJDP, 1999, p. 89). Transfer provisions in 45 states make it easier (as California's Proposition 21 does) to transfer juvenile offenders from the juvenile justice system to the criminal justice system. State sentencing authority was changed through laws to expand juvenile and criminal courts, and sentencing options were changed in 31 states. Finally, laws on confidentiality were modified or traditional juvenile protections were removed, making records and proceedings more open in 47 states.

Overall, the 1980s and 1990s have seen significant changes in terms of treating more juvenile offenders like criminals (Snyder & Sickmund, 1999, p. 89). These changes reflect a movement back to a time when juvenile delinquents were viewed and treated exactly the same as adult criminals. In fact, prior to 1899, juveniles and adults were handled in the same courts and were processed under the same rules. As

(text continues on page 41)

TABLE 2.2 California Juvenile Felony Arrests, 1988-1998, by Crime Category

	Violent Offenses			Property Offenses			Drug Offenses			Other Offenses		
Year	Number of Arrests	Rate per 100,000	Percentage Change From Prior Year	Number of Arrests	Rate per 100,000	Percentage Change From Prior Year	Number of Arrests	Rate per 100,000	Percentage Change From Prior Year	Number of Arrests	Rate per 100,000	Percentage Change From Prior Year
1998	19,646	521.8	-5.0	38,308	1017.5	-8.0	7,392	196.3	-11.5	10,758	285.7	-0.4
1997	21,002	549.0	-6.8	42,287	1105.4	-8.3	8,484	221.8	4.4	10,975	286.9	-1.1
1996	21,962	589.0	-4.6	44,946	1205.5	-5.5	7,921	212.4	-12.7	10811	290.0	-1.5
1995	22,334	617.5	-3.4	46,135	1275.5	-8.1	8,797	243.2	-8.9	10650	294.4	-9.9
1994	22,429	639.0	1.8	48,720	1388.1	-7.3	9,375	267.1	15.9	11475	326.9	-4.3
1993	21,402	627.5	-2.7	51,058	1496.9	-7.7	7,861	230.5	0.0	11652	341.6	5.7
1992	21,367	644.7	-1.0	53,768	1622.2	-4.7	7,636	230.4	0.6	10713	323.2	1.3
1991	21,016	651.0	0.3	54,952	1702.3	-0.2	7,396	229.1	-11.5	10301	319.1	11.7
1990	20,453	649.0	15.1	53,762	1706.0	-1.3	8,158	258.9	-27.9	9000	285.6	16.2
1989	17,325	563.8	25.2	53,116	1728.7	8.7	11,037	359.2	-4.9	7584	245.7	22.9
1988	13,886	450.2		49,061	1590.5		11,646	377.6		6165	199.9	

SOURCE: State of California, Office of the Attorney General, 2000.

TABLE 2.3   California Juvenile Violent Felony Arrests, 1988-1998, for Specific Violent Crimes

	Homicide			Forcible Rape			Robbery			Aggravated Assault		
Year	Number of Arrests	Rate per 100,000	Percentage Change From Prior Year	Number of Arrests	Rate per 100,000	Percentage Change From Prior Year	Number of Arrests	Rate per 100,000	Percentage Change From Prior Year	Number of Arrests	Rate per 100,000	Percentage Change From Prior Year
1998	308	8.2	-10.9	412	10.9	-6.0	6821	181.2	-13.2	12105	321.5	0.7
1997	353	9.2	-11.5	445	11.6	-10.8	7984	208.7	-10.9	12220	319.4	-3.7
1996	389	10.4	-27.8	483	13.0	10.2	8730	234.1	-7.8	12360	331.5	-1.9
1995	521	14.4	-6.5	427	11.8	-9.9	9186	254.0	-0.4	12220	337.9	-5.0
1994	542	15.4	-14.9	459	16.1	-16.0	8947	254.9	5.5	12481	355.6	1.0
1993	618	18.1	-7.2	532	15.6	-8.8	8243	241.7	-1.7	12009	352.1	-2.8
1992	645	19.5	-9.7	566	17.1	-17.0	8151	245.9	-0.3	12005	362.2	0.0
1991	696	21.6	3.3	665	29.6	3.0	7960	246.6	-0.2	11695	362.3	0.3
1990	658	20.9	20.8	630	20.0	1.5	7786	247.1	23.1	11379	361.1	10.8
1989	533	17.3	37.3	606	19.7	11.9	6168	200.7	27.7	10018	326.0	24.1
1988	398	12.6		543	17.6		4850	157.2		8104	262.7	

SOURCE: State of California, Office of the Attorney General, 2000.

TABLE 2.4 Laws Enacted From 1992 Through 1997 to Make Juvenile Justice Systems More Punitive

State	Changes in Law or Court Rule		
Alabama	Transfer provisions		Confidentiality
Alaska	Transfer provisions		Confidentiality
Arizona	Transfer provisions	Sentencing authority	Confidentiality
Arkansas	Transfer provisions	Sentencing authority	Confidentiality
California	Transfer provisions		Confidentiality
Colorado	Transfer provisions	Sentencing authority	Confidentiality
Connecticut	Transfer provisions	Sentencing authority	Confidentiality
Delaware	Transfer provisions	Sentencing authority	Confidentiality
District of Columbia	Transfer provisions	Sentencing authority	
Florida	Transfer provisions	Sentencing authority	Confidentiality
Georgia	Transfer provisions	Sentencing authority	Confidentiality
Hawaii	Transfer provisions		Confidentiality
Idaho	Transfer provisions	Sentencing authority	Confidentiality
Illinois	Transfer provisions	Sentencing authority	Confidentiality
Indiana	Transfer provisions	Sentencing authority	Confidentiality
Iowa	Transfer provisions	Sentencing authority	Confidentiality
Kansas	Transfer provisions	Sentencing authority	Confidentiality
Kentucky	Transfer provisions	Sentencing authority	Confidentiality
Louisiana	Transfer provisions	Sentencing authority	Confidentiality
Maine			Confidentiality
Maryland	Transfer provisions		Confidentiality
Massachusetts	Transfer provisions	Sentencing authority	Confidentiality
Michigan		Sentencing authority	Confidentiality
Minnesota	Transfer provisions	Sentencing authority	Confidentiality
Mississippi	Transfer provisions		Confidentiality
Missouri	Transfer provisions	Sentencing authority	Confidentiality
Montana	Transfer provisions	Sentencing authority	Confidentiality
Nebraska			
Nevada	Transfer provisions		Confidentiality
New Hampshire	Transfer provisions	Sentencing authority	Confidentiality
New Jersey		Sentencing authority	Confidentiality
New Mexico	Transfer provisions	Sentencing authority	Confidentiality

State	Changes in Law or Court Rule		
New York			
North Carolina	Transfer provisions		Confidentiality
North Dakota	Transfer provisions		Confidentiality
Ohio	Transfer provisions	Sentencing authority	Confidentiality
Oklahoma	Transfer provisions	Sentencing authority	Confidentiality
Oregon	Transfer provisions	Sentencing authority	Confidentiality
Pennsylvania	Transfer provisions		Confidentiality
Rhode Island	Transfer provisions	Sentencing authority	Confidentiality
South Carolina	Transfer provisions		Confidentiality
South Dakota	Transfer provisions		
Tennessee	Transfer provisions	Sentencing authority	Confidentiality
Texas	Transfer provisions	Sentencing authority	Confidentiality
Utah	Transfer provisions		Confidentiality
Vermont			
Virginia	Transfer provisions	Sentencing authority	Confidentiality
Washington	Transfer provisions		Confidentiality
West Virginia	Transfer provisions		Confidentiality
Wisconsin	Transfer provisions	Sentencing authority	Confidentiality
Wyoming	Transfer provisions		Confidentiality

SOURCE: Snyder & Sickmund, 1999.

a result, children throughout the nation were often subject to harsh punishments, including the death penalty, and were confined in jails and prisons with adult offenders (Schwartz, 1993, p. 50).

In 1899, the first juvenile court in the country was established by the Chicago Bar Association in an attempt to treat juveniles not as criminals but as wards of the state. The intent in creating the court was to enable children to "receive the care, custody, and discipline that are accorded the neglected and dependent" (cited in Albanese, 1993, p. 68). By 1920, every state in the country, including California, had established a juvenile court based on the Chicago principles of rehabilitation. Still, several juvenile justice historians cite data revealing that "a wide gap remains between the theory and practice of progressive ideas" (Albanese, 1993, p. 68). The gap between the philosophy of rehabilitation and the implementation of the model resulted in great variation among states, and this was especially true throughout California's history of juvenile justice.

CALIFORNIA'S HISTORICAL ROOTS OF JUVENILE CRIME AND JUSTICE

In many respects, the early days of California juvenile justice mirrored the experience in other states. What is unique in California's history is the creation of the **California Youth Authority** or CYA (Krisberg, 1993). The CYA model was developed to provide an efficient, rational, and effective way to administer justice in California. The first stated purpose of the CYA model was "to protect society by substituting treatment and training for retributive punishment of young persons found guilty of public offenses" (Lemert & Rosenberg, 1948, p. 49). What distinguishes the CYA from agencies in other states is that juvenile offenders can remain within the CYA until their 25th birthday. Even those juveniles who are tried and convicted as adult offenders may be housed in CYA detention facilities to protect them from the adult population in the state's correctional facilities.

Still, sociologist Edwin Lemert attributed the emergence of the CYA to the growth of an administrative state in the United States (cited in Krisberg, 1993, p. 47) because the centralization of youth correction agencies enabled them to claim the scarce state delinquency prevention funds. Many criminologists had been critical of the psychologically oriented treatment approaches that were introduced into these institutions (Krisberg, 1993, p. 47). And, by the middle of the 20th century, social scientists began to question the concept of enforced therapy, arguing that treatment-oriented prisons might be even more oppressive than more traditional institutions (Mathieson, 1965). Others viewed this as the period when brainwashing techniques were first used on juvenile offenders (Takagi, cited by Krisberg, 1993).

In 1957, public concerns about juvenile treatment were serious enough to move the governor of California to appoint a commission to investigate the operation of the juvenile justice system in the state. The California commission found many serious deficiencies, including an absence of well-defined standards and norms to guide juvenile justice and an excessive detention of children (Brantingham, 1979, p. 263). In response to the negative findings of the California commission, California's juvenile justice system underwent major revision in 1961, resulting in mandated legal counsel for children involved in felony cases and a pretrial diversion process to keep less serious juvenile cases from formal adjudication. In these measures, California again paralleled the national movement in its emphasis on due process for juvenile delinquents.

On a national level, this change in emphasis was termed a **due process revolution** because it ushered in a trend away from a focus on informal treatment of juveniles toward making juvenile justice more formalized or "adult like" (Albanese, 1993, p. 74). Beginning in the 1960s, the U.S. Supreme Court decided a series of landmark cases that dramatically changed the character and procedures of the juvenile justice system. The most important of these was the 1967 case, *In re Gault*. In this decision, the Court ruled that juveniles had the right to notice and counsel, to question witnesses, and to protection against self-incrimination. Prior to the Gault ruling, it was assumed that because the purpose of the Juvenile Court was to treat and not to punish, juveniles did not need protections from treatment. In the Gault case, it became obvious that treatment may indeed be punishment. Gerald Gault, 15, was charged with making prank telephone calls to a neighbor. During the processing of his case, Gault was not given access to an attorney, was not allowed to question witnesses, and was given no protection against self-incrimination. Because of the charges, the judge committed Gault to a training school until he was 18 years old. The maximum sentence for an adult would have been a $50 fine or 2 months in jail. In the Gault case, the Supreme Court ruled that Gault was being punished rather than helped by the juvenile court and that "unbridled discretion, however benevolently motivated, is frequently a poor substitute for principle and procedure" (*In re Gault*, 1967).

In a related case, the Supreme Court asserted that juveniles were actually receiving the "worst of both worlds—neither the protection accorded to adults, nor the solicitous care and regenerative treatment postulated for children" (*Kent v. United States*, 1966). These cases had a significant impact on all states, including California, in moving the juvenile justice system from an informal process to one in which youth are afforded the same due process protections as adults.

Regardless of the process, the reality is that despite steadily declining crime rates, the current fear of gangs and juvenile violence in California guarantees that juveniles will continue to face growing penalties for youth crime and violence. The national juvenile justice system has undergone a massive change in the 1990s, and California appears to be leading the nation in these changes. In an interview for *Time* magazine, Los Angeles District Attorney Gil Garcetti described what he said were needed changes in California's philosophy and treatment for juveniles. Of these changes, Garcetti said, "We need to throw out our entire juvenile justice system and replace it with one that protects society from violent juvenile criminals and effectively rehabilitates youth who can be saved—and can differentiate between the two" (Lacayo, 1994, p. 61).

Although this comment may not be representative of the ways in which law enforcement officials across California view juveniles, it is notable as an example of a **triage model of juvenile justice** in the Los Angeles District Attorney's Office. The triage model mandates the diversion of scarce resources from those who are believed to be "beyond help" and to those whom Garcetti describes as salvageable. Still, the question remains, how are Los Angeles County prosecutors and judges able to differentiate between those juveniles who can be saved and those who cannot? In Los Angeles and other California cities, gang affiliation appears to be gaining greater importance in determining arrest, processing, treatment, and sentencing.

CALIFORNIA'S URBAN YOUTH GANGS: SOCIAL PROBLEM OR MORAL PANIC?

The current California fear of youth crime and urban gangs is given more meaning when viewed historically. Latino and African American gangs have significantly influenced the ways in which law enforcement and the media have viewed California's juvenile criminal activity since the earliest days of gang identification.

Latino gangs. In her study of the evolution of California gangs over several decades, sociologist Joan Moore views the current **moral panic** surrounding juvenile crime as mirroring similar citizen fears in the 1940s. But, even before then, the stage was set in California for Latino gangs during the economic boom of the 1920s, when thousands of Mexican immigrants were drawn to the state for work. These young Mexicans were primarily rural and brought with them a tradition known as *polomilla* (in Spanish, this means a covey of doves); that is that a number of young men in a village would group together in a coming-of-age cohort (Vigil, 1988, p. 118). In the Los Angeles area during the 1920s and 1930s, these young men began to identify with a particular neighborhood or parish. These groups were the earliest forerunners of the modern Chicano gangs of East Los Angeles (Sheldon, Tracy, & Brown, 2000). But Moore views these early groups as falling within the Mexican tradition of male barrio groups. Indeed, in the earliest years of California Latino gangs, neighbors in the White Fence and El Hoyo Maravilla areas of Los Angeles defined the male groups not as gangs but simply as the "boys from the barrio" (Moore, 1991).

The history of the White Fence group is typical of Latino gangs during these early days. The group grew out of young men's sports groups associated with La Purisima Church, which was the focus of community life during the 1930s. These boys and young men were fully integrated into community activities: "They were the pride of the neighborhood ... there were no drugs at that time ... and they were very family oriented" (Moore, 1991, p. 26). Things changed for the White Fence group later as more established Latino gangs, including El Hoyo Maravilla, began to fight over territory or turf issues (Moore, 1991, p. 27). Although most law enforcement officials would argue that today's Latino gangs bear little resemblance to these church-affiliated groups, most sociologists have found that the strong sense of neighborhood continues in today's gangs, although the willingness to use deadly violence to defend the neighborhood has escalated (Sanders, 1994).

Movement toward today's more territorial Latino gangs escalated during the 1930s. Indeed, during the Depression, when the economy imploded and work became scarce, thousands of Mexican immigrants were repatriated and deported.

Racist polices and widespread discrimination against Mexican Americans set in (Sheldon et al., 2000, p. 7). This culminated in the **Zoot Suit riots** of 1943. Moore (1991) believes that the first moral panic surrounding California's Chicano gangs was a response to the famous 1943 Zoot Suit riots, in which more than 200 uniformed servicemen chartered 20 cabs and charged into the heart of the Mexican American community in East Los Angeles. The servicemen were responding to what they had identified as gang violence against military personnel. Those wearing zoot suits (a distinctive style of men's suit consisting of an oversize, long suit coat with wide lapels and trousers with wide legs) were targeted during the violence and were beaten by the mob. Nine sailors were arrested, but none were charged with any crime. In fact, the servicemen were portrayed in the local press as heroes stemming the tide of the "Mexican crime wave." More than 600 Chicano youth were arrested during the riots. Police called their arrests "preventive action." Police then began to engage in a series of gang sweeps within gang areas. Joan Moore (1991) says these events brought youths closer together and transformed informal youth groups or boy gangs into delinquent gangs. In regard to the Zoot Suit riots, those Chicanos who fought the marauding sailors in East Los Angeles were seen by their younger brothers as "heroes of a race war" (Moore, 1978, p. 62). An important outcome of the Zoot Suit riots was to strengthen the bonds within the Mexican American community and, in many ways, to create even stronger gangs, thus increasing the panicked response to Mexican American youths.

Moore (1991) describes these moral panics as "beginning with reports from law enforcement personnel and reinforced by newspapers and other media" (p. 2). The response to such panics in the 1940s, and the response today by lawmakers and the juvenile justice system, has been to increase the surety and severity of punishment for juveniles. Indeed, the history of California juvenile crime and justice reveals patterns and cyclical changes in attitudes toward youth and how they are to be treated. Attitudes about youth gangs and gang activity have changed markedly since the early days when Los Angeles gangs grew from a traditional barrio base. A similar evolution occurred for California's African American gangs.

African American gangs It may be said that what the Zoot Suit riots did to galvanize Chicano gangs, the Watts riots of 1965 did for African American gangs. Although African American youth did not have the polomilla tradition of their Mexican counterparts,

> Given the emasculating circumstances of ghetto life a quarter century ago, it is small wonder that the cocky, dangerous style of the Latino gangs had a strong appeal for African American youths. It responded perfectly to the need for repackaging defeat as defiance, redefining exclusion as exclusivity. (cited in Sheldon et al., 2000, p. 8)

In many ways, African American youth began to adopt the style and defiance of the Chicano *cholo* (gang) youth. And, like the Latino gangs, African American gangs have a long history, especially in Los Angeles.

According to Alonso (1997), African American gangs began to appear in Los Angeles in the 1920s in the downtown area and later progressed to the area between Slauson Avenue and Firestone. Some active African American gangs were the Goodlows, the Kelleys, the Magnificents, the Driver Brothers, the Boozies, and the Blodgettes. By the early 1950s, clubs in the African American community gained in popularity. Alonso (l997) believes that some were actually early attempts at political organizations. Most were territorial, and some were involved in petty theft, robbery, and assaults, but killing was extremely rare. Despite this, the Los Angeles Police Department began to identify these African American youth groups as gangs, even though some of the clubs were automobile related, or "car clubs," and were not involved in disputes with other clubs.

Following the Watts riots, organized political groups surfaced. Many young African American males became members of the U S Organization and the Black Panther Party. For example, Bunchy Carter, a member of the Renegade Slauson Gang from the late 1950s to 1965, later became the leader of the Los Angeles chapter of the Black Panther Party. After the Watts riots, membership in smaller street gangs also increased for younger youth. Raymond Washington, 15, and classmate Tookie Williams began a gang called the Baby Avenues or Avenue Cribs. This new gang was actually a younger clique or set of a larger and highly respected street gang called the Avenue Boys. This newer, younger gang later became known as the Crips. By the mid 1970s, Crips sets were organized throughout the area around Horace Mann Junior High School, Crenshaw High School, Freemont High School and throughout Compton. At the same time, Piru Street youth were organizing with the Crips. However, in 1972, conflict developed between the Compton Crips and the Pirus from Compton. This ended friendly relations, and the Pirus then created a rival organization, the Bloods. Included in this organization were those groups who had been threatened or attacked by the Crips.

Despite a truce in 1992, the Bloods and Crips continue their sometimes violent conflict. Today, the Crips gang dominates in terms of the numbers of members. Yet, San Diego is known as a "Blood Town" because of the large numbers of Blood sets within the city, especially in the Skyline area where the East Side Pirus remain powerful. Oceanside, which is geographically closer to Los Angeles, has many more Crips members.

According to law enforcement sources, gang membership and violent activity have escalated greatly in California. The National Law Enforcement Institute (1999) maintains that there are more than 300,000 gang members in more than 2,000 gangs in California. The institute also reports 150,000 gang members in 900 gangs in Los Angeles alone, more gang members than in any other city in the nation. The institute also notes 10,000 gang members in San Diego and 1,000 gang members in

Imperial County. Although the data are useful, statistical confirmation of significant gang membership increases is difficult because each California city has a different way of defining and documenting gang members. What is known, however, is that gangs continue to flourish in California as urban youth continue to be drawn to them.

THEORIES ON THE CAUSES OF CALIFORNIA GANG DELINQUENCY

Theories differ greatly about why juvenile involvement in crime and gang activity continues in California. Some researchers and much of the current media point to the increasing involvement of youth gangs in the distribution and sales of illegal drugs (Miller, 1975). A classic work of the 1960s (cited by Moore, 1991, p. 42) is characteristic of this view, in which the adolescent cohorts serve as recruiting and training grounds for adult criminal enterprises.

This view builds on the theoretical framework of Robert Merton (1938), who applied the concept of **anomie**, first proposed by sociologist Emile Durkheim,[1] to conditions of U.S. society. According to Merton, the United States is a goal-oriented society, and youth covet wealth and material goods most of all. Yet, access to these goals by legitimate means are determined by class and status. Those with little formal education and few economic resources find that they are denied the ability to acquire money and other success symbols legally (cited in Siegel & Senna, 1994, p. 153). Consequently, some youth develop criminal or delinquent solutions to the problem of attaining goals. Drug sales can fill the void.

Although this may be true for some, it does not hold true for all California gang members. Data from the National Youth Gang Survey (National Youth Gang Center, 1997) demonstrate that when comparing youth gang involvement in drug sales throughout the country, youths from the western region of the country (including California) were the least likely to be involved in the sale of drugs. As Figure 2.1 demonstrates, the Northeast has the highest percentage (50%) of youth gangs involved in drug sales, followed by 45% for the Midwest, 38% for the South, and only 30% for the West. The differences in drug sales by region were found to be statistically significant. Likewise, in her historical analysis of East Los Angeles gangs, Joan Moore 1991, p. 45) asserts that she did not find that the Chicano gangs she studied had become criminal organizations. She acknowledges the disappearance of good manufacturing jobs in East Los Angeles and their replacement by low-wage work, noting that this provided a context in which the rowdiest adolescent groups became rowdier. A major study of youth gang members by the San Diego Association of Governments (SANDAG, 1994) agreed with this perspective. As Table 2.5 indicates,

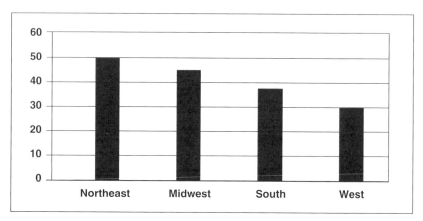

Figure 2.1. Percentage of Youth Gangs Involved in Sales of Drugs in 1997
SOURCE: National Youth Gang Center, 1997.

when each of the more than 200 documented San Diego gang members was asked why he joined a gang, the overwhelming response was because "their friends were in it" or "for the excitement." Drug sales and "to use drugs" were last on a list of more than 20 reasons given for joining the gang, well below the third-most popular reason: "because girls like gang members."

A structural viewpoint of motivations toward gang membership is held by social scientists who believe that gang activity is on the rise because of economic and social dislocation. In her analysis of gangs in postindustrial America, Pamela Irving Jackson (1991, cited by Siegel & Senna, 1994) asserts that gang formation is the natural consequence of the evolution from a manufacturing economy with a surplus of relatively high-paying jobs to a low-wage service economy. This economic viewpoint applies to California, as a state that traditionally required a large population base for its manufacturing and defense-based economy. However, with defense industry cutbacks and a movement to a high-tech economy in California that requires high levels of both skill and education, many are locked out of employment. From this perspective, California youth gangs may be viewed as a response to a drastically changing economy. Several studies of gangs in the Midwest have provided support for this perspective. For example, the increased involvement of African American and Chicano gangs in drug trafficking was tied to increased economic problems in the community (Perkins, 1997). Another study viewed a Midwest Puerto Rican gang as an ethnic enterprise, shifting activities quite consciously toward drug dealing in a community where good jobs had disappeared (Padilla, 1992).

From this economic perspective, gang delinquency is a rational choice made after a careful assessment of legitimate economic opportunities. The **rational choice** view is endorsed by Martin Sanchez Jankowski (1991) in his book *Islands in the*

TABLE 2.5 Reasons Why People Join Gangs, Percentage by Ethnicity (Interview Results, 1991-1992)

Reason	Black	Hispanic	Asian	Other	Total
Friends are in gang	83	90	82	100	87
For the excitement	78	76	91	79	78
Girls like gang members	76	77	77	86	77
It's something to do	61	70	82	57	66
It gives you a family feeling	66	60	77	86	66
To support the neighborhood	58	70	50	71	63
Family member is/was in gang	68	67	27	43	61
Protection from other gang	50	41	82	57	51
To feel important	46	46	59	50	48
Curiosity about gangs	39	45	55	29	43
Important to friends	32	27	82	29	35
Personal survival	37	23	41	43	32
To get material things	34	26	36	43	31
To sell drugs	26	24	18	14	24
To get drugs	16	22	23	14	19

SOURCE: San Diego Association of Governments, 1994.

Street. Jankowski asserts that gangs are made up of adolescents who maintain a defiant individualist character. As Jankowski sees it, the gang offers otherwise unobtainable economic and social opportunities, including both social support for crime and access to parties as well as social and sexual outlets. There is disagreement on the utility of this structural viewpoint, as a historical view provides evidence that youth gangs were present in California in times of economic prosperity as well as economic decline (Moore, 1991). During today's economic upturn in California, with unemployment rates about 3% or 4%, gangs are still flourishing. This points to reasons based on other than pure economics.

In his book, *Race Matters,* Harvard University researcher Cornel West (1994) asserts that the nihilistic thinking of the inner-city black community sets the stage for gang formation. In an environment of despair, West believes that young blacks develop a detached attitude toward others and a self-destructive disposition, a cold and often mean outlook that is damaging to themselves and everyone around them. In his essay "The Killing Fields," recording artist and author Ice-T (1994) echoes this perspective when he says,

> Gangs were born out of this chaos—the inner city. When you grow up in South Central, and you've never had anything in your life you control, you

seek control. Gangs offer you ultimate control to do what you want. Just getting that for a minute is very intoxicating. Gang members are out there trying to control their own little world. It's only a little tiny place. It may not look like much to you—an alley, a street—but it's like a country to them. (p. 147)

Still others support a rational choice model, which holds that potential gang members choose to affiliate to maximize the financial rewards. Research by the Rochester Youth Development Study (Thornberry, Krohm, Lizotte, & Chard-Wierscharn, 1993) supports this rational choice model. These researchers found that before youths joined gangs, their substance abuse and delinquency rates were no higher than those of existing gang members. However, once in the gang, youth crime and drug use rates increased significantly, only to decrease when members left the gang. The Rochester study is important because it concludes that gangs facilitate criminality rather than providing a haven for youth who are already delinquent (cited in Siegel & Senna, 1994). A longitudinal study by Esbensen and Huizinga (1996, p. 238) supports this increased criminality view of gang members. The authors found that, compared to non-gang members, gang members had higher rates of involvement in street crime and other serious forms of offending even before joining the gang. However, the prevalence and individual offending rates were substantially higher during the actual year of membership.

An anthropological viewpoint is offered by those who suggest that gangs appeal not to adolescents' desire for money and fame but to their deep-seated longing for the tribal group process that sustained and nurtured their ancestors (Block & Niederhoffer, 1958, cited by Siegel & Senna, 1994). This approach cites the initiation ceremonies of gang members as marking the death of childhood, as the youth gives up his life as a child and assumes a new adult way of life. Gang names—cobras, jaguars, and kings—are suggestive of totemic ancestors because they are usually symbolic (cited by Siegel & Senna, 1994, p. 335). Contemporary anthropologist James Diego Vigil (1988) points out that current Latino gangs have such traditions as gang initiation rituals in which physical pummeling is used to show that the gang boy is brave and ready to leave his matricentric (mother-dominated) household. For gang members, rituals, hand signals, and graffiti have a tribal orientation and function as a means to bond gang members together.

Although these anthropological perspectives may help to explain why some urban low-income youth join delinquent gangs, they do not explain why others do not. Moore (1991, p. 81) points out that most youth in East Los Angeles never become involved with gangs. Many regions of California and the nation have experienced devastating job losses and conditions of extreme poverty, yet they have not experienced gang formation. Indeed, in his study of San Diego gangs, sociologist William Sanders (1994) asserts that "the problem with most current sociological theories of gang formation is the vast number of low-income urban areas in the United States and the dearth of gangs and gang members relative to this poverty"

(p. 41). Jankowski (1991) attempts to remedy this overprediction of gangs due to low-income lifestyles by suggesting that only those with a character of defiant individualism join gangs. Still, it remains unclear why some youth become defiant individualists and others do not.

An urban setting is an important element in the formation of gangs. At one time, it was thought that the high-density tenement buildings and projects of New York, Boston, and Chicago led to the development of gangs (Sanders, 1994, p. 42). Still, low-density housing in Los Angeles and San Diego also generated gangs (Sanders, 1994, p. 43). Sanders's research on gangs in San Diego revealed that white youth living in the same low-income urban settings as African American and Mexican American gang members simply do not form white gangs. Yet, other researchers (Jankowski, 1991, p. 84) found that East Coast Italian American and Irish American white youth had in fact formed gangs very similar to those found in the Mexican American and African American neighborhoods of the West Coast. San Francisco is experiencing a growth in white Irish American gangs in the Sunset District of the city. These differences in gang membership point to the need to address differences in ethnicity and ethnic identity as important explanatory variables in predicting gang membership and juvenile arrest rates in California as the state's population continues to become more diverse.

ETHNICITY AND JUVENILE CRIME IN CALIFORNIA

The problem with structural explanations of gang formation and violence is that not all cohorts with the same structural situation form gangs. For example, the white youths who live in the South Bay area of San Diego share exactly the same low-income, urban setting as the Southeast Asian, African American, and Mexican American youths. Yet, they do not form white gangs and generally do not join the few mixed ethnic gangs found in the city. Sanders (1994, p. 44) concludes from these ethnic differences that poor white youths in San Diego lack the ideology or the history of being singled out for their membership in a white ethnic group. They are unlike the Italian American youth of New York's Bensonhurst, who saw themselves as needing to protect their neighborhoods from the adjacent African Americans. They are even more unlike the Irish American gangs of South Boston, who saw themselves as oppressed by the white, Brahmin upper class. Boston's "Southies" responded violently to those who enforced undue hardship on them through forced busing to achieve racial integration in the schools (McDonald, 1999).

Instead, most of California's white youth lack the white ethnic group identity of these East Coast Italian and Irish American youth gangs (Sanders, 1994, p. 49).

Cornel West and Ice-T might say that they lack the anger, the nihilism, and the need for control that comes from oppression and years of degradation. Despite a continuing presence of racist skinhead gangs in the state, and an evolving presence of "Straight Edge" gangs populated by working-class and middle-class white youth, California's white youth are less likely to form ethnic gangs than those on the East Coast. The one exception, however, is in San Francisco where Irish American youth gangs may be responding to what they view as the encroaching Asian gangs in the Sunset District, much as the Irish Americans who formed gangs in South Boston.

Most gangs in California and elsewhere are racially and ethnically homogeneous (Siegel & Senna, 1994). Most inter-gang conflict appears to be among groups of the same ethnic and racial background (Klein, 1969). The ethnic distribution of gangs corresponds to their geographic location. In Los Angeles, despite the media presentation of African American gangs including the Bloods and Crips, Latino gangs predominate. This is true in San Diego also, yet Asian gangs are growing quickly. In contrast, San Francisco's gang population is primarily Asian, with a small percentage of Latino gangs and an evolving Irish American presence. This geographic distribution by ethnicity holds true throughout the country, as shown in the most recent *National Assessment of Law Enforcement, Information Resources Report* (National Law Enforcement Institute, 1999).

Ethnic differences in gang membership and activity in California have implications for the future as the trend continues toward greater non-white makeup of the population. Changing California demographics and increasing gang membership are coupled with increased targeting of gang members by law enforcement agencies. The result of this focus is a significant increase in incarceration rates among minority group juveniles in California and elsewhere (Snyder & Sickmund, 1999).

INCREASED MINORITY CONFINEMENT RATES IN CALIFORNIA AND THE NATION

As shown in Table 2.6, a disproportionate number of juvenile minority group members were in residential placement in 1997 in nearly all states. This is especially true in California. About 81% of those committed to California's public residential facilities in 1997 were members of minority groups; minorities accounted for 59% of the juvenile population in general. Under the **disproportionate minority confinement rate requirement** in the Juvenile Justice and Delinquency Prevention Act, states like California must determine whether the proportion of minorities in confinement exceeds their proportion in the population. If such overrepresentation is found, states must demonstrate efforts to reduce it. Although overrepresentation does not necessarily imply discrimination, there are concerns in California because

TABLE 2.6　Percentage of Minorities in Population and in Residential Placement in 1997, by State

	Minority Proportion			
State	1997 Juvenile Population	Public Institution	Private Institution	Detained
U.S. total	34	67	55	62
Alabama	35	69	58	60
Alaska	35	47	67	57
Arizona	43	63	45	56
Arkansas	25	62	56	67
California	59	81	70	70
Colorado	28	56	56	51
Connecticut	26	83	59	77
Delaware	31	75	79	77
District of Columbia	87	100		100
Florida	40	58	63	64
Georgia	40	70	68	70
Hawaii	76	89		
Idaho	13	25	12	4
Illinois	36	70	52	78
Indiana	14	41	31	38
Iowa	7	42	23	27
Kansas	17	52	32	49
Kentucky	11	40	24	38
Louisiana	44	81	74	76
Maine	3	5		7
Maryland	40	68	75	73
Massachusetts	22	64	59	60
Michigan	23	56	57	61
Minnesota	12	46	42	59
Mississippi	47	70		62
Missouri	18	40	34	64
Montana	13	29	19	
Nebraska	14	40	45	44
Nevada	35	50		39
New Hampshire	4	12	12	
New Jersey	37	88		79
New Mexico	62	81		82

(continued)

TABLE 2.6   Continued

	Minority Proportion			
State	1997 Juvenile Population	Public Institution	Private Institution	Detained
New York	41	87	51	81
North Carolina	33	68	36	60
North Dakota	11		29	31
Ohio	18	49	38	51
Oklahoma	26	49	51	60
Oregon	16	29	28	23
Pennsylvania	18	63	66	51
Rhode Island	18	63	38	49
South Carolina	40	69	58	67
South Dakota	17	43		46
Tennessee	24	52	52	51
Texas	53	78	73	77
Utah	12	34	33	28
Vermont	3			
Virginia	32	64	63	66
Washington	21	41	44	41
West Virginia	5	28	27	26
Wisconsin	15	60	39	36
Wyoming	12	27	15	

SOURCE: Snyder & Sickmund, 1999.

minorities are overrepresented at all stages of the juvenile justice system, compared with their proportion in the population. Table 2.7 demonstrates that minority youth are much more likely to be arrested than any other group. These data, coupled with a recent national study sponsored by the Justice Department that documents sharp racial disparities (cited by Glasser, 2000), give reason for alarm. The 1999 Justice Department data demonstrated that among juveniles never before jailed, Latinos were three times more likely to be incarcerated, and blacks were six times more likely, than whites. For drug offenses, black youths are 48 times more likely than whites to be sentenced to juvenile prison. The study shows that minorities also receive longer sentences.

There is substantial evidence of widespread disparity in juvenile case processing. Data available from most jurisdictions across the country show that minority youth

TABLE 2.7 California Juvenile Felony Arrests by Race, 1988 and 1998

Race/Ethnic Group	1988				1998				Percentage Change in Rate, 1988-1998
	Percentage of Population	Number of Arrests	Percentage of Total Arrests	Rate per 100,000	Percentage of Population	Number of Arrests	Percentage of Total Arrests	Rate per 100,000	
White	46.3	25,946	32.1	1,816.8	44.2	21,948	28.8	1,319.3	-27.4
Hispanic	35.2	27,309	33.8	2,518.0	35.8	32,028	42.1	2,374.6	-5.7
Black	8.0	22,018	27.3	8,896.6	7.7	15,335	20.2	5,289.6	-40.5
Other	10.5	5,485	6.8	1,690.5	12.3	6,793	8.9	1,468.4	-13.1
Total	100.0	80,758	100.0	2,618.1	100.0	76,104	100.0	2,021.4	-22.8

SOURCE: State of California, Office of the Attorney General, 2000.

are more likely to be placed in public secure facilities whereas white youth are more likely to be housed in private facilities or diverted out of the juvenile justice system. Furthermore, there is evidence that minority youth are treated differently than white youth at the earliest stages of arrest and detention within the juvenile justice system. In a review of existing research by Pope and Feyerherm (cited by Snyder & Sickmund, 1999, p. 194), about two thirds of the studies examined showed that racial and ethnic status influenced decision making within the juvenile justice system. Because these systems are fragmented and administered at the local level, racial and ethnic differences and treatments exist in some jurisdictions but not in others. When such differences occur, they can be found at any stage of processing within the system. Some researchers call this "justice by geography," and differences can indeed occur for local jurisdictions within the same state (Snyder & Sickmund, 1999, p. 195).

SOLUTIONS TO THE PROBLEM OF JUVENILE DELINQUENCY AND GANGS

As this chapter demonstrates, throughout California's history there have been contrasting perspectives on the underlying problem of California juvenile crime and gangs. Each of these views leads to a different solution about what should be done and where resources should be focused. There are three solution strategies, including (a) using deterrence, (b) addressing community causes of crime, and (c) addressing poverty as the root of juvenile crime. The first solution to juvenile crime is the one that currently dominates: the suppression strategies. This solution rests on the assumption that law enforcement and juvenile justice were too lenient in the past and that to deter juvenile crime, youth must be certain that there will be costs.

SOLUTION 1: LAW ENFORCEMENT AND LEGISLATIVE RESPONSES TO YOUTH GANGS

California cities and other urban areas throughout the country have received millions of dollars from local, state, and national funds to address urban gangs. Much of the funding has helped to establish, strengthen, and maintain gang suppression units of city police departments. These specialized gang units have been successful in making arrests and in gaining convictions for gang-related activities. Still, the most recent negative publicity about abuses within the Los Angeles Gang Unit of the Rampart Division has hurt the image of gang suppression units throughout California. The reports of corruption in Rampart's CRASH Unit

(Community Resources Against Street Hoodlums) has reinforced the already strong negative opinions of the police among many residents of neighborhoods with youth gangs (Ice-T, 1994). Although police abuses may exist, the effective targeting of juvenile gang members by law enforcement continues. Four major law enforcement initiatives have been used to combat the gang problem in California:

Multi-agency enforcement strategies. These are cooperative initiatives that draw from local, state, and federal law enforcement agencies for intelligence gathering, enforcement programs, probation searches, and drug and weapons stings in problem neighborhoods.

The Street Terrorism Enforcement Prevention (STEP) Act (186.22 PC). Enacted in 1988, this act provides enhanced penalties for gang members convicted of a gang-related felony. The act was to "combat the onset of violent street gangs whose members threaten, terrorize, and commit a multitude of crimes against peaceful citizens of their neighborhoods" (California Penal Code Section 12022.53[e]). This sentencing enhancement (also called the 10/20/life) provides for additional prison time to convicted suspects who have a gun or shoot the gun during commission of a crime and a life-term enhancement for causing great bodily injury. In a gang shooting, for example, the felony offense plus the STEP Act enhancement and the 10/20/life section mandates an additional 20 years in prison for the gang member. Before January 1, 1998, the punishment for a documented gang member who shoots a gun would have been 6 years in prison. After January 1, 1998, the same crime warranted a 26-year sentence under these enhancements.

Gang injunctions. These sanctions involve a civil process to abate specific gang behaviors. Several California cities have obtained carefully tailored court injunctions based on local nuisance ordinances to prohibit a list of people identified by name from engaging in specified activities that in many cases are perfectly legal for non-gang members. California cities, including Westminster, Burbank, Van Nuys, San Fernando, and San Jose, have instituted various injunctions specific to gang members. These include barring youths identified by the authorities as gang members from sitting in parks, climbing trees or roofs, whistling, making gang hand signals, wearing large belt buckles, or carrying bottles, baseball bats, or flashlights (Mydans, 1995).

This final strategy has met with strong opposition from California's civil libertarians, who have posed several court challenges to some of the injunctions (Mydans, 1995). In 1999, the U.S. Supreme Court struck down a sweeping Chicago anti-gang law aimed at breaking up sidewalk gathering that intimidates residents in crime-infested neighborhoods. The Supreme Court justices ruled unconstitutional an ordinance that allowed a police officer to order loiterers to disperse or be arrested if they stood or sat with no apparent purpose in the presence of a suspected gang member (Epstein, 1999, p. 1). Thirty one states, including California, had supported the Chicago anti-gang ordinance in arguments before the Supreme Court.

Despite the fact that law enforcement can no longer use these injunctions, the increased penalties for documented gang members, coupled with laws like Proposition 21, have been effective in increasing incarceration rates for juveniles throughout the state. The unfortunate outcome has been an overrepresentation of African Americans and Hispanics within the criminal justice system.

The deterrence solution is flawed, some say (Melville, 1995), because it is based on the assumption that youth who commit serious crimes consider the sanctions. The problem is that most serious juvenile offenders are not efficient decision makers and are not likely to be deterred by the prospect of harsher sentences. This is especially true for juveniles who are convinced that they will not get caught or for those who are convinced that they have nothing to lose.

SOLUTION 2: FOCUS ON STRENGTHENING COMMUNITIES

This solution focuses on strengthening communities in fighting juvenile crime. From this perspective, juvenile crime does not result from the absence of consistent punishment but rather is a symptom of the erosion of moral standards in the community, family, schools, and media. Solutions from this perspective focus on strengthening the community and family responses to crime. An example of this perspective is the imposition of curfews for adolescents as a way to help parents and to convey the message that youths should not be out on the streets late at night. Also, cities have adopted "critical hours" afterschool programs for juveniles. These programs address the fact that most juvenile crime is committed during the unsupervised afterschool hours of 3 p.m. to 6 p.m., when working parents are unable to supervise their children. The federal government recently appropriated several million dollars for these critical-hours programs. In addition, mentoring programs and gang mediation efforts are community solutions to address juvenile crime. Both have been shown to decrease juvenile delinquency.

Critics of this approach, however, say that attributing the juvenile crime problem to the moral collapse of society suggests a much broader problem than the one that actually exists (Melville, 1995, p. 30). Instead, gun control is viewed as a way to reduce juvenile violence. Despite the decline in crimes committed by juveniles, there has been a steep rise in juvenile killings related to firearms. A study of the 65% increase in juvenile homicides from 1987 to 1993 indicates that the increase was disproportionate for African American victims, with the growth in the number of African American victims twice that of white victims. Most significantly, nearly all of the growth in juvenile homicides was in the number of older juveniles killed with firearms.

A study conducted by the Centers for Disease Control and Prevention compared the homicide and suicide rates for children under age 15 in the United States with the rates for several other industrialized countries (reported in Snyder & Sickmund, 1999). Findings presented in Figure 2.2 indicate that the number of homicides per 100,000 children under age 15 in the United States was five times the number in the other countries combined. The rate of child homicides involving a firearm, however, was 16 times greater in the United States. A similar pattern was seen in the suicide rates of children under age 15. The U.S. suicide rate was twice the rate for the other countries combined. However, for suicides involving firearms, the rate in the United States was almost 11 times the rate for the other countries combined. Gun control debates continue in the United States, and it is likely that, despite the alarming data on firearms deaths of juveniles, little agreement will be found.

A criticism of the community approach is that it leads to simplistic solutions that divert attention from what government can and should do about social and economic conditions that contribute to criminal behavior. This latter criticism is related to the third solution: attacking juvenile crime at what some believe are its roots—poverty.

SOLUTION 3: FOCUSING ON THE ROOTS OF JUVENILE CRIME BY ATTACKING POVERTY

Compared with the first two solutions, advocates of this third perspective offer a different diagnosis of the problem of juvenile crime, and they favor a different set of solutions (Melville, 1995, p. 33). Supporters of this perspective acknowledge that the California economy has created a harsh environment in which thousands of juveniles grow up with no reasonable prospect of succeeding, even if they work hard and follow the rules. They assert that, because little is done to help these high-risk juveniles, it is not surprising that many resort to criminal violence. To deal with this problem, a serious effort is needed to address its social roots, and far more resources must be devoted to preventive efforts by strengthening schools and social services (Melville, 1995).

Some critics of this third solution assert that it is not at all certain that social engineering efforts will succeed in reducing the crime rate (Melville, 1995, p. 39). Others criticize this solution as undermining the two-parent family and undercutting the conviction that individuals are responsible for their own behavior and their own choices (Wilson, 1975).

Still, the diversity of the proposed solutions to juvenile delinquency points to the complexity of juvenile crime and gangs in California. Each of the three solutions rests on differing assumptions about the origins of juvenile crime. Each of the solutions has been prominent at different periods of California's history, and each represents contrasting assumptions about the causes and consequences of juvenile

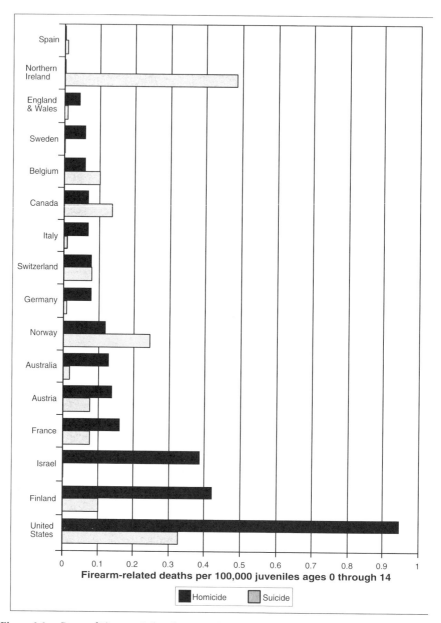

Figure 2.2. Rates of Firearm-Related Homicide and Suicide Among Children in Industrialized Countries

SOURCE: Adapted from Snyder and Sickmund, 1999.

NOTE: If both homicide and suicide rates for a country were 0, that country was not displayed on the graph. Data were provided by Australia, Austria, Belgium, Canada, Denmark, England and Wales, Finland, France, Germany, Hong Kong, Ireland, Israel, Italy, Japan, Kuwait, Netherlands, New Zealand, Northern Ireland, Norway, Scotland, Singapore, Spain, Sweden, Spain, Switzerland, and Taiwan.

crime. What is certain today is that, inspired by California voters, California lawmakers continue to embrace the first solution, which focuses on deterrence and increased punishments for juvenile crime. Indeed, "Lock 'Em Up" continues to be the California response to today's juvenile crime and urban gangs.

NOTE

1. Emile Durkheim was an early French sociologist, who introduced the term *anomie* to indicate a condition of normlessness or confusion over norms.

REFERENCES

Albanese, J. S. (1993). *Dealing with delinquency.* Chicago: Nelson Hall.
Alonso, A. (1997). *Los Angeles gangs* [On-line]. Available: www.bcf.usc.edu/aalonso/gangs/hist.html
Brantingham, P. J. (1979). Juvenile justice reform in California and New York. In E. L. Faust & P. J. Brantingham (Eds.), *Juvenile justice philosophy* (2nd ed.). St. Paul, MN: West.
Demuro, P., Demuro, A., & Lerner, S. (1988). *Reforming the CYA: How to end crowding, diversify treatment, and protect the public without spending more money.* Bolinas, CA: Commonwealth Press.
Epstein, A. (1999, June 11). Anti-gang ordinance is struck down. *San Diego Union-Tribune*, p. A-1.
Esbensen, F.-A., & Huizinga, D. (1996). Gangs, drugs, and delinquency in a survey of urban youth. In D. Rojek & G. Jensen (Eds.), *Exploring delinquency* (pp. 228-242). Los Angeles: Roxbury.
Geist, T., & Pope, V. (1996, February 25). Teens who kill. *U.S. News & World Report,* pp. 29-36.
Glasser, J. (2000, May 8). And justice for some. *U.S. News & World Report,* p. 28.
Ice-T. (1994). The killing fields. In J. Miller, C. Maxson, & M. Klein (Eds.), *The modern gang reader.* Los Angeles: Roxbury.
In re Gault, 387 U.S. 1 (1967).
Jankowski, M. (1991). *Islands in the street.* Berkeley: University of California Press.
Kent v. United States, 383 U.S. 541.86 S.Ct. 1045 (1966).
Klein, M. (1969). *Juvenile gangs in context.* Englewood Cliffs, NJ: Prentice Hall.
Krisberg, B. (1993). Youth crime and its prevention. In I. Schwartz (Ed.), *Juvenile justice and public policy.* New York: Lexington Books.
Lacayo, R. (1994, February 7). Lock 'Em Up! *Time,* pp. 50-53.
Lemert, E., & Rosenberg, J. (1948). *The administration of justice to minority groups in Los Angeles County.* Berkeley: University of California Press.
Mathieson, T. (1965). *Defense of the weak.* London: Tavistock.
McDonald, M. (1999). *All souls: A family story from southie.* Boston: Beacon.

Melville, K. (1995). *Kids who commit crimes.* New York: Basic Books.

Merton, R. (1938). Social structure and anomie. *American Sociological Review, 3,* 672-682.

Miller, W. B. (1975). *Violence by youth gangs and youth gangs as a crime problem in major American cities.* Washington, DC: Department of Justice, National Institute for Juvenile Justice and Delinquency Prevention.

Milliken, J. (2000, February 27). A better way to combat juvenile crime. *San Diego Union-Tribune,* p. G-3.

Moore, J. (1978). *Homeboys.* Philadelphia: Temple University Press.

Moore, J. (1991). *Going down to the barrio: Homeboys and homegirls in change.* Philadelphia: Temple University Press.

Mydans, S. (1995, June 11). New anti-gang weapon bars everyday conduct. *The New York Times,* p. 1.

National Law Enforcement Institute. (1999). *National assessment of law enforcement, 1999, Information resources report.* Santa Rosa, CA: Author.

National Youth Gang Center. (1997). *National youth gang survey.* Washington, DC: Office of Juvenile Justice and Delinquency Prevention.

Office of Juvenile Justice and Delinquency Prevention. (1999). *Juvenile offenders and victims: 1999 national report.* Washington, DC: Author.

Padilla, F. (1992). *The gang as an American enterprise.* New Brunswick, NJ: Rutgers University Press.

Perkins, U. E. (1997). *Explosion of Chicago's black street gangs, 1990 to present.* Chicago: Third World Press.

Sanders, W. (1994). *Gangbangs and drive-bys: Grounded culture and juvenile gang violence.* New York: Aldine de Gruyter.

San Diego Association of Governments. (1994). *Down for the set.* San Diego, CA: Author.

Schwartz, I. (1993). *Juvenile justice and public policy.* New York: Lexington Books.

Sheldon, R., Tracy, S., & Brown, W. (2000). *Youth gangs in American society* (2nd ed.). Belmont, CA: Wadsworth.

Siegel, L. J., & Senna, J. (1994). *Juvenile delinquency, theory, practice, and law.* St. Paul, MN: West.

Snyder, H., & Sickmund, M. (1999). *Juvenile offenders and victims: 1999 national report.* Washington, DC: Office of Juvenile Justice and Delinquency Prevention.

State of California, Office of the Attorney General. (2000). *Report on juvenile felony arrests in California, 1998* (Criminal Justice Statistics Center Report Series, Vol. 2, No.1). Sacramento: Author.

Thornberry, T., Krohm, M., Lizotte, A., & Chard-Wierscharn, D. (1993). The role of juvenile gangs in facilitating delinquent behavior. *Journal of Research in Criminal Delinquency, 30,* 55-87.

Vigil, J. D. (1988). *Barrio gangs.* Austin: Texas University Press.

West, C. (1994). *Race matters.* New York: Vintage.

Wilson, J. Q. (1975). *Thinking about crime.* New York: Basic Books.

CHAPTER 3

RACE AND ETHNIC RELATIONS

Phylis Cancilla Martinelli

Race relations in California follow broader patterns of relations between Anglo Americans and non-whites in America as well as global trends. However, some unique patterns in the state result from the particular demographic, economic, and historical factors in a state that is bigger in size and population than many independent nations.

THE HISTORY OF RACE RELATIONS IN CALIFORNIA

Much of early sociological research on U.S. race relations was focused on African American/Anglo problems, dealing with the issues and ideologies relating to slavery and its post-Civil War consequences. Yet, California, where initial contact was between Spaniards and Native Americans, had a different set of social groups involved in racial contact and competition. Large-scale contact with Anglos who had moved west began in the mid 1800s, when a significant number of Asians also arrived in the state. Also entering then were a smaller number of African Americans and some European groups, such as the Portuguese and Italians, who were often perceived to be racially different. Anti-Semitism also has a long history in the state. So, the stratification system initially must be seen from a multiethnic historical perspective that foreshadows the current situation.

Moreover, California's patterns of racial separation and suspicion of immigrants are also found globally. **Racism** is not an ancient concept. European racism formed out of persistent anti-Semitism and colonial expansion throughout much of Africa, Asia, and Latin America. As explorers encountered non-white and non-Christian peoples, they began to define them as the Other. Attempts to explain differences, such as technological and cultural contrasts, led to the formation of ideas that tried to claim race as a legitimate category. By the 19th century, the idea of racism had developed further and was based on pseudoscientific theories that ranked Europeans as inherently superior to people in the rest of the world (Wieviorka, 1995). These ideas spread to other colonizing cultures. The Japanese, as an example, had long coveted the rich lands of the Ainu, an indigenous people of northern Japan. As Western ideas spread in the 1800s, the Japanese began to label the Ainu as an inferior race, by their nature unable to become more sophisticated and to develop their lands economically. Like other native peoples, such as Hawaiians, Native Americans, and the Maori of New Zealand, the Ainu were relegated to a lower status in society, as the Japanese, in relentless expansion of power, subjected them (Siddle, 1997).

Throughout the world, contact among divergent populations expanded after World War II, increasing opportunities for ideologies of racial difference to catch on. As many colonial empires collapsed and new nation states (e.g., Algeria, India, and Vietnam) emerged, a global economy expanded, and travel became easier and cheaper. For example, in the 1950s and '60s, when French colonies in North Africa and Southeast Asia collapsed, a special class of protected immigrants from these areas began to move to France in large numbers. They were only semi-citizens, however, and their exact rights were not spelled out. Their non-white status elicited previously unspoken racism, which grew into anti-immigrant views, challenging both French liberalism and national identity (Cornelius, Martin, & Hollifield, 1994).

As Cornell and Hartmann (1998) note, racial divisions can happen in most countries affected by modernity and the simultaneous growth of industrialization, urbanization, and rationalization, along with the unprecedented movement of refugees and immigrants across continents.

In California, the first interracial conflict emerged between 1769 and 1800, as the Spanish established missions and forts, coming in contact with a variety of Native American tribes and several major linguistic groups, with an estimated population in 1770 of 133,000. These divisions among Native Americans meant that tribes were unable to unify to oust the invaders; in fact, old enemies were used by the Spanish to fight each other. (Kroeber, 1964). The Spanish goal was to convert tribes to Catholicism and mold them into *gente de razon,* or people of reason, so they could work as serfs on Spanish ranchos. This motive did not, however, prevent soldiers from raping women and killing men or deter missionaries from forcibly bringing converts into the fold. However, under the system of Spanish law, which Mexicans continued to administer, Native Americans were considered human beings with rights and privileges (Cook, 1972).

This viewpoint changed radically with the large-scale movement of Anglo Americans into California in the 1840s. The latter sought to enforce a system of apartheid (the former system of South Africa, which completely separated racial groups). Native Americans were seen as almost subhuman. They were called "Digger Indians," suggesting they were so primitive that all they could do was dig roots and eat insects, placing them below the Plains tribes grudgingly admired for their hunting and riding skills (Almaguer, 1994). For some Anglos, the only fate acceptable for tribes was extermination and removal from their land. The term *genocide* should not be used casually, but the progressive decimation of the native population during this time was not accidental. By 1880, only 20,000 Native Americans remained. Soldiers and civilians killed Native Americans with impunity, children and adults were enslaved, and most of their land was declared public domain. Entire tribes, such as the Yana and Yani of the Sierra foothills, were wiped out (Cook, 1972; Kroeber, 1964). Others, such as the Pomo, survived and today are involved in issues such as cultural preservation, political self-determination, and economic advancement. As of 1990, the number of indigenous Californians was roughly 86,000. The additional Native American population was 236,000, with Cherokees the largest group, followed by other non-California tribes such as the Apache, Choctaw, Navajo, and Sioux. Many of the indigenous Californians live in urban settings, in cities such as Oakland where they have community centers and events like powwows to keep their cultures alive (George & Just, 1992).

For African Americans, California held out promise of more equality. However, California's decision to enter the Union as a free state was spurred by fear on the part of some whites about economic competition from slave labor rather than a concern about injustice. Slaves brought to California by their owners were freed; however, African Americans who escaped from southern plantations could be reclaimed, and runaway slaves were publicly sold. Freed African Americans faced discrimination stemming from the ideology developed to justify slavery (Fredrickson, 1988), and legislation was frequently proposed to prevent free African Americans from settling in the state (Anderson, 1991). And, like Native Americans and Asians, they were forced to contend with legal restrictions. For example, African Americans faced discrimination in public accommodations and were relegated to separate schools. They could not testify in court until 1863, and they were prohibited from intermarrying with whites (Anderson, 1991). In Sacramento, San Francisco, and Stockton, African Americans were segregated into separate neighborhoods and the lowest end of the job market. They did better, however, in Oakland and Los Angeles, where manufacturing jobs were available, a plus for working-class African Americans, and they had the opportunity to own property. This led to a strong middle-class base of African Americans in Los Angeles (Broussard, 1993; Sonenshein, 1993).

Immigrant groups faced different ideologies in the Golden State. Ideologies of Anglo superiority to conquered peoples was not appropriate. However, a nativist ideology, aimed mostly at Asians, focused on three areas: economic competition, cultural differences, and racial differences. The Chinese and Japanese were the first groups to face the hostility of white nativists. The Chinese arrived in large numbers during the Gold Rush era, when they were sought out as cheap, docile labor for the mines and railroads. Soon, the Chinese began economic enterprises of their own (Takaki, 1989).

Repeated acts of violence toward the Chinese were common, starting in 1849 and continuing for more than half a century. In Los Angeles, crimes against Chinese included arson, hangings, and stabbings. A mob of some 10,000 whites overran San Francisco's Chinatown, and Chinese were expelled from 35 California communities early in 1886. Legal assaults were waged as dozens of ordinances and laws were passed at the local, state, and national level (Gonzales, 1990).

Localized expulsions and restrictions did not satisfy nativists, who were able to push through the Chinese Exclusion Act in 1882. The act, which forbade Chinese laborers from entering the United States and prohibited naturalization for 10 years, was extended and not repealed until 1943. California's economic growth led to further urbanization, industrialization, and modern large-scale agriculture. Owners' desire for cheap labor and the continued resentment of Anglo laborers meant newer Asian groups who replaced the Chinese were also harassed. Thus, anti-Japanese, Korean, and Filipino legislation sought to retain the economic dominance of Anglo workers well into the 20th century (Takaki, 1989). With Asian assimilation, color replaced culture as the most important barrier to movement into mainstream society. In contrast, by the second generation, European groups blended in more easily and began to achieve mobility. Ironically, the designation of Japanese Americans as racially suspect led to the internment of 120,000 in concentration camps during World War II, while 10,000 Italian Americans in California were taken from their homes, an action rationalized in part because of their lack of racial visibility (Scherini, 2000).

Mexicans were in a unique place in California's stratification system. The initial Anglo takeover of the state left many of the landed aristocracy with property and political power. Even intermarriage was sanctioned. Because the upper classes were perceived as primarily European, such marriages did not violate Anglo taboos against racial mixing. However, the lifestyle of the elite, based on conspicuous consumption and leisure activities, was judged to be wasteful from the perspective of the Protestant work ethic (Monroy, 1990). The upper echelon Mexicans gradually lost their lands and power in legal battles contesting Spanish land grants (Pitt, 1971).

In contrast, average Mexicans were viewed with hostility and treated more like Native Americans in the rigid racial classification system of the time. California's constitution, forged in 1849, was the legal basis of Mexicans' in-between status. Although the Treaty of Guadalupe Hidalgo granted Mexican people in the United States full citizenship, California law distinguished between white Mexicans and "Indian" Mexicans. White Mexican males could vote in California, but Indian Mexicans could not. Deciding which Mexican Americans were white fell to local communities. Some communities, such as Santa Paula, California, established their own standards (Menchaca, 1995). As was the case in nearby Arizona, "racial orders arose as microsystems" (Gordon, 1999, p. 98). The result was that most Mexicans became disenfranchised politically. Moreover, the legal system dealt more harshly with them in the form of unequal justice often experienced by minorities. Fears of a Mexican rebellion, competition over economic resources, for example, in the gold mines, and fear of banditry frequently led to mob invasions of Spanish-speaking areas, mass arrests, and lynchings (Pitt, 1971). Many average Mexicans remained in agriculture, escaping the economic competition that the Chinese and African Americans faced from white workers, but then they fell into a class of minorities viewed as basically suited to rural labor (Menchaca, 1995).

RACE RELATIONS IN CALIFORNIA TODAY

In 1990, California's population was 29.8 million, increasing to an estimated 33.9 million in 2000.[1] The state showed an emerging diversity that had grown by the end of the century. Non-Hispanic whites,[2] long the dominant group, were still the largest group, at 57.2%, in 1990 but fell to 46.7% in 2000. Hispanics moved from 26% to almost a third of the state at 32.4 million, followed by Asian/Pacific Islanders who went from 9.1% to 11.1%. In contrast, the African American population decreased to 6.4% from 7% in 1990, as did Native Americans from .8% to .5%. People who acknowledged heritage of two or more races, who were counted for the first time in 2000, were 2.7% of the state.

Yet, diversity does not automatically translate into social contact or economic and political power for minority groups, because even very small groups can control a society or state. Certainly, non-whites do not have the same clout as whites. Furthermore, whites remain dominant numerically in some areas. Many have retreated to rural areas, such as Calaveras County, where they are 89% of the population, or they live in affluent suburbs such as Marin County where they are 81% of the residents. Yet, trends from 1990 show that a major shift is under way in some areas. Hispanics dominate two counties, Imperial County at 72% and San Benito at 53%, closely followed by Tulare with 45% (U.S. Bureau of the Census, 2000). Based on an examination of major counties in the state (see Figure 3.1), it is evident that diversity is the order of the day. In northern California, Asians are the largest racial minority group, whereas Hispanics dominate in the south. At the city level, 10 cities

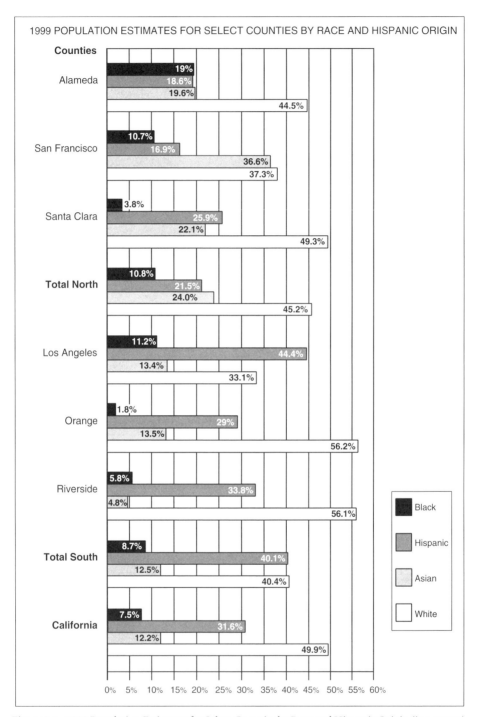

Figure 3.1. 1999 Population Estimates for Select Counties by Race and Hispanic Origin (in percent)
NOTE: Compiled by the author, drawing from data of the 2000 U.S. Census.

showed a multiracial profile, with no majority group: Alhambra, Carson, Daly City, Los Angeles, National City, Oakland, San Francisco, San Jose, Stockton, and Vallejo. Yet, even in these cities, some areas are predominately inhabited by one racial group (California Statistical Abstract, 1999; Clark, 1996).

The above patterns suggest the old stratification system is still present, but change is under way. As often happens during a time of social change, tensions arise. Thus, to use Almaguer's (1994) analogy, California rests on racial fault lines just as it lies on seismic fault lines. Social indicators warn of potential seismic activity. Tremors are often linked to times of economic downturn, a perception of downward mobility on the part of some groups, rising expectations, and a sense of anomie—a feeling that the norms or rules of society do not apply anymore—on the part of minorities. California had these signs in profusion in the early 1990s, and they were major factors resulting in the "rainbow rebellion/riot" of Los Angeles in 1992. Aftershocks continue, with racially polarized reactions to the O. J. Simpson murder trial and a series of propositions targeting minority issues. In 1994, Proposition 187 eliminated services to undocumented immigrants, and in 1996, Proposition 209 ended affirmative action programs meant to promote equality. In 1998, Proposition 227 curtailed bilingual programs in California, which served large numbers of immigrant children, helping them prepare to enter the economic and intellectual mainstream.

The racial fault lines in Los Angeles, globally recognized as a trend-setting postmodern city, merit further examination. As Morrison and Lowry (1994) note, "Territorially based ethnic tension is endemic in California, especially metropolitan Los Angeles" (p. 23). In 1980, the Los Angeles-Long Beach metropolitan area was among 15 metropolitan areas labeled "hypersegregated" (Massey & Denton, 1993).[3] African Americans living in hypersegregated areas had "little direct experience with the culture, norms, and behaviors of the rest of American society . . . African Americans living in the heart of the ghetto are among the most isolated people on earth" (Massey & Denton, 1993, p. 77). By 1990, a new order had emerged. Poor blacks became poorer as they competed for jobs with new immigrants and found themselves living near these groups, for example, in the South Central district of Los Angeles. A Mexican ghetto or mega-barrio emerged, sprawling over 200 square miles on the east side of downtown (Ong & Lawrence, 1992; Waldinger, 1996). By 1990, the number of whites declined, and the African American population dropped slightly. However, the Asian and Latino populations have increased substantially in Los Angeles since 1980, a trend that accelerated during the 1990s (see Figure 3.1) (Morrison & Lowry, 1994). A new type of racial pattern is emerging that cannot be fully explained by older models of ethnic neighborhoods, which are based on older, industrial cities in the East.

The changing dynamics of economics, residence, and race were underpinnings in the 1992 urban chaos in Los Angeles, which was sparked by anger and frustration over the court decision acquitting officers in the assault of Rodney King, a black man whose beating at the hands of Los Angeles policemen was videotaped. For African Americans and Latinos, the explosion was tied to decades of police mistreatment and reflected a more general feeling of discontent and alienation among minority youth.

Certainly, the economic picture played a part in these feelings. By 1990, South Central showed the effects of manufacturing shifts similar to those found elsewhere. Once the core of heavy manufacturing, South Central had lost many good-paying union jobs to nearby areas and Mexico. These jobs were replaced by low-paying positions in the service industries, leaving the district with some extremely impoverished sections. Overall, 33% of area residents existed below the federally established poverty line in 1990, in contrast to 19% of the city's population (Dentler & Hunt, 1992). Countywide, more than 300,000 jobs had been lost since 1990. Welfare rolls grew, as did the number of gang homicides and ethnic tensions (Wildermuth, 1995). Still another factor may have been dissatisfaction in African American neighborhoods regarding the recent influx of Latino and Asian populations into their areas (Bergesen & Herman, 1998).

Earlier riots in Los Angeles had followed different paths. In 1943, the city's Zoot Suit riots involved whites attacking Mexicans. The Watts riot of 1965, similar to uprisings in other U.S. cities, involved African Americans lashing out within their ghettos.

The 1992 events in South Central paralleled the patterns of the 1960s but also reflected changing trends in California's racial makeup, because African Americans, whites, Latinos, and Asians were all affected. As in earlier disturbances, the majority arrested, 90%, were minority males. But unlike the African American riot in Watts in 1965, Latinos were the group with the highest number of arrests and deaths in 1992. Considerable activity during the 1992 disturbances took place in areas where Latinos accounted for 26% to 54% of the population (Navarro, 1993). And although the businesses of whites, African Americans, and Latinos were attacked, Koreans suffered a disproportionate amount of the loss, rather than white merchants as was the case in Watts.

The intergroup dynamics of these areas are important to note. Tension between African Americans and Korean merchants has occurred in a number of cities, including New York, Atlanta, Baltimore, Philadelphia, and Washington, D.C. Koreans followed in the steps of European middle-men minorities, such as Jewish Americans, serving minority clientele when they found other avenues of mobility blocked.[4] The targeting of Korean American businesses during the 1992 Los Angeles uprising reflected these underlying ethnic tensions.[5] African Americans often felt Koreans were unfriendly and overcharged them, but the hostility in Los Angeles really escalated in 1991 when a young African American girl was killed by a Korean store owner in a misunderstanding over a bottle of juice; the shooter received a suspended sentence (Freer, 1994; Oliver, Johnson, & Farrelly, 1993).

Relations between the Korean and Latino communities are also complex. Researchers Cheng and Espiritu (1989), examining ethnic newspapers in Los Angeles between 1972 and 1987, found many instances of hostility between the African American and Korean communities but none between the Korean and Mexican communities. The authors suggest that a common immigrant experience minimized acrimony between Koreans and Mexicans. Other research, however, found tension between Koreans and Latinos, who voiced complaints similar to those of African Americans. Both groups found Korean merchants unfriendly by American cultural standards. Some Latinos believed Koreans employed them because they would work for less than other groups, including Koreans (Bobo, Zubrinsky, Johnson, & Oliver, 1994). This may reflect divisions within the Latino community, because those involved in the uprising were often from Central America and represented the most recently arrived and poorest of the Latino groups. Many established Mexican American politicians stayed away from the fray, and they pointed out that there was little violence in the predominantly Mexican American Eastside area. There is a growing diversity of Latino groups. Mexican Americans, with their longer political history and a stronger voting base in California, have finally begun to gain some political clout. Although they are involved in many broader Latino issues, they still focus attention more on their own interests (Oliver et al., 1993).

In the Korean community, shop owners faced devastating economic losses in 1992, and they found police were slow to respond to their calls, in part because they were protecting middle-class white communities. Koreans felt they were blamed for the economic plight of minorities, when this situation rightfully belonged at the door of the larger white society. Many Koreans realized that numerous African Americans and Latinos saw them as favored "model minorities" and supposedly exploitative merchants. At the same time, they did not receive an upsurge of support from other Asian groups, leaving them feeling isolated. A pan-ethnic identity may be emerging among Asians, but it has not yet coalesced. Some saw the riot as a hate crime toward Koreans. And the community noted that the shooting deaths of some 25 Korean shopkeepers in Southern California since 1990 raised scant attention in the press (Cho, 1993; Kim, 1994)

Another dimension of the 1992 Los Angeles disorder was the systematic nature of the looting. Thus, rather than wanton destruction, activities reflected economic motives, so that some saw the looting as a "commodity riot" (Dentler & Hunt, 1992).

Hostility arises among groups who are all somewhat disadvantaged, in a process of displaced aggression that focuses on adversaries who are in close proximity rather than on the most advantaged groups. As of 1990, demographic profiles in several cities of more than 100,000 showed large minority populations and a lower per capita income than the rest of the state. Thus, Stockton, Inglewood, Oakland, and Fresno are among the major cities with potential for ethnic competition (Hornor, 1992).

RACE RELATIONS IN CALIFORNIA AS COMPARED TO THOSE IN THE UNITED STATES

In terms of race relations, California brings attention to issues of class and interrelationships among racial groups in the nation's increasingly complex stratification system. The economic component leads sociologists to examine the social and economic position of the underclass in minority communities. The recognition of an underclass[6] in the United States was primarily concerned with the development of a core of disadvantaged African Americans left in the inner cities as their middle-class counterparts moved away. The underclass emerged, in part, because of the economic downturn of the Rust Belt manufacturing cities of the East and Midwest as the nation's economy changed from heavy manufacturing to an emphasis on high technology. States like California, which boasts Silicon Valley, seemed immune to recessions in the 1980s. But, South Central, as noted, was hurt by economic recession. Other parts of the Golden State, once seemingly immune to downturn, also suffered the economic consequences of competition for hegemony in the computer industry, the post-Cold War closures of military bases, and natural disasters such as drought, fire, flooding, and earthquakes.

Often, when financial competition escalates, minority groups are the most vulnerable because they are concentrated in the lower social classes. In 1990, U.S. minority groups were overrepresented below the poverty line, whereas whites were underrepresented (U.S. Bureau of the Census, 1992). The same pattern is found in California, where 9.3% of all families, but only 5.5% of whites, were below the poverty line. However, 19% of African American, 15.8% of Native American, 11.4% of Asian, and 18.4% of Hispanic families lived below the poverty line, all proportions that exceed their percentage in the population (U.S. Bureau of the Census, 1993).

In the mid-1990s, a strong economy raised employment levels throughout the country. California's unemployment rate dropped but remained higher than the national average. The state's unemployment rate reached a 30-year low in November 1999 at 4.6%. Remote counties and the Central Valley continued to have the highest unemployment rates. In terms of major metropolitan areas, the San Francisco Bay area fared the best. The Los Angeles area saw lower unemployment only at the end of the century (Ellison & Pochy, 2000).

Furthermore, in 1996, the nation's welfare system was drastically overhauled in an effort to decrease dependency on welfare, often equated with minority households although numerically more whites have used it. The Personal Responsibility and Work Opportunity Reconciliation Act eliminated the Aid to Families With Dependent Children program, in place for decades, and instituted the Temporary Aid to Needy Families (TANF) grant. The emphasis in this program was that benefits would be limited to a 3-year maximum. With a strong economy, the program proposed moving people from welfare into the labor force.

Hopes were raised that poverty would be diminished in the United States as the number of people on welfare dropped significantly to less than 4%, the smallest proportion since 1971. Much of the decline related to the growth of the economy and changes in regulations at the federal and state level. However, in California, poverty rates varied significantly and were higher than in many states and particularly high for children. A study of children living below the poverty line conducted in 13 states showed that California was ranked as one of the three poorest states, next to Mississippi and Texas, in both median income and high inequality. Researchers used 1997 data to calculate both a measure of family income inequality and income differences related to living in a two-parent or one-parent family (Acs & Gallagher, 2000).

One factor in California's poverty standing is the high rate of poverty among immigrants throughout the country. Because California and New York contain 46% of all U.S. immigrants, poverty is a particular problem for both the Golden State and the Empire State. The poverty rate in 1997 was 24.4% for California's immigrants, and this is often associated with high fertility rates and a higher than average number of children among immigrant groups. The state must address the many causes of immigrant poverty, which include racism, lower levels of education, high unemployment rates, jobs with little chance for advancement, poor benefits, and large households (Camarota, 1999).

Furthermore, although many African Americans are justifiably proud of their further economic movement into the mainstream during this robust economy, the plight of poorer African Americans in California cannot be ignored. Any solution to minority poverty must take into account the particular difficulties this group faces. The most disadvantaged minorities can experience a lifetime of discrimination, starting with a pattern of inadequate prenatal care and early health care,[7] an inferior education and career counseling, and less access to job networks. Indirect prejudice and discrimination remain, if blatant, overt practices have generally been eliminated.

CALIFORNIA'S PROBLEM IN GLOBAL CONTEXT

Understandably, some individuals faced with a number of structural barriers to attaining "the good life" are tempted to turn to crime. When they do, they often face a pattern of harsher treatment or **differential justice.** They are more likely to be arrested, pay bail, receive a harsher sentence, and end up on death row. This problem is found throughout the Western world. A recent study showed that although Western countries had different cultures, legal systems, and types of crime (and the studies used different research methods), it was possible to make some generalizations about differential justice (Tonry, 1997). For example, all countries showed higher incarceration rates for some minorities than for the dominant group.

England is one country that showed such problems. A government report on the police in London concluded that the force had a systemic problem with racism so entrenched it had become institutionalized (Hodge, 1999). An examination of incarceration rates nationally showed that in 1994, the rate of British Africans in prison was 1,014 per 100,000 compared to 143 per 100,000 for British whites. The higher rates were linked, in part, to minorities receiving harsher sentences than white criminals (Ethnic Minorities Advisory Committee, 1995).

When looking at crime rates and minorities in a global context, it is important to understand that the United States has an incarceration rate significantly higher than most Western countries, constituting what has been called an "imprisonment binge" (Austin & Irwin, 2001). African Americans and Latinos have been significantly hit by this binge. The War on Drugs and the "three strikes and you're out" efforts have targeted these minority groups. In the War on Drugs, arrests focused on drugs, such as crack, that are used in minority communities and often overlooked the use of similar drugs, such as cocaine, in the white middle-class population.

In 1996, California spent more on corrections than on education. Although African Americans constituted 7% of the state populations in 1996, they accounted for 32% of the prison population. Particularly startling is the rapid increase in the number of African American women going to prison for drug offenses in California. In contrast, whereas 53% of the state's population was white in 1996, only 29% of those who went to prison were white. Latinos fell between the two groups. The arrest rate for young Latino men was 1 in 10, compared to 4 in 10 for African Americans and 1 in 20 for whites (Schiraldi, Kuyper, & Hewitt, 1996).

Concerns about the state's high incarceration rate resulted in a systematic report on how the court system treats minorities (California Judicial Council, 1997). The report looked at a number of factors including the courtroom setup, treatment of minority attorneys, barriers faced by minorities, disproportionate arrest of women of color, family and juvenile laws, the sentencing process, and the jury system.

Findings pointed to the impact of drug laws, but poverty could not be ignored as a factor. In explaining bias in the courts, the report noted that "in misdemeanor cases, where fines are used and where there is a fee for going into a work-service program, the fees and fines are prohibitive for a low-income defendant, who usually goes to jail as a result" (California Judicial Council, 1997, p. 179). Furthermore, it was noted that those in positions of power (judges, public defenders, and attorneys) are predominantly white. This may relate to the public perception that California courts are not as fair to minorities as they are to whites. The report made a number of recommendations, including further education of those in power about the multicultural population that they face in the court system; more research on the status of women of color in the justice system; and increased contact between local courts and minority populations, so that these groups understand their legal rights and responsibilities, particularly if English is not their native language (California Judicial Council, 1997).

The problems of minorities caught in the underclass are not limited to California. They are found throughout the nation and the Western world. While the state's ethnic diversity is more pronounced than the diversity of most other states, other areas will also have a complex ethnic mix similar to that found in California as the U.S. population changes.

ALTERNATIVE EXPLANATIONS FOR RACE RELATIONS IN CALIFORNIA

One view, superficially compelling, claims a difference in the success of racial groups is linked to hereditary traits. Certain social scientists began flirting with the notion of a biological basis for social differences among races in the 1800s. Some tried to justify slavery based on skull size, which supposedly showed racial inferiority, whereas others advocated for immigration restrictions claiming inferior nationalities from southern Europe would weaken the white race. These ideas fit the conservative attitudes of the Progressive Movement, which was led by middle-class whites concerned that social and demographic changes would irrevocably alter the democratic basis of society.

The movement faded with new developments in genetics, psychology, sociology, and anthropology, but it continues to appear under new guises. In the 1960s, the **new hereditarians** advocated separate educational programs for African Americans and whites to accommodate what they claimed was the lower IQ of African American children. Echoing the Progressives, some urged sterilization for African Americans who had low IQs. Their views, predictably, produced a storm of controversy and yet another assessment of the use of IQ tests to measure intelligence. Efforts such as Head Start and the Perry Pre-School Program refuted their views by showing that minority children with low-IQ parents could, over the long run, raise their intelligence when given early assistance (McLemore, 1994).

With re-emerging conservatism, the biological view has appeared again because, for some whites, it supports American values that link success to individual effort in an open, democratic society, rather than to structural advantage in the racial stratification system (Kluegel & Smith, 1986). The recent controversial book *The Bell Curve* by Herrnstein and Murray (1996) again posits the notion that differences in the average IQ can be found among races. The book became a popular best-seller, despite a quick and emphatic response from the scientific community stating that race is not a legitimate biological category. Other authors continue to advance theories that there are biologically based racial differences in brain development and intelligence between groups (e.g., Rushton, 1999).

Instead, most sociologists agree race is a recent social category varying from culture to culture and within one culture over time. Thus, as discussed, sociologists look at structural patterns and ideologies to explain differences created by a lifetime of inequities. This perspective runs counter to the current conservative political views, which attack supposedly excessive privileges gained by minorities, claiming whites are disadvantaged through reverse discrimination. Omi and Winant (1986) see this view as reductionist, a way of seeking to redefine race relations in America from the social level of institutionalized racism to particular cases existing at the individual level. In this way, many whites can feel justified in keeping the status quo and reversing progress made by minorities through programs like affirmative action.

FUTURE PROSPECTS FOR RACE RELATIONS IN CALIFORNIA

Experts disagree on the implications of population trends found in major California cities in 1990. Farley and Frey (1994) are somewhat optimistic and show that two major California cities, San Jose and Anaheim, rank among the least segregated cities among the metropolitan areas they studied. These and some other California cities represent the generally lower segregation found in the West and within cities with substantial new growth in housing after World War II. In addition, Farley and Frey note that Asians and Hispanics tend to be less segregated than African Americans, so that areas with large numbers of these groups and small numbers of African Americans may show less racial division. But deep pockets of segregation still exist in cities such as Los Angeles, San Francisco, and Oakland (Farley & Frey, 1994; Massey & Denton, 1993). Examining the trends in the greater Los Angeles area, Clark (1996) points to two different interpretations of changes evident in the 1990 census. On the one hand, the greater diversity seems to point to a gradual shift toward integration among minority groups, although whites are tending to withdraw. But, continued immigration, which has occurred, can lead to further separation of Hispanics and to less integration in general.

Therefore, racial segregation and the myriad disadvantages that accompany it must be ended. Isolation breeds anomie because the social structure blocks access to the goals society holds out, which this society generally links to occupational status and material success. The wisdom of the older generation in minority communities is now seen as irrelevant. As shown in the neighborhood studied by Anderson (1990), these areas suffer from accelerated moral breakdown. Reactions can be innovation through crime, retreat through drugs and alcohol, or rebellion (Merton, 1968). The "killing zone" in East Oakland, where violence and drugs were institutionalized, showed this pattern among California cities. These reactions will lead to further polarization among the races, further white withdrawal and hostility toward minority concerns, and competition among groups still in central cities.

To combat negative reactions on both sides of the racial divide, Californians must find the will to end segregation, improve the quality of life in poor communities, establish better quality schools, improve health care, and break though the feelings of despair by linking residents to economic opportunity. This can be done when segregation is defined as a critical social problem, especially in communities still resting on racial fault lines. Federal laws exist to combat this problem, but they are not high priorities for many Californians.

SOLUTIONS TO THE PROBLEM OF RACE RELATIONS

Coalition politics is a way to improve the representation of minority groups and work on critical issues. U.S. cities have offered the greatest opportunities for minority groups to become enfranchised. And as Kaufmann (1998) wrote, "coalitional analyses link black and, in some instances, Latino voters with better educated, liberal whites" (p. 2). For such coalitions to work, participants need to organize for a goal, such as decreasing segregation, that transcends securing gains for their own group. This solution is not simple, however, and a number of factors are involved in successful political cooperation, including the size of local minority groups, both in absolute and relative numbers; ethnic group self-identity; the presence of liberal whites; and minority activism at the national and local level. Also important are the amount of intergroup conflict present, economic factors, and state and federal funding available to apply to social problems (Kaufmann, 1998).

Research (Browning, Marshall, & Tabb, 1984) on 10 cities in California from 1970 to 1980 showed varied results in the incorporation of African Americans and Hispanics. In Berkeley and Oakland, large African American populations and unified coalitions aided increasing political incorporation, but in Richmond, the size of the minority group was offset by the lack of a liberal coalition. In San Francisco and Sacramento, fairly small African American communities benefited from multiethnic cooperation, whereas large Latino populations in Stockton and San Jose were hampered by the strength of entrenched coalitions. Finally, the lack of coalitions in Hayward, Vallejo, and Daly City resulted in little minority progress. Sonenshein's (1993) work in Los Angeles parallels the research by Browning et al. As in Berkeley, a highly organized African American group built up an alliance with whites that advanced Tom Bradley, an African American, to the mayor's office, where he remained from 1973 to 1989. The city moved forward on many minority issues, including increased federal aid, better representation on city commissions, and closer control over the police department. However, it was the middle-class and some working-class minorities who benefited. The poor saw little real improvement, because changing federal policies, population shifts, and other factors retarded economic gains. Thus, the city had almost 20 years of a biracial coalition with an African American mayor, but it still erupted in violence in 1992.

Uhlaner's (1991) research on the prospects of coalition building in the state found considerable differences among groups (Asians, Anglos, African Americans, and Mexican Americans) on their perception of structural inequality. Factors such as educational level and immigrant generation cut across simple racial categories. Yet, overall, Uhlaner was cautiously optimistic about the potential for constructing alliances in the future. Research on the Hispanic vote for Proposition 187, which cut back bilingual education programs, is an example of how differences within groups surface during an election. The Hispanic community did not unanimously oppose the proposition. An important factor in voting patterns was the opposition from newer immigrant groups, less likely to be predominantly English-speaking, who saw the proposition targeting them (Newton, 2000).

Furthermore, the dynamics of multiracial coalitions have become increasingly complex as Asian groups move into third place in the state's population. Once holding a low profile in urban politics because of their small numbers and lack of high voter registration, Asians are now a major force in some areas. They, too, are not always unified, as traditional leaders are challenged by highly educated new arrivals with ties to Pacific Rim nations, rather than to local Chinatowns or Koreatowns, whereas poorer immigrant and refugee groups tend to be disenfranchised from the system (Nakanishi, 1991).

How these complex demographic realities played out in California in the 1990s is seen in a case study (Fong, 1994) of Monterey Park, the first suburban Chinatown. Close to Los Angeles, the city was once an all-white bastion with exclusionary racial covenants in deeds. After World War II, it became a stepping stone community for middle-class Mexican, Chinese, and Japanese Americans, although middle-class African Americans were not well received. The Hispanic and Asian populations grew rapidly as whites began to leave, with the two groups being about equal in number in 1980. By 1990, however, whites had declined to 11.7% of the population, Latinos accounted for 31.4%, and Asians, mostly Chinese, made up 56% of the community. The 1992 City Council race found the new Asian immigrant economic elite competing with liberal Chinese Americans, who formed coalitions with other groups. Fong (1994) suggests that policy and political solutions must now embrace "better understanding of the international, multiracial, multicultural, and dynamic class reality exemplified by Monterey Park" (p. 177). Saito (1998), too, found accord between Latinos and Asians in the San Gabriel Valley, where demographic balance aided cooperation. However, he cautions that the alliance was created in part because both groups recognized the need to gain political power to resist the discrimination and prejudice they faced. This coalition could fall apart as the balance of power shifts due to rising numbers of Asian Americans in the San Gabriel Valley.

As community activist Bong Hwan Kim (1994) notes,

> Ultimately, we need a multiracial coalition that includes European Americans, a coalition that supports true equality and enfranchisement. The toughest part will be convincing those with the most that even if a redistribution of power means no gains for them in the short term, the society as a whole will be better for everyone in the long term. (p. 90)

Given the current swing toward the political right, this will indeed be a difficult task. It is hoped that sociology can play a role in directing the course of change by increasing awareness of the continued impact of race for those on the lowest rungs of the ladder in the Golden State. If the public does not recognize continuing patterns of inequality, then social earthquakes will be as much a part of the state's environment as geological ones.

NOTES

1. These numbers are based on the first reports from the 2000 U.S. census.
2. The term *white* will be used to designate non-Hispanic whites, and Hispanic or Latino will be used for those of any race.
3. The index of dissimilarity measures urban segregation. Based on census blocks, it measures the percentage of African Americans or whites who would have to be moved so that each block represented the percentage of those groups found in a particular city. When geographic variation is examined further on five distinct dimensions, a pattern of hypersegregation emerges (Massey & Denton, 1993, pp. 74-78).
4. Middleman minorities are entrepreneurs who serve another minority population that may be ignored by the dominant group (Bonacich, 1972). They are often in a perilous position socially and economically, caught between two groups, because they are generally culturally and racially different from their clientele. They may be attacked by the dominant group for serving minorities, as was the case in the lynching of Italian merchants in southern states in the 1890s, or by the group being served as was the case in South Central Los Angeles. Large-scale immigration from Korea began during the 1970s, as middle-class Koreans seeking economic mobility, which their own nation did not offer them, came to these shores. Often unable to move into their professions due to their immigrant status and discrimination, Koreans turned to small businesses. Some were aided by the system of revolving credit known as a *kye*, which gives members additional capital (Bonacich, 1972).
5. Koreans are not the only Asian group to be seen negatively; research in a San Francisco housing project found that African Americans held unfavorable attitudes, including

cultural stereotyping and economic animosity, toward their Chinese neighbors (Guthrie & Hutchinson, 1995).

6. Ken Auletta's (1983) book, *The Underclass,* helped to define the problem of those poor who are trapped at the bottom of a stratification system in society, where key values emphasize individual success and a belief in freedom of opportunity for all. This work sparked a major controversy and introduced the term *underclass,* which is now commonly used.

7. California (Dumbauld, McCullough, & Stocky, 1994) reflects national patterns of health problems for minorities. One example is that some communicable diseases, virtually eradicated in the white population, are still endemic among minorities. For example, minorities accounted for 80% of the state's tuberculosis cases, with the rate for African Americans, Native Americans, Hispanics, and Asian/Pacific Islanders well above the white rate. Measles showed a similar pattern, with the exception of Native Americans. Although genetics is a factor in some minority health problems, for example, sickle cell anemia, medical experts suggest that socioeconomic issues are also critical. "Preventive care is often a luxury that time rarely affords minority people. . . .ceaseless demands for employment, housing, food, clothing, legal help, public assistance . . . are sufficient to ensure that many illnesses or conditions reach a critical point before health care is sought" (Heckler, 1985, p. 205).

REFERENCES

Acs, G., & Gallagher, M. (2000). *Income inequality among America's children* (Urban Institute, January report) [On-line]. Available: http://newfederalism.urban.org

Almaguer, T. (1994). *Racial fault lines: The historical origins of white supremacy in California.* Berkeley: University of California Press.

Anderson, E. (1990). *Street wise: Race, class, and change in an urban community.* Chicago: University of Chicago Press.

Anderson, S. (1991). Rivers of water in a dry place—Early black participation in California politics. In B. Jackson & M. Preston (Eds.), *Racial and ethnic politics in California* (pp. 55-70). Berkeley, CA: IGS Press.

Auletta, K. (1983). *The underclass.* New York: Vintage.

Austin, J., & Irwin, J. (2001). *It's about time: America's imprisonment binge* (3rd ed.). Belmont, CA: Wadsworth.

Bergesen, A., & Herman, M. (1998). Immigration, race, and riot: The 1992 Los Angeles uprising. *American Sociological Review, 63,* 39-54.

Bobo, L., Zubrinsky, C., Johnson, J., Jr., & Oliver, M. L. (1994). Public opinion before and after a spring of discontent. In M. Baldassare (Ed.), *The Los Angeles riots, lessons for the urban future* (pp. 103-134). Boulder, CO: Westview.

Bonacich, E. (1972). A theory of middleman minorities, the split labor market. *American Sociological Review, 38,* 83-94.

Broussard, A. (1993). *Black San Francisco: The struggle for racial equality in the West, 1900-1954.* Lawrence: University Press of Kansas.

Browning, R., Marshall, D. R., & Tabb, D. (1984). *Protest is not enough: The struggle of blacks and Hispanics for equality in urban politics.* Berkeley: University of California Press.

California Judicial Council Advisory Committee. (1997). *Final report on racial and ethnic bias in the courts.* Sacramento: Author.

California Statistical Abstract. (1999). Sacramento: State of California Printing Office.

Camarota, S. (1999). *Importing poverty: Immigration's impact on the size and growth of the poor population in the United States* [On-line]. Available: http://www.cis.org.articles/poverty_study/findings.html

Cheng, L., & Espiritu, Y. (1989). Korean business in black and Hispanic neighborhoods: A study of intergroup relations. *Sociological Perspectives, 32,* 521-552.

Cho, S. (1993). Korean Americans vs. African Americans: Conflict and construction. In R. Gooding-Williams (Ed.), *Reading Rodney King, reading urban uprising* (pp. 196-214). New York: Routledge.

Clark, W.A.V. (1996). Residential patterns: Avoidance, assimilation, and succession. In R. Waldinger & M. Bozorgmehr (Eds.), *Ethnic Los Angeles* (pp. 109-138). New York: Russell Sage Foundation.

Cook, S. (1972). The destruction of the California Indian. In R. Daniels & S. Olin, Jr. (Eds.), *Racism in California: A reader in the history of oppression* (pp. 13-19). New York: Macmillan.

Cornelius, W., Martin, P., & Hollifield, J. (1994). *Controlling immigration: A global perspective.* Stanford, CA: Stanford University Press.

Cornell, S., & Hartmann, D. (1998). *Ethnicity and race: Making identities in a changing world.* Thousand Oaks, CA: Pine Forge.

Dentler, R., & Hunt, S. (1992). The Los Angeles riots of spring 1992: Events, causes, and future policy. *Sociological Practice Review, 3,* 229-244.

Dumbauld, S., McCullough, J., & Stocky, J. (1994). *Analysis of health indicators for California's minority populations* (No. 180M-5-92). Sacramento, CA: Minority Health Information Improvement Project.

Ellison, B., & Pochy, G. (2000). *California: An economic profile.* Sacramento: California Trade & Commerce Agency.

Ethnic Minorities Advisory Committee. (1995). *Ethnic minorities in Great Britain* [On-line]. Available: http://www.blink.org.uk/reports2/ethic.htm

Farley, R., & Frey, W. (1994). Changes in the segregation of whites from blacks. *American Sociological Review, 59,* 23-45.

Fong, T. (1994). *The first suburban Chinatown: The remaking of Monterey Park, California.* Philadelphia: Temple University Press.

Fredrickson, G. (1988). *The arrogance of race: Historical perspectives on slavery, racism, and social inequality.* Middletown, CT: Wesleyan University Press.

Freer, R. (1994). Black-Korean conflict. In M. Baldassare (Ed.), *The Los Angeles riots: Lessons for the urban future* (pp. 175-204). Boulder, CO: Westview.

George, C., & Just, A. (1992). *Enhanced student outcomes and valuable community resource: Evaluation results about California's Indian Education Centers.* Sacramento: California Department of Education.

Gonzales, J., Jr. (1990). *Racial and ethnic groups in America.* Dubuque, IA: Kendall/Hunt.

Gordon, L. (1999). *The great Arizona orphan abduction.* Cambridge, MA: Harvard University Press.

Guthrie, P., & Hutchinson, J. (1995). The impact of perceptions on interpersonal interactions in an African American/Asian American housing project. *Journal of Black Studies, 25,* 377-395.

Heckler, M. (1985). *Black & minority health* (Vol. 1). Washington, DC: U.S. Department of Health and Human Services.

Hodge, W. (1999, February 23). London's police force riddled with institutional racism. *San Francisco Chronicle*, p. A1.

Hornor, E. (Ed.). (1992). *California cities, towns, & counties: Basic data profiles for all municipalities & counties.* Palo Alto, CA: Information Publications.

Kaufmann, K. (1998). Racial conflict and political choice: A study of mayoral voting behavior in Los Angeles and New York. *Urban Affairs Review, 33*(5), 655.

Kim, E. (1994). Between black and white: An interview with Bong Hwan Kim. In K. Aguilar-San Juan (Ed.), *The state of Asian America: Activism and resistance in the 1990s* (pp. 71-100). Boston: South End Press.

Kluegel, J., & Smith, E. (1986). *Beliefs about inequality: Americans' views of what is and what ought to be.* New York: Aldine De Gruyter.

Kroeber, T. (1964). *Ishi in two worlds: A biography of the last wild Indian in North America* (5th ed.). Berkeley: University of California Press.

Massey, D., & Denton, N. (1993). *American apartheid: Segregation and the making of the underclass.* Cambridge, MA: Harvard University Press.

McLemore, S. D. (1994). *Racial and ethnic relations in America.* Boston: Allyn & Bacon.

Menchaca, M. (1995). *The Mexican outsiders: A community history of marginalization and discrimination in California.* Austin: University of Texas Press.

Merton, R. (1968). *Social theory and social structure.* New York: Free Press.

Monroy, D. (1990). *Thrown among strangers: The making of Mexican culture in frontier California.* Berkeley: University of California Press.

Morrison, P., & Lowry, I. (1994). The demographic setting. In M. Baldassare (Ed.), *The Los Angeles riots: Lessons for the urban future* (pp. 19-46). Boulder, CO: Westview.

Murray, C., & Herrnstein, R. (1996). *The bell curve: Intelligence and class structure in American life.* New York: Free Press.

Nakanishi, D. (1991). The next swing vote? Asian Pacific Americans and California politics. In B. Jackson & M. Preston (Eds.), *Racial and ethnic politics in California* (pp. 25-54). Berkeley, CA: IGS Press.

Navarro, A. (1993). The South Central Los Angeles eruption: A Latino perspective. *Amerasia Journal, 29,* 69-85.

Newton, L. (2000). Why some Latinos supported Proposition 187. *Social Science Quarterly, 81,* 180-193.

Oliver, M., Johnson, J. H., Jr., & Farrelly, W., Jr. (1993). Anatomy of a rebellion: A political-economic analysis. In R. Gooding-Williams (Ed.), *Reading Rodney King, reading urban uprising* (pp. 117-142). New York: Routledge Press.

Omi, M., & Winant, H. (1986). *Racial formation in the United States: From the 1960s to the 1980s.* New York: Routledge & Kegan Paul.

Ong, P., & Lawrence, J. (1992). *Pluralism and residential patterns in Los Angeles.* Los Angeles: UCLA School of Architecture and Urban Planning.

Pitt, L. (1971). *The decline of the Californios: A social history of the Spanish-speaking Californians, 1846-1890.* Berkeley: University of California Press.

Rushton, J. P. (1999). *Race, evolution, and behavior.* New Jersey: Transaction.

Saito, L. (1998). *Race and politics: Asian Americans, Latinos, and whites in a Los Angeles suburb.* Chicago: University of Illinois Press.

Scherini, R. (2000). When Italian Americans were enemy aliens. In F. Iacovetta, R. Perin, & A. Principe (Eds.), *Enemies within: Italian and other internees in Canada and abroad* (pp. 280-306). Toronto: University of Toronto Press.

Schiraldi, V., Kuyper, S., & Hewitt, S. (1996). *Young African Americans and the criminal justice system in California: Five years later.* San Francisco: Center of Juvenile and Criminal Justice.

Siddle, R. (1997). Ainu: Japan's indigenous people. In M. Weiner (Ed.), *Japan's minorities: The illusion of homogeneity* (pp. 17-29). New York: Routledge.

Sonenshein, R. (1993). *Politics in black and white: Race and power in Los Angeles.* Princeton, NJ: Princeton University Press.

Takaki, R. (1989). *Strangers from a different shore: A history of Asian Americans.* New York: Penguin.

Tonry, M. (Ed). (1997). *Ethnicity, crime, and immigration: Comparative and cross-national perspectives.* Chicago: University of Chicago Press.

Uhlaner, C. (1991). Perceived discrimination and prejudice and the coalition prospects of blacks, Latinos, and Asian Americans. In B. Jackson & M. Preston (Eds.), *Racial and ethnic politics in California* (pp. 339-372). Berkeley, CA: IGS Press.

U.S. Bureau of the Census. (1992). *1990 census of population: Social and economic characteristics, U.S.* Washington, DC: Government Printing Office.

U.S. Bureau of the Census. (1993). *1990 census of population: Social and economic characteristics, California* (Vol. 1). Washington, DC: Government Printing Office.

U.S. Bureau of the Census. (2000). *Population estimates for counties by race and Hispanic origin: July 1, 1999* (CO-99-11). Washington, DC: Population Estimates Program, Population Division.

Waldinger, R. (1996). Ethnicity and opportunity in the plural city. In R. Waldinger & M. Bozorgmehr (Eds.), *Ethnic Los Angeles* (pp. 445-470). New York: Russell Sage Foundation.

Wieviorka, M. (1995). *The arena of racism.* London: Sage.

Wildermuth, J. (1995, October 12). The dream dies hard in L.A. *San Francisco Chronicle,* p. 1.

Part II

CHALLENGES TO SEXUALITY AND HEALTH

In this section, we will be looking at the issues of heterosexism/homophobia and the problem of AIDS/HIV in California.

The fourth chapter, "Resisting Heterosexism and Homophobia: The Gay and Lesbian Movement for Equal Rights," by Peter M. Nardi, commences with a review of the history of lesbian and gay oppression and resistance in California. Nardi asserts that the tensions between representatives of the state and gay and lesbian citizens, tensions that exist in all parts of the nation, are often more visibly exhibited and publicly expressed in California. In California, many important national social changes in attitudes and laws toward gays and lesbians began.

Nardi discusses many of the most important first steps in the history of the modern gay social movement in California in the 1950s, from the formation of the Mattachine Society, an organization for homosexual men, to the creation of the first lesbian organization, the Daughters of Bilitis. The author shows how such organizations helped gays and lesbians of that era endure the harassment of politicians, police, and straight citizens alike. Also discussed is the importance of World War II, which brought gays and lesbians from all over the country to major U.S. military installations in San Francisco, Long Beach, and San Diego. Many gay men and lesbians discovered their identities and each other during the war, and they stayed on in California's port cities to join the first gay and lesbian networks and communities. In the years after the war, the growing visibility of gays and lesbians met some resistance, and numerous laws—such as laws prohibiting gay bars—were passed in California to counteract homosexuals. Such anti-homosexual legislation, combined with federal and military purges and local raids, drove gays and lesbians underground during the 1950s. Nardi discusses the gay subculture and movement that began to emerge in the late 1950s in California, particularly in San Francisco. In the early 1960s, many gays and lesbians were entrapped in homosexual bars by undercover police and arrested. Gays and lesbians and their supporters in San Francisco formed numerous organizations to counter this harassment, and a community took form around a shared sexual orientation. These organizations were made of people who started to redefine themselves, not as being sick and perverted, but as sharing an identity oppressed by the dominant culture. By the early 1970s, several core institutions of society—religion, law, education, mental health, police—were challenged by the growing gay movement for equal rights.

Nardi points out how this growing presence of the homosexual community led to a backlash by the New Right in the late 1970s. In California, State Senator John Briggs created Proposition 6, an initiative that would have prohibited gays and lesbians from teaching in public schools. The homosexual community united in force, and the proposition was soundly defeated.

Then in 1978, a gay San Francisco city councilman, Harvey Milk, was assassinated along with San Francisco Mayor George Moscone. This incident galvanized the homosexual community as none had before.

The author points out that the appearance of AIDS in the gay community in the early 1980s caused not only tremendous suffering and grief but also a backlash from the conservative right, blaming gays for the emergence of AIDS. The decade of the 1980s proved to be a tumultuous one for homosexuals in California, as they had to battle AIDS and the rising fear of gay people, who were now synonymous with carriers of terminal diseases. Nardi discusses the growing problem of hate crimes against homosexuals in the early 1990s and the gay community's mobilization of resources to measure the incidence of such hate crimes and to educate the public about gay and lesbian lives.

In his chapter, Nardi shows how the conflict between forces that oppose homosexuality and those that support equal treatment under the law for gays and lesbians has led to a mix of legislative and judicial outcomes. California Assembly Bill 2601, which went into effect in January 1993, bans discrimination against homosexuals in public and private employment, in education, and in public accommodations. This law exemplifies a legislative outcome beneficial to gays and lesbians. Then-governor Pete Wilson's veto of AB 2810, which would have allowed unmarried couples (gay and non-gay) to register their relationships with the Secretary of State and receive some married-couple benefits, represents action that is not beneficial to the gay and lesbian community. Then again, in 1999 and 2000, Governor Gray Davis signed into law policies that would allow same-sex partners to receive the same treatment that heterosexual partners receive in various areas, such as health benefits, hospital visitation rights, and so on. Other new laws have recently been enacted in California to deal with hate crimes and civil rights of homosexuals. The author points out that such legislation is necessary, given the negative events that gays and lesbians continue to encounter, including hate crimes. Nardi points out that the California Department of Justice reported that from 1995 to 1999, there was a 37% increase in hate crimes against gays and lesbians in California.

When the author compares California to the rest of the nation on homosexual rights, California appears relatively more progressive. In addition to California, only 10 other states and the District of Columbia have enacted laws prohibiting discrimination based on sexual orientation. Recent attempts to turn back the clock on homosexual rights in Colorado and Oregon serve to remind the gay community and its supporters that their hard-won rights must be protected. On a more positive note, Nardi discusses the law that passed in Vermont in July 2000, making Vermont the first state to recognize same-sex civil unions, thus giving gays and lesbians the same rights as heterosexuals. Nardi concludes this section by indicating that the climate of the country is changing toward greater acceptance of the diversity of its citizens, including diversity of sexual orientation. The national publicity given to the 1998 brutal slaying of openly gay student Matthew Shepard in Wyoming has led to more calls for extending hate crime legislation to protect gays and lesbians.

In his section on "Lesbian and Gay Issues in a Global Context," the author discusses some gay and lesbian social movements around the world. He asserts that it would be a mistake to assume that globally, people who identify themselves as gay or lesbian do so in the same way and with the same political and social issues as Americans. There is much variation across countries and cultures. Nardi points out that California has provided information, assistance, and cultural images about gay and lesbian lives to many people and groups around the world. Television shows emanating from Hollywood, with gay and lesbian actors, are beamed around the world, serving as a good example of this.

Nardi feels that conflict theory and symbolic interactionist theory are both useful in explaining lesbian and gay oppression and resistance. The history of the lesbian and gay movement is a clear illustration of conflict theory because the competition over whose rules and laws dominate characterizes most of the debates. The author points out that symbolic interactionist theories are especially useful in helping us understand the dynamics of social movements and change. By looking at the everyday world of interactions, symbols, and experiences, we can begin to uncover the ways in which gays and lesbians encounter the world.

Nardi's discussion of future prospects for lesbian and gay oppression and resistance in California looks at several areas: the political, the interpersonal/social, and the cultural. Politically, battles are being fought at both the national and local levels. Many contemporary debates focus on cultural issues, as evidenced by the many conservative candidates who have been running in school board elections. Here, the arguments typically focus on values and the cultural meaning of homosexuality. With regard to changes in the interpersonal/social sphere, research consistently shows that positive and frequent encounters with minorities contribute to a reduction in prejudice. Knowing a gay or lesbian similarly reduces anti-gay attitudes. Gays and lesbians now carry out their daily lives more openly in non-gay environments, and this openness will, in all likelihood, result in increased tolerance of gay lifestyles.

Finally, Nardi discusses possible solutions to the problem of lesbian and gay oppression and resistance. First, even though much of the fight for gay/lesbian rights has been, and will continue to be, fought in California, various federal legislative measures that protect homosexuals need to be taken by the U.S. Congress. Second, changing media images about gays and lesbians so that they are more realistic would help in the resistance of heterosexism and homophobia. Another solution is to work at the state political level by registering gay and lesbian voters, by running openly gay and lesbian candidates, and so on. Also, the criminal justice system must be trained and informed about the issues surrounding hate crimes involving sexual orientation and anti-gay violence. Finally, given the importance of social and interpersonal encounters, it becomes very important for workplaces and other institutions to create climates in which gays and lesbians can more openly lead their lives.

Chapter 5, "HIV/AIDS," by Kevin D. Kelley, starts out by informing us that the state of California contains 16% of all Americans living with AIDS, second only to the state of New York. The author then proceeds to describe the disease and how it is transmitted. Next, Kelley gives a thorough history of AIDS in California, starting with the early 1980s, when physicians in San Francisco and Los Angles began identifying men with infections and cancers they knew to be caused by an immune system deficiency, and ending with the end of the 20th century, which saw a larger and larger share of the epidemic being shared by racial and ethnic minorities. The author then presents sobering statistics that show how the number of AIDS diagnoses and AIDS deaths has grown exponentially from 1990 to 1999. Even more sobering and distressing is the realization that current statistics on AIDS are what Kelley calls "trailing indicators of the epidemic," numbers that include only those people who have been officially diagnosed with the disease. Given the lengthy period of incubation (2 to 15 years), present cases of HIV infection and full-blown AIDS could be just the "tip of the iceberg" of a possibly devastating plague.

Kelley also points out that whereas AIDS in California disproportionately affects males, the growth in AIDS infection in women of color is alarming. This is especially the case for African American females. The author also discusses how Californians become infected with HIV. Whereas men having sex with men remains the dominant way that the HIV virus is transmitted, heterosexual AIDS transmission among intravenous drug users is rising more rapidly than any other category.

In his section on the problem of HIV/AIDS in California, the author demonstrates that the disease is spread disproportionately throughout counties and is clustered in urbanized areas. He also delineates the tremendous direct and indirect economic cost of HIV/AIDS to the state of California and how this burden is being borne by both the private and public sectors. This cost has escalated as new generations of anti-viral drugs, such as protease inhibitors, have become available. The author reports that more than $195,000 is spent in medical treatment costs alone to treat a person with HIV from diagnosis to death. Kelley also deals with the enormous human costs associated with this dread disease.

Comparing California's situation to that of the nation, Kelley finds the trends in California mirroring the nation. As in California, new cases of HIV/AIDS are more likely to be found in communities with low income levels and high proportions of minorities. Also, as in California, women are disproportionately being infected, and heterosexual transmission of HIV/AIDS is the fastest-growing category of transmission.

When Kelley looks at California's problem with HIV/AIDS in a global context, he finds the circumstances around the world to be similar to what is occurring in the Golden State, that is, relatively high rates of HIV/AIDS among men who have sex with men, injection drug users, racial and ethnic minorities, women, and residents in pockets of urban poverty—with relatively lower rates throughout the rest of the population. However, the author finds that circumstances throughout the developing world are significantly more serious, particularly in Africa.

In the section called "Alternative Explanations for the Spread of HIV and AIDS," Kelley uses an extensive quote from Randy Shilts's book, *And the Band Played On*, to show how in 1982 the U.S. government sprang into action and spent enormous amounts of money to deal with the cyanide-laced Tylenol capsule scare in Chicago, while virtually ignoring the AIDS cases that were striking American citizens at the same time. The author points out that the media gave the Tylenol scare story tremendous coverage while giving very little coverage to AIDS. Why all the attention to the Tylenol scare and so little to AIDS victims? The author answers by pointing out that the Tylenol scare was affecting "regular" Americans whereas AIDS was affecting homosexuals. Conflict theory is used to explain why so little interest was shown in combating AIDS in the 1980s. Kelley argues that it was not until the disease started to affect heterosexuals, and not until enough money could be made selling drugs to victims of AIDS, that the drug industry started to develop drugs to combat AIDS. Pharmaceutical companies have been making huge profits from sales of drugs developed to deal with AIDS, at the expense of AIDS victims. The author shows that other companies, such as those that buy up life insurance policies of AIDS victims, also have been making enormous profits from the AIDS epidemic.

The section on future projections of AIDS in California is unnerving. The author presents data from the U.S. Centers for Disease Control showing exponential growth in cumulative HIV infections, living AIDS cases, and cumulative AIDS deaths. Furthermore, minority groups will increasingly account for higher and higher proportions of AIDS infections, and minority children will be increasingly at risk for AIDS.

In the last section of this chapter, Kelley presents seven unique and insightful recommendations to address the AIDS epidemic in California.

CHAPTER 4

RESISTING HETEROSEXISM AND HOMOPHOBIA
The Gay and Lesbian Movement for Equal Rights

Peter M. Nardi

On March 7, 2000, 61% of California voters approved Proposition 22, the initiative sponsored by Republican State Senator William "Pete" Knight. This measure restricts marriage in the state to two people of the opposite sex, thereby prohibiting so-called gay marriages. Yet, according to a poll conducted in April 2000 by Decision Research, 81% of Californians oppose discrimination on the basis of sexual orientation (Warren, 2000). The same poll also found that 42% of respondents believe homosexuality is morally wrong, whereas 54% say that **homophobia** (the irrational fear and hatred of homosexuals) is morally wrong. These views reflect the often ambiguous and contradictory relationship that has existed for some time between the state of California and its lesbian and gay citizens.

Like its ecology, California is a state of many contrasts politically and socially. What can be said about a state that elects a conservative Republican like Ronald Reagan for governor, yet sends a liberal Democrat like Barbara Boxer to the Senate? How can it be home to both San Francisco (the symbolic center of gay America) and Orange County's Traditional Values Coalition, a religious right organization actively campaigning against equal rights for gays and lesbians? How can you categorize a state that includes not only those people who gave birth to the modern American gay movement through the formation in Los Angeles of the Mattachine Society in 1951 but also the person who assassinated Harvey Milk, the first openly gay political official, in 1978, or those who contribute to the rise in hate crimes based on sexual orientation?

These seemingly contradictory events and organizational forces, their interconnections, and the ways they can be understood sociologically are the subjects of this chapter. The attitudes expressed toward gays and lesbians in California cannot be explained simply by individual homophobia. Such attitudes are more clearly understood by uncovering the political, cultural, and social structural arrangements that favor heterosexuality (heterosexism), reproduce institutional homophobia, and legitimize individuals' fear of and violence against lesbians and gay men.

Like racism, classism, and sexism, **heterosexism**—the set of institutional practices and assumptions that privilege heterosexuality—is best explained by looking to sociological theories that focus on analyzing the social organization, conflicts, and meanings embedded in the social system. How California deals with lesbian and gay issues illustrates not only the complexity of our contemporary social system but also the numerous ways diversity based on sexual orientation must constantly be negotiated between the state and its citizens.

A HISTORY OF LESBIAN AND GAY OPPRESSION AND RESISTANCE

In the earliest hours of the New Year, 1967, the Black Cat, a gay bar on Sunset Boulevard in the Silver Lake neighborhood of Los Angeles, was raided by the Los Angeles Police Department. An employee was beaten, and others were arrested there and at a nearby bar down the street. In some ways, this was not unusual. Vice squads regularly harassed people in gay bars and public places in San Francisco, Los Angeles, and other major cities throughout California and the nation in the decades following World War II. For 3 weeks in November 1968, 75 San Diego men were arrested in Balboa Park and the nearby beaches for homosexual activities (see Thompson, 1994, for a historical chronology of the important gay/lesbian events cited here and throughout this chapter). And as far back as 1906, some religious leaders blamed the San Francisco earthquake on the city's tolerance of "deviants."

But the Black Cat raid was unusual in that, within a few weeks, a group called Personal Rights in Defense and Education (PRIDE) organized the largest gay protest march to date in California, 2 years before the infamous Stonewall rebellion at a bar in New York City in June 1969 (the symbolic start of the contemporary gay and lesbian movement). Given the lack of coverage by local newspapers, PRIDE decided to turn its newsletter into a regularly scheduled gay newspaper, called *The Advocate*. By 1968, more than 5,000 copies were circulated throughout Southern California. By the late 1980s, *The Advocate* had become the most widely read gay magazine in the nation, providing an important communications link to the many lesbians and gay men who lived outside of major urban areas.

In many ways, the story of the bar raid at the Black Cat, the emergence of a gay protest group, and the development of a major gay communications medium—all occurring in Southern California[1]—represent many historical and contemporary dynamics of the modern gay and lesbian movement nationally and globally. Tensions between representatives of the state and gay and lesbian citizens, which exist in all parts of the nation, are often more visibly exhibited and publicly expressed in California. In California, many important national social changes in attitudes and laws toward gays and lesbians began.

In fact, some of the most important first steps in the history of the modern gay social movement began in California, in particular, in Los Angeles (see D'Emilio, 1983, for details about this early history). There, in November 1950, Harry Hay, Rudi Gernreich, Chuck Rowland, Bob Hull, and Dale Jennings met in a Silver Lake house to discuss the formation of an organization for homosexual men. By April 1951, a "homophile" group called the Mattachine Society was started, named after guilds of entertainers who wandered in 11th-century southern Europe, using their art in the service of the common people.

The Mattachine Society became an important group, with chapters throughout the country. Another Los Angeles homophile organization, One, Inc., was formed in 1952. It published the first widely circulated gay magazine (*One*) in the country.[2] In 1955, the first lesbian organization, the Daughters of Bilitis, started in San Francisco. These associations served the personal and social needs of those who felt isolated and alienated in society, as well as the political need to organize in resistance to social oppression.

During the 1950s, the medical, psychiatric, and legal systems considered homosexuals to be sick, perverted, and criminal. Although some lived in coupled relationships or were comfortable with themselves as single men and women, many lived closeted lives in unhappy marriages, in lonely alcoholic hazes, risking their careers and reputations merely by entering a bar known to attract homosexuals (see Nardi, Sanders, & Marmor, 1994). In the McCarthy era, known homosexuals (or as the State Department would refer to them, "perverts") were routinely hunted down and fired from jobs, ostracized by their families, and denied basic civil rights. So great was the possibility of entrapment by police that, in 1952 in Los Angeles, the Committee to Outlaw Entrapment was established and successfully fought to have a case dropped (Gorman, 1991).

Numerous military bases were located in California during World War II, with San Francisco, Long Beach, and San Diego serving as major military re-entry ports. Some gay men and lesbians discovered their identities and each other during the war, and they stayed on in the port cities to join the first gay networks and communities (see Bérubé, 1990). By the 1940s, many California cities (in particular, Long

Beach, Los Angeles, Oakland, San Diego, and San Francisco) had both lesbian and gay bars, gay areas, and gay beaches (Gorman, 1991). However, military police often cracked down and cited many of these bars for liquor law violations, declaring them off-limits to military personnel. In 1942 and 1943, for example, a series of vice raids led to the closure of at least six or seven gay bars in San Francisco. Ironically, this resulted in the emergence of a strong underground network, a more public nightlife, and the beginnings of a political sensibility of resistance. As Bérubé (1990) wrote, "By uprooting an entire generation, the war helped to channel urban gay life into a particular path of growth—away from stable private networks and toward public commercial establishments serving the needs of a displaced, transient, and younger clientele" (p. 126).

In the years after the war, the growing visibility of gay and lesbian people met some resistance. Claiming that San Francisco's Black Cat Cafe was a "disorderly house," the California Board of Equalization in 1949 tried to close it. Successful appeals in 1951 to the State Supreme Court finally "led the California legislature in 1955 to pass a law prohibiting the licensing of gay bars because they were 'resorts for sexual perverts' " (Bérubé 1990, p. 356, n. 31). Combined with federal and military purges and local vice raids, gays and lesbians were driven underground during the 1950s, although bars in many cities remained the primary meeting place for many of them. It was in this kind of climate that the first homophile groups in the nation (Mattachine, One, Daughters of Bilitis) were founded in California in the 1950s.

By the end of the decade, a gay subculture and movement began to emerge in California, particularly in San Francisco. The bohemian "beat" subculture of the early 1950s (among such poets as Allen Ginsberg, Robert Duncan, and Lawrence Ferlinghetti) developed an "opposition to the bland conformity and consumerism of the postwar years" through its poetry, bookstores, cafes, and bars (D'Emilio, 1983, p. 177). Coexisting and, at times, overlapping with the homosexual subculture of bars and cafes, the beat generation in San Francisco helped propel the growing gay movement into more protest-oriented directions. When Ginsberg's famous poem, "Howl," was confiscated on obscenity charges in 1957 by the San Francisco police department, primarily because of its homosexual content, a series of gay protests, legal victories, and consciousness-raising began.

San Francisco, and by extension the entire state of California, soon became associated with a more radical, bohemian, alternative culture. During the 1959 mayoral campaign, opponent Russell Wolden accused San Francisco Mayor George Christopher and his police chief, Thomas Cahill, of letting the city become the homosexual capital of the nation. Soon after his re-election, Christopher announced a campaign against gay bars in San Francisco as a way of cleansing the city's allegedly scarred and vice-filled reputation (D'Emilio, 1983).

Then, in December 1959, the California Supreme Court ruled that a gay bar could not lose its license just because homosexuals congregate there; illegal sexual activity on the premises needed to be proved. One result of this ruling was that several bar owners publicly accused the police and the Alcohol Beverage Control departments of taking payoffs, resulting in what the press called a "gayola" scandal in San Francisco (D'Emilio, 1983). But the angry police and city government in 1960 took this opportunity to send plainclothesmen into gay bars to entrap patrons.

As D'Emilio (1983) reports, felony convictions of male homosexuals in San Francisco rose to 29 by the end of 1960 and to 76 by mid-June 1961, from a June 1960 rate of 0. Misdemeanor charges were estimated at 40 to 60 a week, and 89 men and 14 women were arrested in August 1961 at the Tay-Bush Inn. In October 1961, "Every establishment that had made charges against the police during the gayola scandal lost its license" (D'Emilio, 1983, p. 184).

Several significant events followed throughout the 1960s. In the 1961 election for San Francisco's city supervisors, José Sarria, an openly gay man, received almost 6,000 votes. In 1962, the city's gay bars organized into the Tavern Guild, which sponsored voter registration and provided legal assistance for those arrested in the bars. In 1964, the Society for Individual Rights was founded to create a sense of community among gay men and to start a political action committee. In December of that year, liberal Protestant ministers and gay leaders formed the Council on Religion and the Homosexual, which resulted in the religious leaders condemning police harassment of a gay New Year's dance the council sponsored (D'Emilio, 1983; Wolf, 1979).

All these events helped San Francisco to become the "gay capital" of the United States a decade before its more publicly celebrated fame of the 1970s. As D'Emilio (1983) concluded about the qualitatively profound changes occurring in the consciousness of gay San Franciscans, "A 'community' was in fact forming around a shared sexual orientation, and the shift would have important implications in the future for the shape of gay politics and gay identity throughout the nation" (p. 195).

A variety of social, economic, and political forces converged between the end of World War II and the beginning of the 1970s, including the growth of jobs, cities, and industries throughout postwar California; a network of military bases; economic and business associations of gay bars; changes in the laws dealing with sexuality and obscenity due to State Supreme Court rulings; the critique of dominant culture by the beat generation; the rise of the women's and black civil rights movements; and the alliance of religious leaders with gay business and political leaders. These factors, as well as other cultural and economic changes in the larger society, contributed to the organization of a critical mass of people into an identity community (Murray, 1996). San Francisco and other major cities in California began to develop visible, large, and politically powerful gay and lesbian communities made up of people who started to redefine themselves, not as being sick and perverted, but as sharing an identity oppressed by the dominant culture.

By the early 1970s, several other significant events in California occurred. Founded in Los Angeles in 1968, the Reverend Troy Perry's Metropolitan Community Church grew rapidly in size. The Gay and Lesbian Community Services Center, the nation's oldest (1970) and largest gay center, began helping numerous clients and providing invaluable assistance to an underserved population in Los Angeles. In December 1969, the California Supreme Court ruled that a teacher's credentials could not be revoked for charges of homosexuality. And Troy Perry led 250 protesters in a rally seeking reform of the Los Angeles Police Department in January 1970. In short, several core institutions of society—religion, law, education, mental health, police—were challenged by the growing gay movement for equal rights.

Although arrests, raids, arson, beatings, and job discrimination continued against gays and lesbians (what today might be called sexual orientation hate crimes or anti-gay violence and victimization), resistance soon emerged. Significant social, cultural, and political changes appeared. For example, the *Los Angeles Times* dropped its ban against the word "homosexual" in advertising in 1970. In June 1970, a march celebrating the first anniversary of the Stonewall riots was held in Los Angeles, and a "gay-in" rally happened in a San Francisco park. In 1972, a California Superior Court ruled unconstitutional the state law making oral sex a felony, and Hollywood filmed its first positive gay TV movie, *That Certain Summer*. In 1973, the *The Tonight Show* with Johnny Carson was zapped by gay activists protesting "fag jokes." In 1975, California passed a law decriminalizing consensual sex acts, and it was signed into law by Governor Jerry Brown. In the same year, Santa Cruz County became the state's first county government to ban anti-gay discrimination in employment.

As the growing infrastructure of agencies, associations, and organizations took shape, as antiquated laws were overturned, and as people continued to protest and resist, a significant development occurred in the large cities, especially in San Francisco, where an economic and political consolidation of resources became evident. By the mid-1970s, neighborhoods made up of gay residences and gay-owned businesses sprang up in the Castro district of San Francisco, along Santa Monica Boulevard in West Hollywood and Los Angeles, and in the Hillcrest section of San Diego. The Castro district, which was once a working-class section of inexpensive homes, experienced a white (mostly heterosexual) flight to suburbia. Young white gay men bought up the inexpensive homes, renovated them, leased storefronts for businesses, and created a visible economic and political neighborhood with substantial influence in San Francisco's growth. It soon became the national symbol for the rising gay and lesbian social movement. As Castells (1983) stated it, gays "had to organize themselves spatially to transform their oppression into the organizational setting of political power. . . .[The Castro] brought together sexual identity, cultural self-definition, and a political project in a form organized around the control of a given territory" (p. 157).

As these communities became political voting districts with some clout—enough to elect the first openly gay San Francisco city supervisor, Harvey Milk, in 1977—it was not long before the new right formally organized a countermovement (Adam, 1995). Beginning nationally in 1977 with Florida orange juice spokeswoman and former Miss America Anita Bryant, a strident and divisive campaign to limit the civil rights of gay men and lesbians began. Focusing her attention on Dade County, Bryant's attempts to restrict the rights of gay men and lesbians inadvertently gave tremendous impetus to the gay movement and individuals' personal decisions to "come out." Its impact on California gay politics was powerful. By 1977, the Municipal Elections Committee of Los Angeles (MECLA) became the first gay political action committee to create financial resources for candidates supportive of equal rights for lesbians and gays.

Bryant's campaign spawned an important anti-gay crusade in California's history. In 1978, Republican State Senator John Briggs created Proposition 6, an initiative that would have prohibited gays and lesbians from teaching in public schools and anyone from teaching about homosexuality. It was so restrictive of basic civil rights that even former Governor Ronald Reagan openly opposed it. Yet its effect on California (and the nation's) gays and lesbians, not unlike the impact of the Bryant campaign, was to provoke a group of people to mobilize resources strategically, politically, and economically. Working through the gay neighborhoods and their networks of bars, businesses, newspapers, and organizations, lesbian and gay Californians mobilized an effective campaign and learned to resist this external threat to their basic liberties. Proposition 6 went down in defeat with almost 60% of the votes going against it.

Then, San Francisco supervisor Harvey Milk and Mayor George Moscone were assassinated on November 27, 1978, by recently resigned city Supervisor Dan White. That evening, more than 25,000 people marched in candle-lit silence from the Castro to City Hall in an unprecedented public display of honor and grief. Months later, in May 1979, 3,000 people angrily protested in the streets when White's conviction was reduced to voluntary manslaughter. In 1985, he committed suicide after being released on parole after 5 years in prison.

Meanwhile, then-Governor Jerry Brown publicly supported the inclusion of sexual orientation in anti-discrimination legislation. In April 1979, he signed an executive order prohibiting discrimination in state hiring based on sexual orientation. In the same year, the State Supreme Court ruled that public solicitation of a sexual act that is legal in private is not punishable by law. And, in January 1980, hundreds of gay activists marched on Sacramento's state capitol calling for passage of Assembly Bill 1, a civil rights bill protecting gays and lesbians from discrimination.

With the visibility, organizational skills, and economic and political clout garnered in resisting the repeated legal, social, media, and political attacks throughout the 1970s, gays and lesbians were now ready to meet one of their greatest challenges: the appearance of what would later be called human immunodeficiency virus (HIV) and the complications that the virus caused, collectively called Acquired Immune Deficiency Syndrome (AIDS). In June 1981, the Centers for Disease Control announced the discovery that five previously healthy men in Los Angeles were diagnosed with a rare form of pneumocystis carinii pneumonia (PCP).

A few months later, in October, 200 Christian fundamentalists marched to bring San Francisco "back to Jesus." The beginning of a decade-long struggle between gay/lesbian people and religious and conservative leaders was evident in these early years, empowered by the 1980 election of Ronald Reagan to the presidency. The gay and lesbian communities continued to battle not only AIDS but also the rising fear of gay people, who were now synonymous with carriers of a terminal disease. That fear is evidenced by these events: the veto by California Governor George Deukmejian of a state bill to prohibit discrimination in the workplace based on sexual orientation (1984); the appearance of yet another state proposition (64, the LaRouche initiative in 1984), this time to quarantine people with AIDS and to bar people suspected of carrying HIV from certain jobs; heated debates between San Francisco city health officials and the gay community about closing the bathhouses, pitting many gays against each other (1984); the U.S. Supreme Court upholding Georgia's sodomy law in *Bowers v. Hardwick* (1986); and Congressman William Dannemeyer, Republican from Fullerton, testifying before the Presidential Commission on AIDS in 1988 and calling AIDS "God's punishment" for homosexuals.

These appeared to be significant setbacks to equal rights for gays and lesbians in California and the nation, in part in reaction to the growing AIDS epidemic. However, as with all attempts to oppress, gays and lesbians mobilized to resist. Activist groups such as ACT-UP (AIDS Coalition to Unleash Power) organized to deal with AIDS discrimination and called for better health care. Other groups formed to continue to fight for equal rights in the workplace and in housing for gays and lesbians in California by marching in support of Assembly Bill 101 and by working within the political system. West Hollywood incorporated as a city in 1984 and elected a gay majority to its city council. California voters in 1988 once again defeated a Lyndon LaRouche initiative to quarantine people with AIDS (Proposition 69). And in 1988, the Gay and Lesbian Alliance Against Defamation (GLAAD) opened a powerful branch in Los Angeles to fight inaccuracies and negative images of gays and lesbians in the media.

Of growing concern to many gays and lesbians was the rise of **hate crimes** and violence based on actual or perceived sexual orientation. Gays and lesbians had been forced to rely on themselves at first when resisting the AIDS epidemic, by setting up their own social and health networks in the face of government indifference. For some time, gays and lesbians also took it on themselves to measure the discrimination, defamation, and violence committed against them by those who were homophobic. However, one of the first major national victories, which had important outcomes in California, occurred in 1990 when Republican President George Bush signed into law a bill that included the monitoring of hate crimes based on sexual orientation.

Part of the effort to educate the public about gay and lesbian lives was to develop a system to record and measure the kinds of attacks against gays and lesbians and against those who appeared to be gay and lesbian. Although many gay organizations continue to monitor these various hate crimes, several state, county, and federal agencies are now involved in tracking them. The campaign against violence and attempts to codify laws to prevent other forms of social and legal discrimination against gays and lesbians mark much of the contemporary debate in California.

LESBIAN AND GAY OPPRESSION AND RESISTANCE IN CALIFORNIA TODAY

On September 30, 1991, and continuing for an unprecedented 10 days, hundreds of lesbian and gay men took to the streets of Los Angeles protesting then-governor Pete Wilson's veto of State Assembly Bill 101, which would have prohibited discrimination in the workplace and housing based on sexual orientation. Similar protests were happening in San Francisco, San Diego, Palo Alto, and other communities throughout California. People were angry, not only because the governor had promised during his campaign to support the bill but also because, in 1984, former Governor George Deukmejian had vetoed a similar bill (AB 1). Gays, lesbians, and their supporters had had enough and were no longer going to take it quietly.

This conflict between forces that oppose homosexuality and those who support equal treatment under the law for gays and lesbians has led to a variety of legislative and judicial outcomes, as Figure 4.1 illustrates. Beginning in January 1993, when Assembly Bill 2601 went into effect, discrimination against people based on sexual orientation was prohibited in the state of California in public and private employment, in education, and in public accommodations. Also, the California State Labor Code (sections 1101, 1102, and 1102.1) was amended to cover discrimination based on sexual orientation.

Yet in September 1994, with the urging of conservative religious organizations, Governor Pete Wilson vetoed AB 2810, which would have allowed unmarried couples (gay and non-gay) to register their relationships with the secretary of state and receive some married-couple benefits. But in 1999 and 2000, Governor Gray Davis signed into law policies that allow same-sex partners to register as couples and for the partners of state and local public employees to receive health benefits; that give **domestic partners** hospital visitation rights; and that give a domestic partner the right to inherit property if the other partner dies without a will.

California also appears to be the first state to allow a nonbiological co-parent who is the sole wage earner supporting a same-sex partner and child to be allowed to claim "head of household" status on state income tax forms. The State Board of Equalization in November 2000 narrowly approved this ruling. Another bill that was passed in 2000 prohibits the use of sexual orientation as a means to disqualify California citizens from serving as jurors.

- No sodomy law: Consensual sexual acts (same-sex and opposite-sex) are not criminalized in California.

- A domestic partner registry for same-sex couples has been established; equivalent health benefits are available for same-sex partners of state and local public employees; hospital visitation rights for domestic partners have been established.

- Domestic partner has the right to inherit property if other partner dies without a will, to make funeral arrangements, and to be considered a family member for medical treatment issues.

- Hate crime laws include sexual orientation.

- It is unlawful to discriminate in employment and housing accommodations based on sexual orientation.

- All people in California public and postsecondary schools are afforded equal rights and opportunities in the educational institutions of the state regardless of sexual orientation.

- No eligible person shall be exempt from service as a trial juror by reason of sexual orientation.

Figure 4.1. Some of California's Laws Involving Sexual Orientation
SOURCE: Legislative Council of California (http://leginfo.ca.gov).

But are such policies, laws, and ordinances necessary in the state of California? As the events outlined above demonstrated, the history of gays and lesbians in California is a story of oppression, resistance, and mobilization. For decades, gays and lesbians often retreated or silently accepted official harassment. Then, working in ever-increasingly visible and powerful numbers, they slowly organized into a social movement willing to test and resist the oppression. With this visibility and mobilization came equally strong countermovements, often among the agents and organizations of the religious right and conservative Republicans.

Today, such groups as the Traditional Values Coalition in Orange County continue to fight policies and legislation aimed at protecting gay and lesbian citizens in California and elsewhere. With the significant impact of the AIDS epidemic throughout the state (particularly in San Francisco, Los Angeles, San Diego, and Laguna Beach), many of those who oppose homosexuality found a way to justify their hatred. In addition to these groups that foster organized oppression, individuals who continue to put their homophobia into action have increased their physical attacks on lesbians and, more often, gay men.

A look at the reported figures on hate crimes in the state illustrates the effects of homophobia and heterosexism on people's lives. Hate crimes are defined as "acts directed at an individual, institution, or business expressly because of race, ethnicity, religion, sexual orientation, gender, or disability status" (Los Angeles County Commission on Human Relations, 1995, p. 23). They include graffiti (with epithets or symbols), vandalism, obscene or threatening phone calls, and assaults without other apparent motives.

According to figures recorded by the Criminal Justice Statistics Center at the California Department of Justice (2000), there has been an increase of about 37% in hate crime events targeting people based on their actual or perceived sexual orientation, as well as gay-related institutions. As Figure 4.2 shows, events against gay men are second only to those against African Americans (the data do not reflect any hate crime events committed against those who are both gay and black). Combining the incidents against gay men with those against lesbians and transgendered people, more than 22% of all reported hate crime events in California in 1999 were motivated by sexual orientation.

Hate crimes based on sexual orientation include criminal threats, intimidation, assault or battery, and property vandalism against buildings or organizations. In most hate crimes, the perpetrators are males under the age of 30, and many incidents involve two or more assailants, a higher number of offenders per incident than hate crimes committed against other groups (Community United Against Violence [CUAV], 1995).

In a survey of more than 2,200 lesbians, gay men, and bisexuals in the Sacramento (California) area (Herek, Gillis, & Cogan, 1999), about 20% of the women and 25% of the men experienced bias-related criminal victimization at some point since the age of 16. About 12% of the women and 17% of the men were victims of hate crimes based on sexual orientation within the previous 5 years. They also were less likely to report hate crimes to police than other kinds of nonbias crimes.

Most hate-crime statistics reflect the underreporting of actual incidents to police (see Nardi & Bolton, 1997). For example, in San Francisco, for every offense classified anti-gay by the police department, the local community agency recorded 3.43 events (CUAV, 1995). The most common reason for not reporting these incidents to the police is fear of secondary victimization: "Victims fear an insensitive or hostile response by police, physical abuse by police, and public disclosure of their sexual orientation" (CUAV, 1995, p. 13).

But what is most significant about the monitoring and reporting of anti-gay incidents is the emergence of anti-violence projects which, of necessity, must engage in coalition building with racial and religious groups (Jenness, 1995). In many ways, the growth of organizations developed to document anti-gay violence—in California and nationally—represents an attempt to deal with institutional heterosexism by bringing together various governmental, social, religious, and racial groups united against intolerance.

102 ■ CHALLENGES TO SEXUALITY AND HEALTH

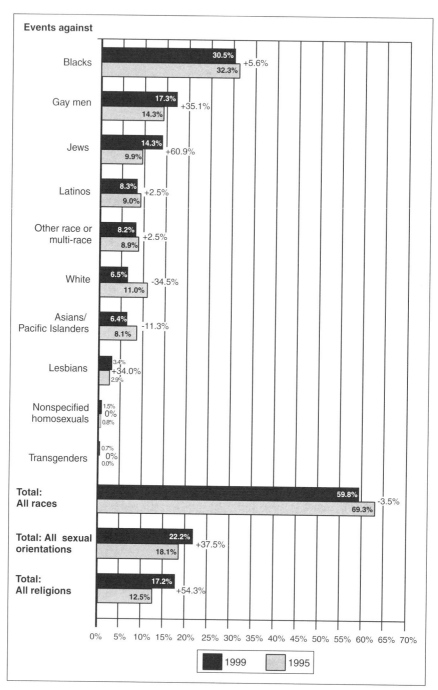

Figure 4.2. Hate Crimes in California (in percent)
SOURCE: California Department of Justice, 2000.

LESBIAN AND GAY OPPRESSION AND RESISTANCE IN CALIFORNIA AS COMPARED WITH THE UNITED STATES

Attitudes toward lesbians and gay men vary widely in California, from the tolerance of large cities and their gay neighborhoods to the intolerance evidenced in the attempts of various groups to eliminate what they wrongly call "special rights" for lesbians and gay men. Nevertheless, compared to much of the rest of the country, California appears to be relatively progressive. The state is not alone, however, in its attempts to fight heterosexism and homophobia. The history of resistance to police and state control of gay men's and lesbians' lives could probably be documented for every state, and certainly for every region of the country. A great variety of state and local legislation and ordinances protects the rights of lesbian and gay citizens.

In addition to California, as of July 2001, 11 other states (Connecticut, Hawaii, Maryland, Massachusetts, Minnesota, Nevada, New Hampshire, New Jersey, Rhode Island, Vermont, and Wisconsin) and the District of Columbia have enacted laws prohibiting discrimination based on sexual orientation. In addition, 27 states and the District of Columbia have hate crime laws that include sexual orientation (18 other states have hate crime laws that do not explicitly include sexual orientation, and 5 have no hate crime laws of any kind). Sodomy laws are still in effect in 15 states 11 of which apply to acts of both same-sex and opposite-sex couples (see www.ngltf.org for current information).

One of the most infamous debates over a state's civil rights law was the battle in Colorado. In November 1992, voters passed Amendment 2, which would have repealed all anti-discrimination protections for lesbians, gay men, and bisexuals throughout the state, including a 1990 state Executive Order as well as city ordinances and human rights laws in Boulder, Denver, Telluride, and Aspen. When the amendment was challenged in court as unconstitutional, a friend-of-the-court brief supporting Amendment 2 was filed by seven states; California's brief was filed by then-state Attorney General Daniel Lungren. However, in October 1994, the Colorado Supreme Court upheld a lower court ruling that Amendment 2 was unconstitutional. And in a 1996 landmark 6 to 3 decision (*Romer v. Evans*), the U.S. Supreme Court ruled Amendment 2 in violation of the equal protection clause of the Fourteenth Amendment.

Whenever such campaigns against lesbians and gay men begin, an increase in the reports of anti-gay violence inevitably follows. After attempts in 1992 in Oregon (Ballot Measure 9) and in Colorado to deny civil rights to gay men and lesbians, "both states experienced an explosion in anti-lesbian/gay violence, including the bias-motivated homicides of a gay man and a lesbian in Oregon" (CUAV, 1995, p. 22). Often, this is a result of a variety of videos, pamphlets, and other fund-raising media used by the radical right to incite disgust and anger at lesbians and gay men. The organizers of these state anti-gay initiatives

> stoke their campaigns with the groundless myths that gay men and lesbians seek "special rights," that extensions of civil rights protections will require "hiring quotas for homosexuals," that gay men and lesbians are sexual predators seeking to "convert" children, and that gays and lesbians do not deserve civil rights protections because being gay or lesbian is a "choice." (CUAV, 1995, p. 22)

Continuing attempts to defeat pro-gay bills or propose anti-gay ones are evident. In Oregon, Portland's 1992 gay rights ordinance was upheld by a two-to-one margin in the November 2000 election. Maine voters decided against a measure that would make discrimination based on sexual orientation illegal in employment, housing, public accommodation, and credit.

One of the most contested areas of debate focuses on the issue of civil unions for same-sex partners. The November 2000 election resulted in landslide victories in Nevada and Nebraska supporting proposed state constitutional amendments denying legal recognition to same-sex marriages. In response to a possible extension of the marriage law in Hawaii, many state legislators and conservative religious leaders have expressed concern about the possibility that same-sex marriages will become legal and the impact on their states. As a result, since 1995, 34 states have passed laws banning same-sex marriage either performed in their own state or in other states. In 1996, Congress passed, and Democratic President Bill Clinton signed, the Defense of Marriage Act, limiting the federal definition of marriage to exclude same-sex couples.

On the other hand, in July 2000, Vermont became the first state to recognize same-sex civil unions, which have all the same benefits, protections, and responsibilities under Vermont law that are granted to spouses in a marriage. This landmark law can be a model for other states to ensure equal protection and rights for same-sex couples without altering the legal definition of marriage. Whether a "separate but equal" law will survive court cases has yet to be determined, but in the November 2000 election, which was seen as a referendum on same-sex relationships, Vermont voters re-elected Govenor Howard Dean, the Democratic governor who supported state-sanctioned same-sex relationships. Most of the pro-civil union state senators survived the election, but many House members lost their seats due to their support of the law.

Another area of litigation and civil rights activism centers on gay and lesbian youth. Although the state Supreme Court of New Jersey unanimously ruled in 1999 that the Boy Scouts was discriminatory in its anti-gay policies, the U.S. Supreme

Court in 2000 (*Boy Scouts of America v. Dale*) overturned that decision in a 5 to 4 vote, thereby allowing the Boy Scouts to discriminate legally against gay youth and adults. How to deal with America's youth, in an era of "coming out" to live more openly as gay and lesbian, has become a major social issue not only in California but around the nation and globe.

Continuing efforts by some schools districts (such as at El Modena High School in California's Orange County Unified School District) to eliminate "gay-straight alliance" clubs for students have received unusual opposition from many conservative religious organizations, which see such attempts to silence school groups as leading eventually to the elimination of religious clubs on school campuses. These gay-straight school-based clubs were formed as a way of providing support for gay and lesbian youth, many of whom experience ongoing verbal and physical abuse on today's high school campuses.

A 1996 landmark case in Wisconsin (*Nabozny v. Podlesny*, U.S. Court of Appeals for the Seventh Circuit) for the first time made school officials legally and financially responsible for failing to protect gay and lesbian students from harassment, and many schools are now more attentive to the urgency of providing safe spaces for students who are different. In the November 2000 election, voters in Oregon once again defeated an anti-gay measure (also called Measure 9), this time one that would have barred schools and community colleges from instruction that encourages, promotes, and sanctions homosexual and bisexual behavior.

Clearly, the climate of the country is changing toward greater acceptance of the diversity of its citizens. Unprecedented nationwide attention was given to the brutal 1998 slaying of openly gay college student Matthew Shepard in Wyoming. This attention led to more calls for extending hate crime legislation and for creating protection of gay men and lesbians from housing and employment discrimination (the state of Wyoming has declined to pass a hate crimes bill). A federal Hate Crimes Prevention Act that includes sexual orientation as a category was supported by a majority of members in the House and Senate in the summer of 2000, but congressional Republicans dropped it as a provision of the defense authorization bill in the fall of 2000. Despite some setbacks in civil rights and other legislation (such as the popular support of Proposition 22 in California and the Supreme Court's Boy Scout decision), the public's attitudes toward gays and lesbians are becoming more supportive of nondiscrimination.

A nationwide *Los Angeles Times* poll in June 2000 indicated that 66% of those surveyed favor laws protecting gays against housing discrimination, and 68% favor laws against job discrimination (54% said the latter should apply even for Boy Scout leaders). Also, 64% said they can accept "two men or two women living together like a married couple," and 50% support gay couples receiving "the same type of rights or benefits that married couples currently enjoy" (such as inheritance rights and health care benefits). However, only 34% support legal gay marriage, and 64% still think that sexual behavior between two people of the same sex is wrong (Rubin, 2000). The legal support of discrimination against gay and lesbian citizens remains low, while the moral disapproval about sexual behavior stays high, illustrating that the ambiguity Californians feel toward sexuality and diversity is reflected nationally.

LESBIAN AND GAY ISSUES IN A GLOBAL CONTEXT

Internationally, gays and lesbians are also seeking equal rights, developing economic and political communities, and defining identity in a variety of complex ways (see Duyvendakl, Krouwel, & Adam, 1998). In the early 1970s, gay liberation groups first appeared in such countries as Australia, Britain, Canada, Germany, France, the Netherlands, Mexico, Argentina, and Italy.[3] By the 1980s, a growing international network of gay media diffused ideas, symbols, and issues around the globe. Tourists to many of the major gay centers in America, especially San Francisco, brought many of the political and social values of gay life back home to their various countries. It is not unusual to find rainbows and pink triangles (symbols often used to indicate gay and lesbian political and social identities) in many urban spaces around the world, along with the work of activists struggling for equal rights for gay men and lesbians.

However, it would be incorrect to assume that globally, people who identify as gay or lesbian do so in the same way and with the same political and social issues as Americans (see Murray, 2000). Cultural variations in the conceptions of sexuality and gender preclude a globalization of standardized gay and lesbian roles (Herdt, 1997). Any shared causes and symbols among different cultures might be explained more by "common urban and ideological pressures" than by cultural backgrounds, according to Altman (1997, p. 424). In his view, what has changed in the developing and developed Asian countries he studied is not the denial of race and nationality identities but the emergence of a wider range of meanings for homosexuality, including "a definable group of self-identified homosexuals" who see themselves as part of a common global gay community (Altman, 1997, p. 424).

Miller (1992) listed four preconditions that must be present in a society for a modern gay and lesbian identity and community to emerge in other societies and become a sociopolitical movement:

> A modicum of personal freedom and social tolerance; a level of economic development that offered some degree of independence and social mobility; a relatively high status for women; and a decline of the power of the family and

religious institutions in defining and determining every aspect of an individual's life. (p. 360)

Such changes have been evident, for example, in many of the countries that form the European Union.

Both the Parliamentary Assembly of the Council of Europe and the European Parliament have been some of the most progressive voices in equal rights for gays and lesbians. In 1981, the Assembly accepted a recommendation that member states (a) ban all laws and practices that criminalize or medicalize homosexual acts between consenting adults, (b) make the age-of-consent laws the same for both homosexual and heterosexual acts, (c) eliminate discrimination in the workplace based on sexual orientation, (d) prohibit medical acts aimed at changing sexual orientation, (e) not restrict custody or visitation rights by parents based solely on sexual orientation, and (f) introduce measures to reduce prison violence (van der Veen, Hendriks, & Mattijssen, 1993).

The founding of the International Lesbian and Gay Association (ILGA) in 1978 in Coventry, England, was also an important moment in the global recognition of gay issues. ILGA does not attempt to standardize gay lives and movements. Rather, it seeks to unite through politics the societal pluralism that exists among the many lesbian and gay communities worldwide. According to ILGA's constitution (as quoted in Holtmaat & Pistor, 1988, p. 35), its aims "are to work for the liberation of lesbian/gay women and gay men from legal, social, cultural, and economic discrimination" by providing information about applying political pressure to governments and international bodies for gay rights, by coordinating international political actions, by promoting unity and cooperation among gay people throughout the world, and by exchanging information about gay liberation.

California has provided information, assistance, and cultural images about gay and lesbian lives to many people and groups around the world. One of the key organizations, the International Gay and Lesbian Human Rights Commission (IGLHRC), is located in San Francisco. According to its mission statement, the group is devoted to protecting and advancing

> the human rights of all people and communities subject to discrimination or abuse on the basis of sexual orientation, gender identity, or HIV status.... IGLHRC responds to such human rights violations around the world through documentation, advocacy, coalition building, public education, and technical assistance. Our overarching commitment is to defend the rights of people worldwide to define their own sexualities and gender identities. We support the efforts of individuals and groups to organize to create societies free from heterosexism and homophobia. (www.iglhrc.org)

Besides political support, California has also inadvertently provided images that have been used to resist equal rights for gays and lesbians. In the fall of 1998 at the Interpride Conference in Los Angeles, a resolution was presented and approved (as it had been in other cities around the world) that, in July 2000, Rome would be the host city for the annual "world day of gay pride." The International Association of Lesbian and Gay Pride coordinators had agreed to hold the events in Rome, but many religious and political leaders were resisting the idea. They wanted to stop the World Pride march because Rome would be hosting a religious jubilee celebrating the Christian year 2000. As a way of stopping the events, some sensationalized videos of pride marches in San Francisco were distributed, along with warnings about how these images and behaviors could also occur in Rome if the festival and political conferences took place. Despite attempts to block the World Pride event, 200,000 participants marched with the motto, "Freedom, Respect, Equality, Enjoyment" in all their rainbow finery.

Global gay rights movements and marches like the one in Rome develop in various forms and styles, incorporate local customs and identities, and build on the expertise and knowledge of other countries' experiences. They are not simply appropriations of the American way of doing things. Many differences exist. For example, in 2000, the Netherlands passed the first bill anywhere in the world giving gay men and lesbians full family rights including marriage, adoption, and divorce. Same-sex couples had been able to register their relationships since 1998, and they may now trade their registrations for actual marriage certificates. On the other hand, the House of Lords in England has been reluctant to lower the age of consent for same-sex partners to 16, the age currently approved for consenting opposite-sex sexual relationships.

California is often at the forefront in the creation of gay and lesbian images and global communications networks, such as television shows sent around the world and the Internet with its numerous Web sites, like the San Francisco-based *PlanetOut.com*. Yet, each society interprets and uses information according to its own history and cultural customs. Uncovering and resisting the social inequalities linked to privileging heterosexuality and subordinating gays, lesbians, bisexuals, and transgendered people are some of the common goals of lesbian and gay people throughout the world, just as they have been and continue to be the objectives for gays and lesbians throughout California and the rest of the United States.

ALTERNATIVE EXPLANATIONS FOR LESBIAN AND GAY OPPRESSION AND RESISTANCE IN CALIFORNIA

What becomes clear when studying history and contemporary debates over equal rights for lesbian and gay citizens is the struggle for social, cultural, and political power among competing interest groups. In addition, meanings and values about homosexuality and "traditional lifestyles" are routinely contested in symbolic and actual battles. Analyzing these dimensions of a social movement and its countermovements is made easier by considering various sociological explanations and theories.

For example, conflict theory posits that social systems are temporary arrangements of competing forces in which struggles over power and resources are the norm (Hess, Markson, & Stein, 1996). The history of the lesbian and gay movement is a clear illustration of conflict theory because, as described above, the competition over whose rules and laws will dominate characterizes most of the debates. In many of those struggles, the debates have centered on the identities of lesbian and gay people and who gets to create meanings for their lives. Considering symbolic interaction theories can further our understanding about how gays and lesbians make sense of their identities and lives. Let's look at both of these sociological perspectives and see how they contribute to an analysis of heterosexism and homophobia.

Explaining why some people engage in anti-gay violence and others use ballot propositions or school board elections to control lesbians' and gay men's lives can too easily be reduced to psychological theories about homophobia and insecurities over gender identity. Uncovering the structural and systemic heterosexism that exists in many of the legal, political, social, and cultural organizations of society may be more useful. It is important to see the lesbian and gay movement's attempts at fighting anti-gay violence and discrimination in terms of a power struggle between those who control and those who are controlled.

Conflict theory directs us to look at the tension that arises among different groups who hold various amounts of power. When the police raided gay bars and arrested patrons, it was a means of asserting power over those whom a system defined as perverted, sick, and criminal. A goal of a social movement is to mobilize material and symbolic resources so that those who are disenfranchised can exert some influence over social change and alter the power structure. Conflict theory explains these mobilizations and the techniques used by those in power to control the challenges and changes. Rather than viewing society as a stable structure with anomalies of disruption, conflict theorists begin with the premise that the norm is an ongoing struggle over power and resources. Social order results from the conflicts among groups and the ways the powerless resist the influence of those in control. The history of gays and lesbians is the story of their attempts to resist the political, religious, medical, social, and cultural control of their lives.

The best example of how this theory works is to analyze the ways various groups compete to pass or defeat legislation that protects gays and lesbians from hate crimes and discrimination in the workplace and housing. Generating a list of supporters of equal rights legislation for lesbians and gays requires playing political power games, mobilizing financial and political resources to assist in the passage of the bills, and building coalitions with other groups who have experienced some form of discrimination. Including sexual orientation in a hate crimes law involves convincing some religious and racial organizations to join in support. This means mobilizing several groups against those who traditionally hold the power to make decisions about the lives of some less powerful groups.

Part of the process involves changing people's concepts and meanings about what homosexuality is. This is where symbolic interactionist theories can help us understand the dynamics of social movements and change. A central question posed by symbolic interactionists is how do people make sense of their lives, define themselves, and affect others' meanings and values (Hess et al., 1996). By looking at the everyday world of interactions, symbols, and experiences, we can begin to uncover the ways in which gays and lesbians encounter the world from their perspectives. For decades, religious, psychiatric, and criminal justice systems have imposed their meanings and definitions of homosexuality. But with the rise of a lesbian and gay social movement, the voices of those who are marginalized have become more evident and are heard more clearly.

Thus, by producing their own Web pages, newspapers, and magazines—for example, when *The Advocate* came into existence in 1967 after the mainstream newspapers failed to communicate the events of a bar raid from viewpoints other than the official police and establishment ones—gays and lesbians began the first stages in redefining for themselves their experiences and their identities. The history of the lesbian and gay movement is a story of the creation and re-creation of identity, the symbols that are associated with it, and the infrastructure that has evolved to support it.

Take, for instance, the way lesbians and gays throughout the United States and in many international countries have appropriated the rainbow to represent gay identity and community. Originally depicted as a rainbow flag, the colors now appear on a variety of objects, including jewelry and clothing, and as triangles which, when pink, also symbolize the oppression homosexuals experienced in the Holocaust. A symbolic interactionist would evaluate how, over time, meanings have changed about these objects and about lesbians and gay men, from the breakdown of the media's negative depictions and stereotypes, to the move in 1974 to get the American Psychiatric Association to declassify homosexuality as a mental disorder, to the inclusion of sexual orientation as a legitimate category of classification in hate crime legislation.

In other words, a sociological study of lesbian and gay oppression and resistance would involve a complex analysis of the competing sites of political power struggles using conflict theories, as well as the arenas in which cultural meanings and social encounters have been debated and negotiated using symbolic interactionist theories.

FUTURE PROSPECTS FOR LESBIAN AND GAY OPPRESSION AND RESISTANCE IN CALIFORNIA

Given what we know about conflict theory, vigilance is required in attempts in California to revoke legislation that prohibits discrimination based on sexual orientation or to put propositions on the ballot restricting equal rights for gays and lesbians. Several areas are regularly being contested when it comes to lesbian and gay issues: the political, the interpersonal/social, and the cultural.

The political battles are both national and local. The debates surrounding Vermont's adoption of civil unions for same-sex partners had a profound impact on the successful passage of California's Proposition 22. Court cases, however, including those about high school gay-straight clubs, child adoption and custody by same-sex parents, and domestic partner benefits, have increasingly supported equal rights for lesbians and gay men. The more such legal precedents are set, the more difficult it will become to undo these kinds of changes. Because the struggle is over power, favorable court rulings establish some control for formerly powerless groups.

But many of the contemporary debates focus on cultural issues, as is evident from the debates over the role of gay youth and leaders in the Boy Scouts. Many issues focus on so-called traditional family values and the cultural meanings of homosexuality. Should our schools be teaching children about alternative family models and use books that discuss homosexuality as a variant way of life? How should gays and lesbians be depicted in the media? Are accurate gay images destroying the family and inhibiting the free speech of religious people? The regular appearance of lesbian or gay characters in television sitcoms and dramas is clearly a significant cultural change that will become more difficult to undo. Given the centrality of Hollywood to the production of media images, California undoubtedly will play an important role in defining and structuring a good deal of the national discussion about meanings and symbols involving lesbians and gays.

Finally, during the past several decades, an important change has occurred socially and interpersonally. Research consistently demonstrates that positive and frequent encounters with minorities contribute to a reduction of prejudice. Knowing gays or lesbians similarly reduces anti-gay attitudes (Herek, 1991). Many gays and lesbians now carry out their daily lives more openly in non-gay environments. It is no longer unusual to see, especially in the larger cities in California, gays and lesbians openly mingling with heterosexuals at parties, restaurants, workplaces, and social venues. Numerous mainstream Internet sites are open to gays and lesbians in cyberspace. It is no longer unusual to see textbooks like this one including chapters on gay and lesbian issues.

Given this visible presence of gays and lesbians in California, it would be extremely difficult to turn back the clock to the pre-1970s style of oppression discussed above. The laws and ordinances have made it easier for many to live more openly and honestly. This, in turn, makes it more difficult to push people back into more hidden lives. Yet, complacency can be the oppressors' ally. Court cases, legislation, and protection against discrimination, hate crimes, and defamation are still not easily won or guaranteed. The large infrastructure of gay and lesbian organizations in California allows for a more organized resistance when repressive measures become mobilized. But resources can easily be drained when they are called on to fight numerous fires simultaneously, including AIDS, hate crimes, and various forms of discrimination.

Change certainly has happened in California and nationally with lesbian and gay issues. When President Bill Clinton made gays in the military a topic of concern in 1992, and when Proposition 22 initiated a debate about gay relationships in 2000 (despite the failure to win those battles), issues of importance to gays and lesbians became part of the national political and social agenda in ways they never had been before. Numerous other issues loom on the horizon that will guarantee a continuation of the debates, and California will more than likely remain at the center of many of them.

SOLUTIONS TO THE PROBLEM OF LESBIAN AND GAY OPPRESSION AND RESISTANCE

When debates develop over contested issues involving gays and lesbians (such as gay marriage, child adoption and custody, domestic partner benefits, gay youth groups on high school campuses, workplace discrimination), California will be an important site. Because it is a state in which class, ethnic, and racial diversity also characterizes the lesbian and gay population, many of the national arguments centering on these issues will have their roots in California. However, working at the local level to pass protective measures based on sexual orientation is not always the most efficient method. Politically, several federal legislative measures need to be taken by Congress, including passage of a federal bill prohibiting employment and housing discrimination on the basis of sexual orientation and authorization for the attorney general of the United States to investigate civil rights violations that involve sexual orientation. Political leaders in states with sodomy laws must work to repeal them, because they are inevitably used to justify discrimination against lesbians and gay men in employment, housing, child custody cases, and related legal challenges.

But local work has to continue in California, as well, despite its relatively progressive policies. Californians must not assume that efforts to introduce and maintain equal legal protection for gays and lesbians will come easily, as the historical events chronicled above warn us. Therefore, several steps should be taken to protect the changes that have occurred and to prevent challenges from those who feel the changes have gone too far.

One method that has experienced some success in attacking defamation and in educating the public about lesbian and gay lives has been changing media images (Nardi, 1996). Because images and meanings are often at the center of cultural debates over equal rights for gays and lesbians, controlling media information becomes an important project, especially in media-centered California. Herein lies the power struggle: Advertisers, corporate owners of networks and publications, and religious organizations are all involved in the fight to maintain certain values and images. Media will continue to be a salient site in resisting cultural heterosexism and homophobia.

Pressure from audiences to present accurate and balanced depictions of gays and lesbians will involve letter-writing campaigns, demonstrations, and negotiations with those who control the media (see www.glaad.org). For example, throughout the spring and summer months of 2000, E-mail campaigns and protests against sponsors of Dr. Laura Schlessinger's syndicated radio show resulted in several major companies pulling their advertising and withdrawing sponsorship of her television show. Her on-air comments about gays and lesbians being "biological errors" and gay men being "predatory" mobilized many to take action. Demonstrations in February 2001 against what many see as anti-gay and anti-women lyrics in the songs of rap performer Eminem raised national debates about the role music and other media play in contributing to a climate of hate against lesbians and gay men.

Another solution is to work at the state political level by registering gay and lesbian voters, by running openly gay and lesbian candidates, by encouraging political leaders to appoint gays and lesbians to judicial and other political positions, and by lobbying State Senate and Assembly leaders to enact protective legislation that includes sexual orientation. It is important to establish legal precedents to prevent successful attempts to reverse laws and ordinances.

Personnel in the criminal justice system—from the police to prosecutors to judges—must be trained and informed about the issues surrounding sexual orientation hate crimes and anti-gay violence. Too often, even in states that include sexual orientation in hate crime legislation, prosecutors may not push very hard when a gay man has been attacked, and judges may dismiss the case because they blame the victim for having brought on the bashing. Having anti-gay violence legislation or ordinances does not guarantee that the public, the media, and the criminal justice system will understand the issues involved.

Finally, given the importance of social and interpersonal encounters, it becomes important for workplaces and other institutions to create climates in which gays and lesbians can lead their lives more openly. Many corporations now have gay employee groups that lobby management to include domestic partner benefits and other policies. These can make it easier for lesbian and gay employees to be themselves at work, thereby reducing prejudice and the spread of misinformation and myths about sexual orientation.

Attempts to restrict the lives of gay men and lesbians over the past 50 years in California have met with various types of responses, from retreat to revolution. The rise of lesbian and gay movements has been one powerful reaction to systemic heterosexism and homophobia. The more recent history is a collection of events in which gays and lesbians have effected important political, social, legal, and cultural changes. However, forces of oppression against people who try to challenge the hegemony of those in control are growing more vocal and visible. Although significant guarantees of equal rights for lesbians and gay men have been won, the struggles continue throughout California, the nation, and the world.

NOTES

1. Similar stories can be told about the growing gay subculture in San Francisco of the 1950s and 1960s. About two dozen gays picketed San Francisco's Federal Building in July 1968 to demand an end to discrimination against homosexuals and the legalization of consensual sex practices in private.

2. In June 1947, a Los Angeles lesbian whose psuedonym was Lisa Ben produced on her typewriter nine editions of *Vice Versa,* the first known publication for lesbians. Using carbon paper, she was able to produce 10 copies each and distribute them among her friends. See Marcus, 1992, for an interview with Lisa Ben.

3. Portions of this section previously appeared in Nardi (1998).

REFERENCES

Adam, B. (1995). *The rise of a gay and lesbian movement* (rev. ed.). New York: Twayne.

Altman, D. (1997). Global gaze/global gays. *GLQ: A Journal of Lesbian and Gay Studies, 3*(4), 417-436.

Bérubé, A. (1990). *Coming out under fire: The history of gay men and women in World War Two.* New York: Free Press.

Bowers v. Hardwick et al., 478 U.S. 186 (1986).

Boy Scouts of America v. Dale, 530 U.S. (2000).

California Department of Justice. (2000). *Hate crime in California, 1999.* Sacramento: Bureau of Criminal Information and Analysis, Criminal Justice Statistics Center.

Castells, M. (1983). *The city and the grassroots.* Berkeley: University of California Press.

Community United Against Violence. (1995). *Anti-Gay/Lesbian Violence in 1994: San Francisco and the United States.* San Francisco: Author.

D'Emilio, J. (1983). *Sexual politics, sexual communities.* Chicago: University of Chicago Press.

Duyvendakl, J. W., Krouwel, A., & Adam, B. D. (Eds.). (1998). *The global emergence of gay and lesbian politics: National imprints of a worldwide movement.* Philadelphia: Temple University Press.

Gorman, E. M. (1991). The pursuit of the wish: An anthropological perspective on gay male subculture in Los Angeles. In G. Herdt (Ed.), *Gay culture in America* (pp. 87-106). Boston: Beacon.

Herdt, G. (1997). *Same sex, different cultures: Gays and lesbians across cultures.* Boulder, CO: Westview.

Herek, G. (1991). Stigma, prejudice, and violence against lesbians and gay men. In J. Gonsiorek & J. Weinrich (Eds.), *Homosexuality: Research implications for public policy* (pp. 60-80). Newbury Park, CA: Sage.

Herek, G., Gillis, J. R., & Cogan, J. C. (1999). Psychological sequelae of hate crime victimization among lesbian, gay, and bisexual adults. *Journal of Consulting and Clinical Psychology, 67*(6), 945-951.

Hess, B., Markson, E., & Stein, P. (1996). *Sociology* (5th ed.). Boston: Allyn & Bacon.

Holtmaat, H., & Pistor, R. (1988). Ten years of international gay and lesbian solidarity. In International Lesbian and Gay Association (Ed.), *The second ILGA pink book: A global view of lesbian and gay liberation and oppression* (pp. 33-45). Utrecht, the Netherlands: Interfacultaire Werkgroep Homostudies.

Jenness, V. (1995). Social movement growth, domain expansion, and framing processes: The gay/lesbian movement and violence against gays and lesbian as a social problem. *Social Problems, 42,* 145-170.

Los Angeles County Commission on Human Relations. (1995). *Hate crimes in Los Angeles County, 1994.* Los Angeles: Author.

Marcus, E. (1992). *Making history: The struggle for gay and lesbian equal rights, 1945-1990.* New York: HarperCollins.

Miller, N. (1992). *Out in the world: Gay and lesbian life from Buenos Aires to Bangkok.* New York: Vintage.

Murray, S. O. (1996). *American gay.* Chicago: University of Chicago Press.

Murray, S. O. (2000). *Homosexualities.* Chicago: University of Chicago Press.

Nabozny v. Podlesny, No. 95-3634 (7th Cir., July 31, 1996).

Nardi, P. M. (1996). Changing gay and lesbian images in the media. In J. Sears & W. Williams (Eds.), *Combating heterosexism: Strategies that work.* New York: Columbia University Press.

Nardi, P. M. (1998). The globalization of the gay & lesbian socio-political movement: Some observations about Europe with a focus on Italy. *Sociological Perspectives, 41*(3), 567-586.

Nardi, P. M., & Bolton, R. (1997). Gay-bashing: Violence and aggression against gay men and lesbians. In P. M. Nardi & B. E. Schneider (Eds.), *Social perspectives in lesbian and gay studies: A reader.* London: Routledge.

Nardi, P. M., Sanders, D., & Marmor, J. (1994). *Growing up before Stonewall: Life stories of some gay men.* London: Routledge.

Romer v. Evans, 517 U.S. 620 (1996).

Rubin, A. (2000, June 18). Public more accepting of gays, poll finds. *Los Angeles Times,* p. A-3.

Thompson, M. (Ed.). (1994). *Long road to freedom: The advocate history of the gay and lesbian movement.* New York: St. Martin's.

van der Veen, E., Hendriks, A., & Mattijssen, A. (1993). Lesbian and gay rights in Europe: Homosexuality and the law. In A. Hendriks, R. Tielman, & E. van der Veen. *The third pink book: A global view of lesbian and gay liberation and oppression* (pp. 225-246). Buffalo, NY: Prometheus.

Warren, J. (2000, June 14). Gays gaining acceptance in state, poll finds. *Los Angeles Times,* p. A-3.

Wolf, D. (1979). *The lesbian community.* Berkeley: University of California Press.

CHAPTER 5

HIV/AIDS

Kevin D. Kelley

The Centers for Disease Control (CDC) estimates that 650,000 to 950,000 Americans were living with HIV infection in the United States at the end of 1999, and it is estimated that 40,000 new infections occur during each year (Simon et al., 1999, p. 987). A significant percentage of those infections occur in people under the age of 25; "in the United States, approximately 2 young people become infected with HIV every hour" (Bende & Johnston, 2000, p. 526). As of December 1999, about 734,000 people had been diagnosed with AIDS, and nearly 433,000 of those people had died; California is home to about 16% of these people, giving it the second-highest AIDS population in the United States, behind only New York (CDC, 1999b).

THE BASIC SCIENCE

AIDS (acquired immune deficiency syndrome) is a disease caused by the human immunodeficiency virus (HIV), which damages the infected individual's immune system. In its official literature, the CDC created the acronym AIDS for acquired immune deficiency syndrome during 1982. Because that term was used in CDC reports, it became the conventional name for the disease (Grmek, 1990; Nussbaum, 1990). HIV is a retrovirus that replicates itself in essential cells of the immune system. As it replicates and multiplies, it destroys those cells and thereby degrades immune system function (Young, 1994).

As the HIV virus destroys T-4 cells during replication, the immune system weakens to the point that common infections can become life threatening (Hombs, 1992;

Morse, 1992). Death from AIDS is caused not by the viral infection itself but by opportunistic infections such as recurrent pneumonia, pulmonary tuberculosis, cancers, and infections of the nervous system that attack the weakened immune system (Hombs, 1992).

HIV is transmitted in various ways, none of which are casual. As Hombs (1992) stated,

> Only four routes of HIV transmission have been identified: (1) unprotected sexual contact, (2) unscreened transfusions of blood or blood by-products, (3) sharing of intravenous needles, and (4) congenital or perinatal transmission from a woman to her fetus or newborn. (p. 7)

The virus is fragile and dies when exposed to air (Johnston & Hopkins, 1990). Unbroken skin acts as a barrier to infection, and HIV is not transmitted by normal everyday activities. Razel (1998) reported,

> Sharing household facilities such as kitchen, bath, shower, and toilet; sharing household items such as bed, razor, toothbrush, comb, towel, clothes, eating utensils, plates, glasses . . . interactions with an infected individual such as shaking hands, hugging, kissing on the cheek, kissing on the lips, bathing with him, sleeping with him, or giving him injections; contact with infected individuals through mosquitoes, bedbugs and other insects; . . . none of these leads to transmission of HIV. (p. 115)

The designation HIV positive is commonly used to describe people who have tested positive for antibodies for the virus. This indicates that the person has been exposed to and infected with HIV. AIDS describes a combination of symptoms that result from a damaged immune system, such as the presence or history of certain opportunistic infections or the simple lack of T-4 cells in the blood (Hombs, 1992). Everyone with AIDS is HIV positive; however, not everyone who is HIV positive has AIDS.

There is reason for cautious optimism today, as new treatments for HIV infection are prolonging lives and several vaccines are under development. Bende and Johnston (2000) write that "more than 60 Phase 1 (initial) trials of approximately 30 candidate vaccines have been conducted in uninfected volunteers worldwide" (p. 529). At the same time, new treatment options have proven to be relatively effective. Beginning in 1996, for the first time since the epidemic began, there have been declines both in progression from infection to AIDS disease and in related deaths (CDC, 1999a). The treatments, known collectively as protease inhibitors, "cause profound and sustained suppression of viral replication, reduce morbidity, and prolong life in patients with HIV infection" (Flexner, 1998, p. 1281). However, there is still no vaccine to protect unexposed people from being infected, nor is there a cure for people already exposed. Treatments for HIV infection are still limited to slowing or disrupting the replication of the virus in cells. In addition, drugs taken on a maintenance basis may lower the risk of contracting life-threatening opportunistic infections (Hombs, 1992; Kakuda, Struble, & Piscitelli, 1998). Significant evidence suggests that resistance to current-generation HIV drugs may be an obstacle in the future (Boden et al., 1999).

HISTORY OF HIV/AIDS IN CALIFORNIA

As Hombs (1992) noted,

> The Human Immune System disorder known as Acquired Immune Deficiency Syndrome, or AIDS, was first identified in the United States in 1981, when physicians in New York City and San Francisco were confronted with the mysterious deaths of a growing number of young men who had illnesses usually held in check by the body's natural defenses. (p. 1)

In 1981, physicians in San Francisco and Los Angeles began to identify men with infections and cancers they knew to be caused by a deficiency of the immune system. Initially, the doctors related the deaths to viruses and bacterial infections the men were exposed to by being sexually active. These men had a history of sexual activity with other men, and in the beginning, the explanation for the disease had much to do with its identification with homosexuality. Until it was discovered that the disease was transmitted through blood and blood products, the new disease was named GRID (gay-related immune disorder) (Hombs, 1992; Shilts, 1987).

From its initial discovery, HIV infection and AIDS disease spread rapidly throughout the state and nation, and by December 1985, the California Department of Health Services (DHS) had reported 4,983 cases of AIDS and 4,077 AIDS-related deaths. A total of 11,417 cases were reported throughout the United States. Just 5 years later, in 1990, California DHS reported 36,969 AIDS cases and 32,159 AIDS-related deaths; U.S. AIDS cases totaled 45,022. By 1995, about 80,000 Californians were living with AIDS, and 50,000 deaths were attributed to the disease (California DHS, 1995b, p. 12).

Throughout the 1990s, women and racial and ethnic minorities made up a larger and larger portion of the epidemic, and unprotected heterosexual sex as the vector for transmission accounted for more and more of all new infections (Haverkos, Turner, Moolchan, & Cadet, 1999; Klevens, Fleming, Neal, & Li, 1999; Padian, 1998; Padian, Shiboski, Glass, & Vittinghoff, 1997; Wortley & Fleming, 1997). Wortley and Fleming (1997) wrote that "between 1981 and 1995, the number of women diagnosed as having acquired immune deficiency syndrome (AIDS) increased by 63%, more than any other group of persons reported as having AIDS, regardless of race or mode of exposure" (p. 911). Haverkos et al. (1999) report that African Americans and Hispanics have a "significantly higher" rate of HIV/AIDS exposure (p. 17), and according to Neal, Fleming, Green, and Ward (1997),

> Heterosexual contact was the most rapidly increasing mode of HIV exposure. The average annual percent increase in the number of AIDS cases was 28% for heterosexual contact, 12% for injection drug use, 6% for men who have sex with men, 5% for men who have sex with men and inject drugs, and 2% among persons with hemophilia or coagulation disorders. . . . From 1988 through 1995, heterosexually acquired AIDS accounted for 4% of all AIDS cases among men and 44% among women. (p. 468)

COURSE OF HIV INFECTION IN CALIFORNIA

Figure 5.1 illustrates the cumulative rise of both AIDS cases and deaths in California, using the years 1990, 1995, and 1999 as reference points. By December 1999, it is estimated that 117,568 Californians were living with AIDS, and 71,794 deaths related to HIV infection had been documented since record keeping began in 1980 (California DHS, 2000a). It must be understood, however, that AIDS cases and AIDS deaths are only *trailing indicators of the epidemic,* recognizing only those people who HIV infection in the state are only estimates; California has anonymous HIV antibody testing (CDC, 1994), making it ineffective to track the number of positive tests. Currently, the California DHS (1999) estimates that an additional 200,000 Californians are living with HIV infection, and according to current estimates, it can take from 2 to 15 years before HIV infection becomes symptomatic (Coffin, 1996, p. S75).

Although females are the fastest-growing category of new infections (Wortley & Fleming, 1997), AIDS cases by gender in the state remain disproportionately male; only 7% of the AIDS cases statewide are female (California DHS, 1999). However, when the proportions of people infected are broken down by ethnicity and compared to the percentage represented in the population, the rate of infection in women of color is significant and alarming. Figure 5.2 illustrates the percentages of AIDS infection in women by ethnicity, compared to the percentage each group represents in the general population of California. The number of AIDS cases is growing faster among African American women than among any other group (Haverkos et al., 1999).

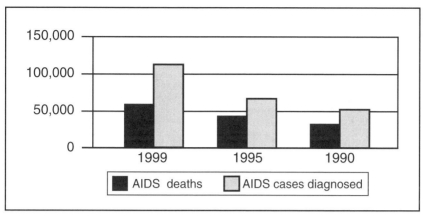

Figure 5.1. AIDS Cases Diagnosed, AIDS Deaths in California, 1990-1999
SOURCE: California Department of Health Services, 2000a.

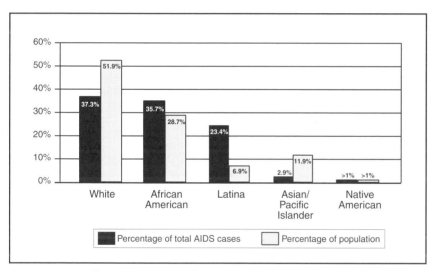

Figure 5.2. Female AIDS Cases in California, by Race/Ethnicity, Compared to Percentage of Female Population, December 1999
SOURCE: California Department of Health Services, 2000b.

Figure 5.3 shows how Californians are being infected with HIV. Although the category "men who have sex with men" continues to be the most common avenue of transmission of the HIV virus, as indicated earlier, heterosexual transmission is the fastest-growing category. Following established trends, "heterosexual transmission currently accounts for more than 4% of all infections in men and more than 44% of infections in women" (Neal et al., 1997, p. 468; Padian, 1998, p. 9). Finally, in

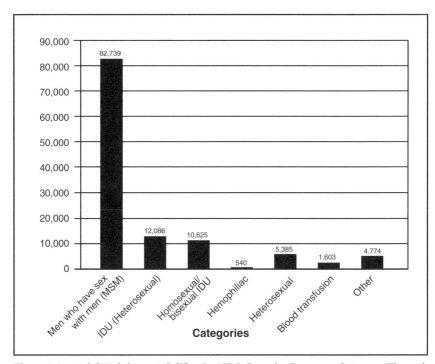

Figure 5.3. Adult/Adolescent California AIDS Cases by Exposure Category, Through December 2000
SOURCE: California Department of Health Services, 2000a.

California, injection drug users account for nearly 20% of infections prior to December 1999.

■ THE PROBLEM OF HIV/AIDS IN CALIFORNIA TODAY

As of December 1999, it is estimated that California had nearly 118,000 persons living with AIDS disease and just under 72,000 related deaths since tracking of the disease began in 1980 (California DHS, 2000a). With an average of 1.7 HIV infections for every case of AIDS, the state would, therefore, have nearly 200,000 additional individuals infected with the HIV virus, leaving about 1 in every 100 Californians infected (U.S. Bureau of the Census, 1999). Mirroring the situation throughout the United States, in California, the disease continues to be spread disproportionately throughout counties that include urbanized areas (Stephenson, 2000b). As illustrated by Figure 5.4, AIDS has substantially affected some cities and counties in California, while leaving others relatively unaffected.

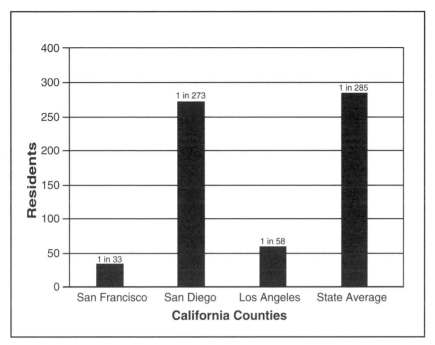

Figure 5.4. Estimated AIDS Prevalence Rates, Counties in California
SOURCE: California Department of Health Services, 2000a.

IMPLICATIONS FOR CALIFORNIA

The development of new treatments for HIV infection and AIDS disease has increased direct costs of HIV/AIDS to the state of California substantially. Analyzing these costs, Holtgrave and Pinkerton (1997) had this to say:

> The development of new generations of anti-viral drugs, including protease inhibitors and nonnucleoside reverse transcriptase inhibitors; combination drug therapies; and improved techniques for monitoring disease progression and therapeutic effectiveness have radically altered how HIV disease is perceived. Despite their promise to prolong survival and improve quality of life of persons infected with HIV, the new therapeutic regimes are much more costly than their predecessors. Moreover, as persons live longer, they consume greater health care resources, driving overall health care costs associated with HIV even higher. (p. 54)

It is estimated that it currently requires more than $195,000 in medical treatment costs alone to treat a person with HIV from diagnosis to death (Holtgrave & Pinkerton, 1997); these costs continue to be carried both by the private and public sectors. Some people infected with HIV are able to maintain private medical insurance, whereas others lose their coverage and migrate to public programs such as Medicare and Medi-Cal (California's version of Medicaid). Still others don't have private coverage from the outset of the disease. The burden of caring for people without private insurance is necessarily carried by the public sector. For the 2000 to 2001 fiscal year, Medi-Cal projected spending more than $258 million to pay for medical treatment of people with AIDS (State of California, 2000). As of July 1999, about 50% of all Californians being treated for HIV/AIDS are receiving Medi-Cal benefits (Feinstein, 1999). There is no reason to believe this trend will improve. In fact, there is evidence that the trend toward more public financing of HIV/AIDS treatment will continue (State of California, 2000).

Future public and private sector medical expenditures for the current number of HIV-positive Californians can be estimated by multiplying the number of people either projected to be HIV positive or with an AIDS diagnosis by the estimated costs for the course of the disease. In California, with its current HIV prevalence, this amount is more than $62 billion, and this figure represents only the medical costs associated with treating HIV and AIDS in California if the disease were to spread no further.

Additional indirect costs will occur for a variety of reasons. There will be costs to both the public and private sectors associated with the loss of productivity due to the progression of the disease in workers and overall losses to wages earned, saved, spent, and taxed. When affected individuals become unable to work and provide for themselves, it often falls to the public sector to pick up the costs. Housing and care for disabled individuals is necessary, and families that may have depended on these individuals for support also frequently need public aid.

Beyond economic costs, there are enormous human costs associated with the AIDS epidemic. AIDS is a long-term, disabling, highly debilitating, ultimately terminal condition. It has destructive effects not only on the people with the disease but also on caregivers, family members, and their extended social networks. A woman left this account:

> When I was diagnosed with AIDS in December of '89, I was only 21 years old. It was the shock of my life and my family's as well. I have lived to see my hair fall out, my body lose over 40 pounds, blisters on my sides. I've lived to go through nausea and vomiting, continual night sweats, chronic fevers of 103-104 that don't go away anymore. I have cramping and diarrhea. I now have convulsions and forgetfulness. I have endured trips twice a week to Miami for 3 months only to receive painful IV injections. I've had bone marrow biopsy. I cried my heart out from the pain of the biopsy. I lived through the fear of whether or not my liver has been completely destroyed by ddI and

other drugs. It may very well be. I lived to see white fungus grow all over the inside of my mouth, the back of my throat, my gums, and now my lips. It looks like white fur and gives you atrocious breath. Isn't that nice? I have tiny blisters on my lips. It may be the first stages of herpes . . .

Have you ever awakened in the middle of the night soaking wet from a night sweat—-only to have it happen an hour later? Can you imagine what it's like to realize you're losing weight in your fingers and that your body may be using its muscles to try to survive? Or do you know what its like to look at yourself in a full-length mirror before you shower—and you only see a skeleton? Do you know what I did? I slid to the floor and cried. Now I shower with a blanket over the mirror. (Arno & Feiden, 1992, pp. 11-12)

COMPARISON OF CALIFORNIA'S SITUATION TO THE UNITED STATES

Assuming national trends resemble those found in California (about 1 AIDS case for every 1.7 HIV infections), it is possible to estimate that, by the end of 1999, slightly more than 1,100,000 to 1,615,000 Americans either were HIV-infected or had developed AIDS. From 1993 to 1996 (before the availability of protease inhibitor treatments), the CDC (1999a) listed AIDS as the leading cause of death for Americans between the ages of 25 and 40. Throughout California, and throughout the nation, the disease is spread disproportionately, with significant clusters in urbanized areas (Stephenson, 2000b).

In California and throughout the United States, **socioeconomic status** is associated with rates of HIV infection (Moss et al., 2000, p. 123): Racial and ethnic minorities and women make up the preponderance of new infections (Rotheram-Borus, Gillis, Reid, Fernandez, & Gwadz, 1997; Wortley & Fleming, 1997). As discussed earlier, women in general, both at the state level and nationally, are being infected at greater and greater rates. Infection rates among women increased 63% from 1981 to 1995 (Wortley & Fleming, 1997, p. 911).

Nationally, young people, especially homeless and gay youth, are also at significant risk for infection. Rotheram-Borus et al. (1997) write that "certain subgroups of adolescents (e.g., homeless, gay, and bisexual youth) [remain] at increased risk for human immunodeficiency virus (HIV) owing to their sexual and substance use behaviors" (p. 217).

Nationally, women in general are the fastest-growing category of new infections (Wortley & Fleming, 1997), but when HIV and AIDS cases are examined by gender in the nation, males still account for the majority. Only 17% of current AIDS cases

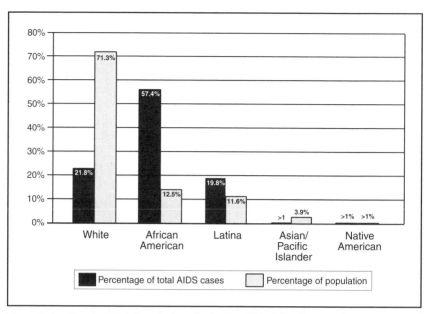

Figure 5.5. Female AIDS Cases in America by Race/Ethnicity, Compared to Percentage of Female Population, December 1999
SOURCE: Centers for Disease Control, 2000; U.S. Bureau of the Census, 2000.

throughout the United States occur in women. Reflecting recent trends, women make up 28% of all people living with HIV throughout the country (CDC, 1999b). However, as in California, when national rates are broken down by ethnicity and compared to the population percentages, the current HIV infection rate in racial and ethnic minority women is serious. Figure 5.5 illustrates the percentages of AIDS infection in women by race or ethnicity, compared to the percentage they represent in the general U.S. population. Nationally, as in California, the number of AIDS cases is growing faster among African American women than among any other group (Haverkos et al., 1999).

Figure 5.6 shows the ways in which Americans are becoming HIV infected. As in California, the category "men who have sex with men" continues to be the main route of transmission of the HIV virus nationally, but heterosexual transmission is the fastest-growing category (Wortley & Fleming, 1997). And finally, Carroll and Ferreboeuf (1999) note that injection drug users are central in the spread of HIV in the United States, as about a third of AIDS cases can be attributed to high-risk drug use. Similarly, Vlahov and Junge (1998) reported that "in the United States, injection drug users . . . account for nearly one-third of the cases of acquired immune deficiency syndrome (AIDS), either directly or indirectly (heterosexual and perinatal cases where the source of infection was an [injected drug user]" (p. 75).

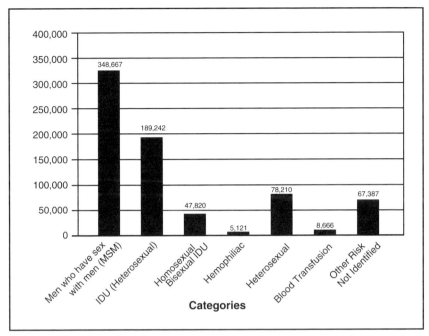

Figure 5.6. Adult/Adolescent U.S. AIDS Cases by Exposure Category, Through December 2000.
SOURCE: Centers for Disease Control, 2000; U.S. Bureau of the Census, 2000.

CALIFORNIA'S PROBLEM IN A GLOBAL CONTEXT

Circumstances throughout the rest of the so-called developed world of Western Europe, Australia, New Zealand, and Japan are similar to those found in the United States: relatively high rates of HIV/AIDS among men who have sex with men, injection drug users, racial and ethnic minorities, women, and residents in pockets of urban poverty, with relatively lower rates throughout the rest of the population. However, circumstances throughout the developing world are significantly more serious. In December 1999, the Joint United Nations Programme on HIV/AIDS estimated 34.3 million people were living with HIV/AIDS worldwide. Infection rates run from more than 20% to as high as 36% of the general population in some African countries, and parts of Asia are experiencing a surge in infections (Joint United Nations Programme on HIV/AIDS, 2000; Stephenson, 2000a). The U.N. further estimates that more than 15,000 new infections occur every day throughout the world (Bende & Johnston, 2000).

Stephenson (2000a) writes,

According to projections from a new study commissioned by the U.S. Agency for International Development (USAID), by 2003, Botswana, South Africa, and Zimbabwe will be experiencing negative population growth, and five other countries will be experiencing a growth rate of nearly zero [because of shortened life expectancy due to the effects of AIDS]. (p. 556)

Significantly, heterosexual intercourse is the "primary mode of HIV transmission world wide" (Neal et al., 1997, p. 465).

Fournier and Carmichael (1998) cite a "growing body of evidence that in developing countries and among areas of poverty in the United States, socioeconomic status clearly correlates with risk and influences survival, both generally and specifically with regard to HIV infection" (p. 214). Predictably, the overwhelming majority of infected people in developing countries, because of their **poverty**, are lacking access not only to the new generation of HIV/AIDS drugs but also to basic medical care. Infected individuals in these countries generally die more quickly, as was true early in the U.S. epidemic (Panos, 2000).

In its *AIDS Epidemic Update,* the Joint United Nations Programme on HIV/AIDS (2000) stated the following:

- The total numbers (as of December 1999) of AIDS orphans worldwide since the beginning of the AIDS epidemic is 8.2 million. By the year 2020, an estimated 40 million orphans younger than age 15 will exist in the 23 countries most affected by HIV.

- An estimated 2.6 million AIDS-related deaths occurred during 1999 worldwide, more than ever before in a single year.

- More than 95% of all HIV-infected people now live in the developing world, which likewise has experienced 95% of all deaths to date from AIDS.

- Worldwide, women represent 46% of all people older than 15 living with HIV or AIDS.

- One tenth of newly infected people were younger than age 15; 1,600 new infections each day are among children younger than 15.

In 1998, Stover and Way wrote that "the U.S. Bureau of the Census included the impact of AIDS in its 1994 and 1996 global population projections and its forthcoming 1998 report. . . ." For the 1998 projections, the U.S. Census Bureau will report the following:

1. AIDS will reduce the projected population of 21 countries in sub-Saharan Africa by 11.8% (59 million people) by 2005 and by 16.6% (120 million people) by 2025.

2. The most severely affected country, Zimbabwe, will have 21% (3.1 million) fewer people in 2005 than there would have been without AIDS. The projected populations of Botswana, Malawi, Namibia, and Swaziland in 2005 are also more than 15% less than they would be without AIDS.

3. By 2010, life expectancy in Botswana will be reduced by 42% from what it would have been in the absence of AIDS (from 66 to 38 years). For all 21 countries, life expectancy will be reduced by 27%, from 62 to 45 years.

Trends worsen quickly. Just 2 years before the Joint United Nations Programme on HIV/AIDS made its predictions regarding negative population growth, Stover and Way (1998) wrote that "population growth rates will remain positive in all countries but will be reduced significantly due to AIDS. For all 21 countries, the annual rate of growth from 1990 to 2005 will be 2.2%, rather than the 2.6% that would be projected without AIDS" (p. S30).

ALTERNATIVE EXPLANATIONS FOR THE SPREAD OF HIV AND AIDS

As we discussed previously, the course of HIV/AIDS in California and the United States is directly related to both the social and the economic status of the people who are most commonly infected. Bradley-Springer (1999) points out that "epidemics do not exist in isolation; they are integrated into the social fabric of the communities in which they exist" (p. 50). She goes on to note that epidemics are

> subject to social, cultural, economic, political, religious, and technologic patterns and controlled by the rules, customs, and mores of a society... meanings of specific diseases vary according to preexisting social agendas, and the community's emotional response to a disease is a major consideration in how the disease progresses. (p. 50)

In the United States, the AIDS epidemic first emerged in so-called **deviant subcultures,** appearing first primarily among men who have sex with men, followed by IV drug users. HIV and AIDS became permanently associated with these groups, and commonly held assumptions about their **subcultures** shaped and slowed initial social, governmental, and cultural responses to HIV and AIDS.

Arguably, the United States, more than any other nation, possessed the resources to significantly slow or stop progression of the disease early. Instead, as it emerged, the disease was allowed to race relatively unchecked until infection reached a numeric critical mass beyond which it was too late to stop movement into other populations. After the initial population of men who have sex with men, the disease moved to and continued to spread disproportionately among other less powerful subcultures (those in poverty generally, ethnic/racial minorities, and IV drug users). This characteristic continues to affect the social and scientific response in California and the United States. And arguably, once again, the infection rate in the state and nation has reached the point of critical mass, moving from subcultures into the general population, and again, it is too late to stop it.

Worldwide occurrence is similar, as both the likelihood of infection and the social response (locally and globally) have been directly tied to social and economic status. Early infections worldwide were also associated with so-called deviant subcultures, associated with men who have sex with men and IV drug users, and allowed to race unchecked until significant mass infection occurred. Infection rates, primarily in the developing world, have been allowed to reach critical mass (years too late to stop). In the developing world, the rate of infection has reached crisis proportions that literally threaten world stability.

The following quote, taken from Randy Shilts's (1987) history of the AIDS epidemic, *And the Band Played On*, is useful in examining the unfolding of the HIV/AIDS crisis in California, the United States, and the world:

> The discovery of cyanide in Tylenol capsules occurred in those same weeks of October 1982 [about the same time as AIDS deaths were being reported in large numbers]. The existence of the poisoned capsules, all found in the Chicago area, was first reported on October 1. *The New York Times* wrote a story on the Tylenol scare every day for the entire month of October and produced twenty-three more pieces in the two months after that. Four of the stories appeared on the front page. The poisoning received considerable coverage in the media across the country, inspiring an immense government effort. Within days of the discovery of what proved to be the only cyanide laced capsules, the Food and Drug Administration issued orders removing the drug from store shelves across the country. Federal, state, and local authorities were immediately on hand to coordinate efforts in states thousands of miles from where the tampered boxes appeared. No action was too extreme and no expense too great, they insisted, to save lives.
>
> Investigators poured into Chicago to crack the mystery. More than 100 state, federal, and local agents worked the Illinois end of the case alone, filling twenty-six volumes with 11,500 pages of probe reports. The Food and Drug Administration had more than 1,110 employees testing 1.5 million similar capsules for evidence of poisoning and chasing down every faint possibility of a victim of the new terror, according to the breathless news reports of the time. Tylenol's parent company, Johnson & Johnson, estimated spending $100 million in the effort. Within five weeks, the U.S. Department of Health

and Human Services issued new regulations on tamper-resistant packaging to avert repetition of such a tragedy.

In the end, the millions of dollars for CDC Tylenol investigations yielded little beyond the probability that some lone crackpot had tampered with a few bottles of the pain reliever. No more cases of poisoning occurred beyond the first handful reported in early October. Yet the crisis showed how the government could spring into action, issue warnings, change regulations, and spend money, lots of money, when they thought the lives of Americans were at stake.

Altogether, seven people died from the cyanide-laced capsules; one other man in Yuba City, California, got sick, but it turned out he was faking so he could collect damages from Johnson & Johnson.

By comparison, 634 Americans had been stricken with AIDS by October 5, 1982. Of these, 260 were dead. There was no rush to spend money, mobilize public health officials, or issue regulations that might save lives.

The institution that is supposed to be the public's watchdog, the news media, had gasped a collective yawn over the story of dead and dying homosexuals. In New York City, where half the nation's AIDS cases resided, *The New York Times* had only written three stories about the epidemic in 1981 and three more stories in all of 1982. None made the front page. Indeed, one could have lived in New York, or in most of the United States for that matter, and not even been aware from the daily newspapers that an epidemic was happening, even while government doctors themselves were predicting that the scourge would wipe out the lives of tens of thousands. (p. 191)

Shilts demonstrates that during the emergence of HIV/AIDS in the United States, another public health emergency briefly appeared. His example describes how governmental and social agencies and the media, with sufficient motivation, responded rapidly to a relatively minor threat and reached a resolution. Occurring simultaneously, the HIV/AIDS crisis was not effectively addressed for several years.

In circumstances defined as a crisis, a concerted, rapid effort by government agencies, businesses, and the public would be generally accepted as a normal response. A crisis definition of HIV/AIDS in the early years would have generated immediate measures to contain and eradicate the disease. Shilts proposes that HIV/AIDS was not treated as a public health emergency in the early years because it initially struck gay men, followed by IV drug users.

By establishing a pattern of concealment and relative paralysis for the government in the early years of the epidemic, public policy makers promoted an impression that HIV/AIDS was a relatively insignificant and inconsequential risk to the general public, not a public health emergency. It took 5 years and 11,000 AIDS cases before an American president spoke the word *AIDS* publicly (Hombs, 1992, p. 36). Public health policy and intervention strategies have been and continue to be shaped and affected by similar political and moralistic barriers. By examining reactions and adaptations of the private and public sectors to the opportunities and challenges presented by HIV and AIDS, the notion of AIDS as a component of society is contrasted with the emergency response the epidemic merits.

Looking back at Shilts's comparison of governmental response to the cyanide Tylenol scare and the response to early AIDS deaths, it is clear that social definitions of the populations believed to be at risk for contracting HIV/AIDS prevented and continue to prevent an adequate governmental and social response to the epidemic. This alternative explanation for the state of HIV/AIDS in California, the country, and the world demonstrates that the expanding epidemic may have been preventable.

FUTURE PROSPECTS: AIDS PROJECTIONS IN CALIFORNIA

Projections for the extent of both HIV infection and cumulative AIDS deaths in California are calculated for the year 2005. Assuming that the state will maintain its current share of about 16% of the nation's infections (CDC, 1999b), and using projections for trends in sexual and drug use behavior, California in 2005 can expect to have more than 1.5 million HIV infections, more than 80,000 living AIDS cases, and a cumulative total of more than 300,000 dead from AIDS.

According to Greeland et al., "Data on the current HIV transmission patterns in Los Angeles County indicate that minorities, women, and young gay males are groups that are being increasingly affected and represent the 'leading edge' of the epidemic" (cited in Sorvillo, Kerndt, Odem, & Castillon, 1999, p. 151). Figure 5.7 shows how different impacts of AIDS in racial and ethnic groups continue to grow. This table illustrates that AIDS in California affects minority children at a disproportionately higher rate than white children. AIDS in children is closely related to needle sharing of one or both parents during injection drug use (CDC, 1995, p. 5). Because California has been politically unable to accept wide-scale public health measures to address this situation (e.g., clean needle and syringe exchange programs, improving access to drug treatment), these figures are unlikely to change or improve in the near future. The future, in the absence of these kinds of demonstrated effective interventions, will continue to exhibit an increase of minority pediatric AIDS.

Of the 650,000 to 950,000 estimated AIDS cases in the United States as of December 1999, 52% were among African Americans (CDC, 1999b, p. 11). To date, about 60% of all AIDS cases have occurred among African Americans and Latinos,

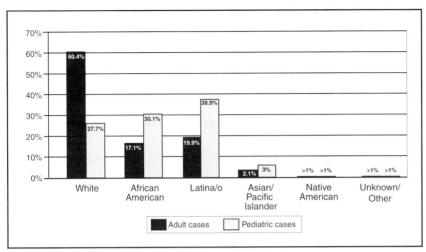

Figure 5.7. Comparison of Adult to Pediatric AIDS Cases by Ethnicity, California, Through October 31, 2000
SOURCE: California Department of Health Services, 2000b.

although they make up less than 21% of the overall population. Haverkos et al. (1999) wrote that by 1999, African Americans were 4.7 times more likely than whites to contract AIDS, African American women were up to 13.5 times more likely than white women to contract AIDS, and African American children were 3.6 times more likely than white children to contract AIDS. At the same time, Hispanics were 2.5 times more likely than whites to contract AIDS, Hispanic women were up to 8 times more likely than white women to contract AIDS, and Hispanic children were 3.2 times more likely than white children to contract AIDS. It is clear that HIV infection and AIDS continue to strike minority communities at a greater and greater rate. With these facts in mind, the inability or unwillingness of California's government to address the rise of HIV infection in the heterosexual population will affect minorities in California at a much greater level.

POSSIBLE SOLUTIONS TO AIDS IN CALIFORNIA

In projecting the future course of the epidemic, researchers have used two main criteria: projected advances in medical science and the levels of sustained behavioral change required. In consideration of current information and projections available, the following recommendations to address the AIDS epidemic in California have been developed:

RECOMMENDATIONS TO ADDRESS THE AIDS EPIDEMIC IN CALIFORNIA

1. *Expanded and comprehensive AIDS education in schools at age-relevant levels.* As we have discussed, young people make up a significant majority of new HIV infections (Bende & Johnston, 2000); in the United States, of the newly infected, "the CDC estimates that over half are younger than 25 and are infected sexually" (Fauci, 1999, p. 1047). Aggressive strategies need to be developed and expanded to give young people the information they need to protect themselves from HIV (Schoeberlein, 1995). Risk behaviors rather than risk groups need to be described to young people to help them identify the behavioral changes needed to guard against HIV transmission. Funding for HIV education and prevention needs to be significantly expanded in California. Prevention is demonstrably less expensive than treatment for the infection and disease.

2. *Change state drug policy to allow state-of-the-art public health interventions to be used.* Injection drug users are directly or indirectly responsible for about one in three HIV infections throughout the country (Vlahov & Junge, 1998). The human and economic costs associated with HIV transmission through injection drug use and associated pediatric transmission require an innovative and direct response. Changes in drug policy would include statewide decriminalization of needle and syringe possession and Department of Health Services-sponsored needle-exchange programs in metropolitan areas, to include aggressive outreach and education. International evidence demonstrates needle- and syringe-exchange programs work to lower transmission of HIV. Decriminalization of possession of injection equipment is needed to provide those unable to access needle- and syringe-exchange programs the means to protect themselves from infection (National Commission on AIDS, 1993). As part of a total harm reduction strategy, expanded culturally appropriate drug treatment programs available on demand are required. These should include aggressive outreach programs to identify those in need of treatment. There should also be programs for those who are not injection drug users but are still at risk for HIV infection, such as prostitutes and people who engage in other kinds of sexual activities that occur with drug and alcohol dependence.

3. *Fund expanded research to eliminate researchers' dependency on profit.* Fauci (1999) writes that the "solution to the HIV pandemic is the development and availability of a safe and effective vaccine against the infection" (p. 1049). Most AIDS treatment research is financed by drug companies with an economic stake in the outcome of the research (Arno & Feiden, 1992), and the development of an AIDS vaccine is currently hampered by a dependency on profit. Bende and Johnston (2000) note that "when the costs and the time frame of vaccine development and the chances for success are weighed against the potential profit, there appears to be

little incentive for companies to aggressively pursue HIV vaccine development" (p. 536). By more aggressively funding research, California and the United States can provide insight into solutions that are currently neglected. Because California contains about 16% of U.S. AIDS cases, our population would be well served by a state initiative to fund expanded research.

4. Assist advanced biotechnology industries already located in the state by sponsoring a national Manhattan Project for HIV/AIDS. Again, as California currently represents 16% of national AIDS cases, our state would benefit from sponsorship of advanced biotechnology industries already in place in the state to address an HIV/AIDS solution.

5. Develop, in concert with affected communities, expanded and culturally relevant AIDS education programs. Rather than producing an effort *for* at-risk communities, good public health policy suggests developing an effort *with* these communities (Jenkins, Lamar, & Thompson-Crumble, 1993; Valdiserri & West, 1994). Using this strategy, solutions will be culturally appropriate, and the community members involved will have a personal investment in their own success.

6. Provide comprehensive health care and education to all who require it. A California-wide health care initiative would work to reduce risk behaviors on the whole, provide early intervention and treatment to HIV-infected Californians, and support healthy behavior changes on a statewide level (National Commission on AIDS, 1993). Providing comprehensive health care to all who need it is an important investment in California's future.

7. Work to end poverty, racism, sexism, and homophobia. Evidence clearly shows that social and governmental response to the AIDS epidemic at every level has been and continues to be influenced and shaped both by contemporary assumptions about different social groups with relatively high rates of infection and by relative levels of social authority, status, wealth, and power. Prejudice with regard to sexual orientation, race and ethnicity, gender status, and poverty have allowed the epidemic to reach the crisis proportions experienced today, and work on changing the course of the epidemic must necessarily embrace a new response to these issues.

REFERENCES

Arno, P. S., & Feiden, K. L. (1992). *Against the odds: The story of AIDS drug development, politics, and profits.* New York: HarperCollins.

Bende, S., & Johnston, M. I. (2000). Update: Search for an AIDS vaccine. *The AIDS Reader, 10,* 526-537.

Boden, D., Hurley, A., Zhang, L., Cao, Y., Guo, Y., Jones, E., Tsay, J., Ip, J., Farthing, C., Limoli, K., Parkin, N., & Markowitz, M. (1999). HIV-1 drug resistance in newly infected individuals. *Journal of the American Medical Association, 282*, 1135-1141.

Bradley-Springer, L. A. (1999). The complex realities of primary prevention for HIV infection in a "just do it" world. *Nursing Clinics of North America, 34*, 49-70.

California Department of Health Services. (1995a). *Estimates of HIV prevalence and AIDS incidence—California, 1994.* Sacramento, CA: Office of AIDS.

California Department of Health Services. (1995b). *HIV-AIDS reporting system surveillance report.* Sacramento, CA: Office of AIDS, HIV/AIDS Epidemiology Branch.

California Department of Health Services. (1999). *California HIV/AIDS update.* Sacramento, CA: Office of AIDS.

California Department of Health Services. (2000a). *Acquired Immunodeficiency Syndrome (AIDS) HIV-AIDS reporting system surveillance report for California—August 31, 2000.* Sacramento, CA: Office of AIDS. On-line: http://www.dhs.ca.gov/org/ooa/Statistics/

California Department of Health Services. (2000b). *HIV/AIDS among racial/ethnic groups in California.* Sacramento, CA: Office of AIDS.

Carroll, A. M., & Ferreboeuf, M. (1999). *Self-perceived risk for HIV/AIDS and partner risk assessment among drug users in the Bay Area, California* [On-line]. Available: http://www.cdc.gov/hiv/conferences/hiv99/abstracts/277.pdf

Centers for Disease Control, Advisory Committee on Prevention of HIV Infection. (1994). *External review of CDC's HIV prevention strategies: Summary of the final report.* Atlanta, GA: U.S. Department of Health and Human Services.

Centers for Disease Control. (1995). *HIV/AIDS surveillance report: U.S. HIV and AIDS cases reported through December 1994.* Atlanta, GA: U.S. Department of Health and Human Services.

Centers for Disease Control. (1999a). *CDC guidelines for national human immunodeficiency virus case surveillance, including monitoring for human immunodeficiency virus infection and acquired immunodeficiency syndrome.* Atlanta, GA: U.S. Department of Health and Human Services.

Centers for Disease Control. (1999b). *HIV/AIDS surveillance report, December 1999 year end edition.* Atlanta, GA: U.S. Department of Health and Human Services.

Centers for Disease Control. (2000). *HIV/AIDS surveillance report, June 2000.* Atlanta, GA: U.S. Department of Health and Human Services.

Coffin, J. (1996). HIV viral dynamics. *AIDS, 10*, S75-S84.

Cohen, J. (1993). A "Manhattan Project" for AIDS? *Science, 259*, 1112-1114.

Fauci, A. S. (1999). The AIDS epidemic: Considerations for the 21st century. *The New England Journal of Medicine, 341*, 1046-1050.

Feinstein, D. (1999). *Senator Feinstein urges Governor Davis to increase Medi-Cal reimbursement rate: Increased reimbursement needed to head off fiscal collapse* [Press Release]. Sacramento, CA: Author.

Flexner, C. (1998). HIV-protease inhibitors. *The New England Journal of Medicine, 338*, 1281-1292.

Fournier, A. M., & Carmichael, C. (1998). Socioeconomic influences on the transmission of human immunodeficiency virus infection. *Archives of Family Medicine, 7*, 214-217.

Grmek, M. D. (1990). *History of AIDS: Emergence and origin of a modern pandemic.* Princeton, NJ: Princeton University Press.

Haverkos, H. W., Turner, J. F., Moolchan, E. T., & Cadet, J.-L. (1999). Relative rates of AIDS among racial/ethnic groups by exposure categories. *Journal of the American Medical Association, 91,* 17-24.

Holtgrave, D. R., & Pinkerton, S. D. (1997). Updates of cost of illness and quality of life estimates for use in economic evaluations of HIV prevention programs. *Journal of Acquired Immune Deficiency Syndroms and Human Retrovirology, 16,* 54-62.

Hombs, M. E. (1992). *AIDS crisis in America: A reference handbook.* Santa Barbara, CA: ABC-CLIO.

Jenkins, B., Lamar, V. L., & Thompson-Crumble, J. (1993). AIDS among African Americans: A social epidemic. *Journal of Black Psychology, 19,* 108-122.

Johnston, W. B., & Hopkins, K. R. (1990). *The catastrophe ahead: AIDS and the case for a new public policy.* New York: Praeger.

Joint United Nations Programme on HIV/AIDS. (2000). *AIDS epidemic update* [On-line]. Available: http://www.unaids.org

Kakuda, T. N., Struble, K. A., & Piscitelli, S. C. (1998). Protease inhibitors for the treatment of human immunodeficiency virus infection. *The American Journal of Health-System Pharmacy, 55,* 233-254.

Klevens, R. M., Fleming, P. L., Neal, J. J., & Li, J. (1999). Is there really a heterosexual AIDS epidemic in the United States? Findings from a multisite validation study, 1992-1995. *American Journal of Epidemiology, 149,* 75-84.

Morse, S. S. (1992). AIDS and beyond: Defining the rules for viral traffic. In E. Fee & D. M. Fox (Eds.), *AIDS: The making of a chronic disease* (Vol. 99, pp. 49-83). Berkeley: University of California Press.

Moss, N., Lemp, G., Bluthenthal, R., Goldsmith, M., Haubrich, R., & Kahn, J. (2000). From the cell to the community: AIDS research in California. *Western Journal of Medicine, 173,* 119-124.

National Commission on AIDS. (1993). *AIDS: An expanding tragedy* (Final report of the National Commission on AIDS). Washington, DC: Author.

Neal, J. J., Fleming, P. L., Green, T. A., & Ward, J. W. (1997). Trends in heterosexually acquired AIDS in the United Sttates, 1988 Through 1995. *Journal of Acquired Immune Deficiency Syndromes and Human Retrovirology, 14,* 465-474.

Nussbaum, B. (1990). *Good intentions: How big business and the medical establishment are corrupting the fight against AIDS.* New York: Atlantic Monthly Press.

Padian, N. S. (1998). Recent findings about heterosexual transmission of HIV and AIDS. *Current Opinion in Infectious Diseases, 11,* 9-12.

Padian, N. S., Shiboski, S. C., Glass, S. O., & Vittingoff, E. (1997). Heterosexual transmission of human immunodeficiency virus (HIV) in Northern California: Results from a ten-year study. *American Journal of Epidemiology, 146,* 350-357.

Panos. (2000, July). *Beyond our means? The cost of treating HIV/AIDS in the developing world* (HIV/AIDS Briefing No. 7) [On-line]. Available: http://www.panos.org.uk/aids/BeyondOurMeans.pdf

Razel, M. (1998). Casual household transmission of human immunodeficiency virus. *Medical Hypotheses, 51,* 115-124.

Rotheram-Borus, M. J., Gillis, J. R., Reid, H. M., Fernandez, M. I., & Gwadz, M. (1997). HIV testing, behaviors, and knowledge among adolescents at high risk. *Journal of Adolescent Health, 20,* 216-225.

Schoeberlein, D. (1995). Redefining HIV/STD education. *SIECUS Report, 23*(2).

Shilts, R. (1987). *And the band played on: Politics, people, and the AIDS epidemic.* New York: St. Martin's.

Simon, P. A., Thometz, E., Bunch, J. G., Sorvillo, F., Detels, R., & Kerndt, P. (1999). Prevalence of unprotected sex among men with AIDS in Los Angeles County, California, 1995-1997. *AIDS, 13,* 987-990.

Sorvillo, F., Kerndt, P., Odem, S., & Castillon, M. (1999). Use of protease inhibitors among persons with AIDS in Los Angeles County. *AIDS Care, 11,* 147-155.

State of California. (2000). *Governor's budget summary.* Sacramento: California Department of Health Services, Public Health Section.

Stephenson, J. (2000a). Apocalypse now: HIV/AIDS in Africa exceeds the experts' worst predictions. *Journal of the American Medical Association, 284,* 556-557.

Stephenson, J. (2000b). Rural HIV/AIDS in the United States: Studies suggest presence, no rampant spread. *Journal of the American Medical Association, 284,* 167-168.

Stover, J., & Way, P. (1998). Projecting the impact of AIDS on mortality. *AIDS, 12,* S29-S39.

U.S. Bureau of the Census Bureau. (1999). *State population estimates and demographic components of population change: July 1, 1998, to July 1, 1999* [On-line}. Available: http://www.census.gov.population/estimates/state/st-99-1.txt

U.S. Bureau of the Census. (2000). *Resident population estimates of the United States by sex, race, and Hispanic origin: April 1, 1990 to July 1, 1999, with short-term projection to October 1, 2000.* Washington, DC: Population Estimates Program.

Valdiserri, R. O., & West, G. R. (1994). Barriers to the assessment of unmet need in planning HIV/AIDS prevention programs. *Public Administration Review, 54,* 25-30.

Vlahov, D., & Junge, B. (1998). The role of needle exchange programs in HIV prevention. *Public Health Report, 113* (Suppl. 1), 75-80.

Wortley, P. M., & Fleming, P. (1997). AIDS trends in women in the United States. *Journal of the American Medical Association, 278,* 911-916.

Young, J.A.T. (1994). The replication cycle of HIV-1. In P. T. Cohen, M. A. Sande, & P. A. Volberding (Eds.), *The AIDS knowledge base: A textbook on HIV disease from the University of California, San Francisco, and San Francisco General Hospital* (pp. 3.1.1- 3.1.7). Boston: Little, Brown.

Part III

CHALLENGES TO CALIFORNIA INSTITUTIONS

The chapters in this section cover problems with families, with poverty and inequality, with the poor and welfare reform, and with the continuing crisis in higher education in California.

Chapter 6, "Problems of Families," by Robin Wolf, Robin Franck, and James A. Glynn, deals with the problems being faced by California's families. The trends in divorce, nonmarital childbearing, and teenage pregnancy are discussed, as are the effects of these trends on children. During the 1980s, California's rising divorce rate reversed direction and started to drop, then leveled off in the 1990s. The leveling off is due, in part, to the growing proportion of California's population that is of either Hispanic or Asian origin, segments of the population that have historically had lower divorce rates than the general population.

The authors also point out that nonmarital childbearing in California and the United States is increasing. Although there is little historical information on nonmarital childbearing in California, U.S. data show how unusual childbirth outside of marriage was as recently as 1960 and how common it is today. In 1960, only 4% of all children in one-parent families were living with a never-married mother. By 1998, 4 of 10 children in one-parent families in the United States lived with a never-married mother. The latest data from California show the same tendency. Comparing data on nonmarital births, 34% of all births in California are to unmarried mothers, compared to 26% of all U.S. births. The authors discuss the debate over whether nonmarital childbearing leads to poverty or poverty leads to nonmarital childbearing. They conclude that both are true. They argue that the high rate of nonmarital childbearing and divorce in California has led to a feminization of poverty: Low-income families tend to be headed by single women.

Compared to divorced mothers, never-married mothers are much less likely to receive child support from the fathers of their children. This helps explain the above-mentioned link between nonmarital childbearing and poverty.

Wolf, Franck, and Glynn also delve into teen childbearing. They argue that the history of teen childbearing in California suggests that the significant development over the last 40 years is that teens who become pregnant are opting not to get married. The authors point out that although it was not unusual in the 1950s for teens to become pregnant outside of marriage, weddings generally followed the pregnancies. The authors point out that teenage marriage—with or without pregnancies—was easier in California in the 1950s than it is today, because inexpensive housing and plentiful well-paying jobs were available, neither of which required great amounts of education. In 1998, California's teen pregnancy rate ranked 11th among states. In 1999, teen pregnancies resulted in 59,000 California births. Latina teens in California have the highest birth rates, followed by African American, white, and Asian teens. Teenage mothers who try to raise their children by themselves find that they are, for the most part, destined to a life of poverty. In most cases, their educational and occupational training have been cut short as a result of their becoming pregnant and having children.

Wolf, Franck, and Glynn also discuss the impact of abortion on teen childbearing. Whereas California ranks first in the nation in teen pregnancy rates, it ranks 28th in the nation in teen births as a percentage of all births in the state. The increased choice of abortion among California teens accounts for this differential. However, the choice of abortion is more common among middle-class teens compared to lower-class teens, who are more likely to see their pregnancies come to term. Another issue touching on teen births is the fathers. The data show that the fathers of teenage girls who give birth are much more likely to be older males rather than teenagers themselves. In fully two thirds of births to teenagers in California, the fathers are age 20 or older. Yet another issue has to do with the relationship between sexual abuse and teen births. There is growing evidence that the majority of teens who give birth have a personal history that includes prior sexual abuse.

Wolf, Franck, and Glynn offer a number of alternative explanations for the aforementioned trends. First, a high divorce rate weakens the institution of marriage. As the divorce rate rises, the general willingness of individuals to commit themselves fully to marriage declines. Second, the sexual revolution of the 1960s and 1970s resulted in a more accepting attitude toward sex outside of marriage. Third, the time between sexual initiation and marriage has been increasing. Fourth, there has been a significant rise in the number of cohabiting relationships, which are less stable than marriages and are generally of a shorter duration. Nonetheless, women in cohabiting relationships are exposed to considerable pregnancy risk. Fifth, research suggests that individuals raised in one-parent families are more likely to drop out of high school and are more likely to bear a child as a teenager than are individuals raised in two-parent families. Finally, the massive economic restructuring that occurred in the United States and California since the mid 1970s has resulted in a tremendous loss of good-paying jobs. The economic insecurity resulting from this restructuring means that fewer young adult men are willing to make a commitment to marriage even if the girlfriend becomes pregnant.

When the authors look at the family from a global perspective, they see some interesting differences between California and other parts of the world. Comparing California to developing countries, it becomes clear that families in developing countries face many more difficulties than families in California. For example, extreme poverty and health problems in underdeveloped countries result in many children being orphaned. Comparing California to the developed regions of the world, the authors point out that the percentage of all births to unmarried women is even greater in many developed countries than it is in the United States and California.

The authors do not have any reason to think that the trends discussed in their chapter will change radically in the future. They do offer some suggestions on how we might deal with the problems of divorce, nonmarital childbearing, and teenage pregnancy. First, they suggest that it is time to shift attention to fathers. Because some evidence shows that an employed father is more likely than an unemployed father to marry the mother of their child, society ought to do everything possible to put people to work, even if it means in jobs such as the Works Progress Administration provided in the 1930s.[1] Also, government agencies can use new techniques to locate fathers who fail to support their children. Adult men should also be discouraged from having sexual relationships with young teens. Second, pregnancy prevention programs among teens should be given much more attention. The authors report that meaningful pregnancy prevention programs in the schools have been shown to be effective in reducing teen pregnancy and should be emulated and expanded in California. Such programs emphasize assertive communication in sexual situations, stress abstinence, inform youth about the risk of sexually transmitted diseases, and provide contraceptive information. Also, teens who have been sexually abused need to be offered counseling, and the adults responsible for such abuse need to be prosecuted. Another suggestion is to encourage three-generation families. Requiring teenage recipients of Temporary Aid to Needy Families to live with responsible adults in a family setting could benefit the children of the teenagers. The last suggestion offered by the authors is an increase in classes in high school and college that focus on problem solving and communication skills.

Chapter 7, "Inequality and Poverty," by Robert Enoch Buck, states that although some inequality is inevitable in complex societies such as California, there are four conditions under which it constitutes a serious social problem: (a) where wealth confers privilege and esteem and poverty confers low respect; (b) where the poor are numerous and their material quality of life sinks below societal standards for health, nutrition, housing, medical care, and personal safety; (c) where employment security erodes to the point that the possibility of experiencing poverty over the life course becomes widespread; and (d) where government efforts to prevent poverty or mitigate it through the provision of income supports and social services are unsuccessful in overcoming its human costs. The author points out that all four conditions exist, both in California and the United States as a whole, and have been generally worsening since the 1970s.

In fact, Buck points out that if California were an independent nation, it would have the fifth richest economy in the world. It would also be, by far, the most unequal of all advanced industrial societies. Buck suggests that it is as if California is dividing, with economic growth lifting one group, and the rest falling further and further behind. He also shows that although inequality is growing in the United States as a whole, it is growing much more rapidly in California. The author points out that it is significant that slightly less than 10% of the population of California experienced major income gains during the economic boom that started in the late 1990s, with about 70% losing ground during this period of time. Buck discusses the inequality ratio, which is calculated by dividing the percentage of all personal income that is received by the richest fifth of the population by the percentage received by the poorest fifth. He shows that when using the inequality ratio as a measure, the rate of inequality in California increased by 83% in only 18 years.

The author points out that the inequality of wealth in California is increasing even faster than is inequality of income. Buck defines wealth as people's net worth, which is the value of all of their assets minus all of their debts. The author shows that if we use constant 1999 dollars, the number of billionaires in California has skyrocketed from 2 in 1983 to 68 in 1999, an astounding 3,300% increase.

When focusing on inequality of income, Buck points out that the primary cause for this increase in inequality has been the deterioration of the bottom end of the labor market. Californians are aware of the role played by changes in the occupational structure due to the rapid decline of the aerospace industry during the late 1980s. However, the author points out that this deterioration has been going on since the mid-1970s, partly as a result of increased competition in the world economy and partly due to changing business practices with regard to both increasing the percentage of part-time and temporary jobs and exporting labor-intensive jobs to foreign countries with lower wage rates. The deterioration of opportunities for workers with relatively little education is continuing as a result of ongoing patterns of occupational growth and shrinkage. Buck points out that the problem is further exacerbated by increasing returns on education, which are widening the pay gap between those with a college education and those without one. The government has also played a role in labor market deterioration, both by choosing to not enforce the National Labor Relations Act rigorously and by allowing business deregulation, which resulted in a massive wave of mergers that led to large-scale layoffs of workers.

Another way that government has been responsible for increased inequality is through change in taxation policy. A series of tax reforms during the 1980s, including much lower marginal tax rates on high incomes and a tax break on income from capital gains (that is, income deriving from investment rather than income deriving from employment), greatly reduced the federal income tax rate for the richest Americans. It also increased payroll taxes, such as social security, which effectively shifted more of the tax burden to the middle class. At the same time, many of what formerly were federal responsibilities were passed on to the states.

The author points out that people falling through the cracks of the economy is not a new phenomenon. It is as old as industrial society, when people first entered the cities and became dependent on the labor market for their living. Buck points out that industrial societies have recognized this for nearly a century and have, in varying degree, created systems of social services to ameliorate the misery it causes. The author states that cutbacks in the provision of such services since the early 1980s, and especially welfare reform, which began in 1997, have also contributed to increases in inequality and poverty in the Golden State. This takes us to the next chapter, which is also written by Buck.

In Chapter 8, "Poverty and Welfare Reform," by Robert Enoch Buck, the author discusses the change in the welfare system in California and the effectiveness of the welfare-to-work program. Before discussing welfare programs in California, the author points out that there are two basic kinds of social programs run by welfare states: social insurance programs, in which all can participate, such as social security; and means-tested programs, which provide services to individuals on the basis of their need, such as welfare. Buck points out that unlike Europe, the United States has shown a greater preference for social insurance programs over welfare and, as a result, erected a far less complete social safety net for the poor. The author also points out that from the beginning of the welfare state until welfare reform began in 1997, it was assumed that people have a right to welfare, although not a right to assistance sufficient to lift them above the poverty level.

Welfare reform began with the Personal Responsibility and Work Opportunity Reconciliation Act of 1996 (PRWORA). This law ended the right to welfare, replacing Aid to Families with Dependent Children (AFDC) with Temporary Aid to Needy Families (TANF), which required 32 hours of work activity per week to receive assistance and established a 60-month lifetime limit on aid. TANF funds are distributed through large block grants to each state, allowing states to establish and administer programs based on their own rules and requirements for eligibility, so long as they meet broad federal guidelines. Under the new law, if a state cuts its welfare expenditures, it gets to keep the rest of the money.

Buck points out that California was one of the last states to enact a welfare-to-work program, when the California Work Opportunity and Responsibility to Kids Act (CalWORKs) was signed into law in August 1997. The author points out that CalWORKs added several requirements beyond those established by the federal government, most notably shorter time limits for welfare-to-work participants, 24 months for those on welfare when the program began and 18 months for those entering after that time. Under this law, counties are allowed to provide a variety of programs to help individuals prepare for work, do a job search, and get a job. Participants not holding unsubsidized jobs after their welfare-to-work time expires must do 32 hours of community service per week to continue to receive TANF until their federal 60-month limit is reached. Buck points out that CalWORKS gives counties a strong fiscal incentive to cut costs, allowing them to keep 75% of all savings from their grant from the state.

How well has welfare reform fared in the Golden State? Buck states that welfare reform has been widely heralded as a success in the political arena as a result of the decrease in caseloads since the program began. However, he points out that critics have questioned whether welfare reform has done anything to alleviate poverty or whether it has just gotten poor people off of the welfare rolls. If we look at the poverty rate in California, it does not appear that welfare reform has really worked. Despite its great wealth, California has a poverty rate of 16.3%, a figure that is 23% higher than the national average. Indeed, during the long economic boom beginning in the early 1990s, the poverty rate in California increased by 30% at the same time that the U.S. poverty rate was declining.

Buck questions how the welfare rolls have been reduced in California. He presents convincing evidence that the decline in the number of people on welfare in California is not due to CalWORKs getting people good jobs. Buck points out that the number of people on welfare started to decline before CalWORKs was initiated, during fiscal year 1996-1997, coincident with the takeoff of California's economic recovery. He also points out that the reduction in caseloads is due to a reduction of recidivism, shorter episodes of being on welfare, and barriers being created that prevent those who become poor from getting on the rolls. Indeed, the number of applications for welfare relief in California plummeted because an increasing number of applications are being denied or withdrawn.

How are those leaving welfare doing? In California, the answer is not very well. Those who work are employed at low-wage jobs that pay between $5.50 and $7 per hour. Buck points out that given the cost of child care and the very high cost of housing in California, these wages are not enough to lift people above the poverty line.

Buck then discusses various continuing problems with welfare reform in California and puts forth some policy suggestions. One major concern is policy making that focuses on reducing the number of people on welfare, treating problems faced by participants after the fact and not dealing with the causes of poverty and the means to lift the poor to economic self-sufficiency. Buck discusses four areas of policy that are critical for California. First is the state's effort in terms of funding and programs, and the degree to which those programs address the causes of the problems in question and reach those in need. Buck shows that the Golden State does poorly here. In fact, California does not spend all of its annual federal block grant of $3.73 billion. By the end of the 1999 fiscal year, it had not spent $1.61 billion, 43% of that year's funding. States generally hold back obligated grant money to create a "rainy day fund" to be used in case of a recession. Nevertheless, the amount of unspent funds could easily address most of the problems facing California's poor. Buck points out that although California has developed a wide variety of programs for the poor, the poor seem to be losing ground. The percentage of families with incomes less than half the poverty level is increasing. The poorest of the poor are getting poorer. The percentage of the poor receiving food stamps, MediCal, and child care assistance is declining. Housing assistance remains extremely low.

The second area of policy is how the state addresses the fundamental issue of poverty: improving income and employment. In Buck's opinion, the state of California is not doing well in this area. He points out that the primary reason that only a third of CalWORKs participants are employed in unsubsidized jobs, the great majority of them low-wage and part-time, is that participants lack the human capital (education, training, and experience) to qualify for higher-paying jobs. The state needs to pay much more attention to these needs.

The third area of policy has to do with housing, especially given the shortage of housing and rising costs in California at the beginning of the 21st century. Buck shows the seriousness of this problem by pointing out that the typical CalWORKs family is a single working mother with two children and a monthly income from all sources, including TANF, of about $1,000, whereas the median rent for a two-bedroom apartment in California is about $900. If 90% of a person's income is going for rent, there is not enough money left to buy food and other necessities. The author shows that the rapid increase in the number of homeless families in California is a direct result of the above mismatch between income and cost of housing. Buck suggests that the state could provide rent vouchers to help ease the problem, especially because only 56.6% of its TANF block grant was spent during the last fiscal year. A pool of $1.6 billion is available to fund state rent vouchers.

The fourth policy question concerns the welfare of children, the intended prime beneficiaries of TANF and other programs. Buck thinks that child care could be the most crucial element in the process of welfare reform. First, it is extremely expensive, so much so that forcing a single mother to work at a minimum-wage job is not cost-effective. Second, the supply of child care is inadequate to meet the needs of all who are eligible. Third, just as there are time limits on receiving income assistance, there are even shorter limits on child care assistance. Fourth, the child care program for low-income families in California is divided among several different organizations, with the result that many families eligible for assistance fall through the administrative cracks. Finally, and perhaps most important, child care is a crucial element in preventing the transmission of poverty to the next generation. Buck then gives a number of suggestions on how child care could be improved in California.

In his conclusions, Buck states that real solutions to the problem of poverty are readily available. But until the politics of deception can be replaced by a public honesty about the nature of poverty, little will be done about it. He states that although many factors influence poverty, and a host of personal characteristics affect the likelihood that specific individuals will be poor, the fundamental cause of poverty is the labor market. A lasting solution to poverty must start from this fact.

Chapter 9, "The Continuing Crisis in Higher Education: Threats to California's Master Plan," by James L. Wood and Lorena T. Valenzuela, starts by showing how the crisis in higher education has changed from a financial problem in the early 1990s to a problem of who controls higher education at the start of the 21st century. To fully appreciate the work of Wood and Valenzuela, students need to know about California's Master Plan. Throughout the nation, it has been assumed for most of the past century that every child should attend school up to a certain age (usually defined by each state). However, 40 years ago, Californians extended that entitlement to postsecondary education (meaning education beyond high school). In 1960, Californians approved a broad policy called the Master Plan for Higher Education, often simply referred to as the Master Plan. The Master Plan was forged by educators, legislators, and the state's governor, and it virtually guaranteed higher education to any Californian who desired to continue studying beyond high school. If fact, it produced the "most affordable, accessible, diverse, and highly respected system of higher education in the United States" (California Citizens Commission on Higher Education, 1998, p. ix).

The dream of higher education begins with the California community college system. By the year 2005, it is projected that nearly 408,000 additional students will be attending the state's 108 two-year community colleges.

The second part of the education Master Plan is composed of the California State University (CSU) system, with 23 campuses that stretch from San Diego State University not far from the Mexican border to Humboldt State University in Arcata, not too far from the Oregon border. These universities are really multi-universities, offering a wide variety of courses and majors. Students can often earn not only the 4-year bachelor's degree but, in selective fields, the master's degree, usually involving a couple of years of additional study and independent work. In a few areas, the doctorate degree, which requires an additional 3 to 6 years of work, is also offered.

The final segment is the University of California (UC), currently a nine-campus system. The tenth campus is now under construction in Merced and is scheduled to open in 2003. The UC was intended to be the advanced research division (although much research is also done at the CSUs), granting bachelor's and master's degrees along with the highest educational degrees (for example, Ph.D., Ed.D.) and professional degrees (for example, the M.D. for physicians, the J.D. for lawyers, or the Pharm. D. for pharmacists). Nearly 36,500 students will be added to the UC system by 2005.

In the early 1990s, higher education was plagued by insufficient resources. However, the situation changed from 1995 to 2000 in California when a series of "compacts" (non-legally binding agreements) were worked out between Governors Pete Wilson and Gray Davis on the one hand and the heads of the three branches of California higher education on the other. The compacts called for a series of 4% yearly increases in the budgets of the three systems of California higher education, an infusion of funding that significantly improved the operation of each of the branches.

According to the authors, the problem that higher education in California faces now is a series of strategic plans presented to and often imposed on the faculty, students, and staff by university administrations, plans that often entail different views of governing the university, as well as different views of what constitutes quality higher education. The authors also assert that the above plans are often borrowed from the world of business, and they argue that this world cannot be equated with the world of higher education.

The authors discuss distance learning as an example of a type of instruction that emerges from the business model. Distance learning uses modern computer technology to transmit higher education from a central location to many students, who sit in front of computers or television screens at physical distances from the academic point of origin, many times with minimal interaction with faculty and other students. A recent conference in Sonoma, California, keynoted by an administrator from Britain's Open University, purported to show participants how to teach 5,000 students per class section. The authors show that powerful outside interests, often connected to corporations involved with selling modern computer technology, are attempting to influence academic decisions by helping devise these new strategic plans for colleges and universities. This is an influence that can negatively affect the quality of higher education provided to California students.

The authors point out that a central problem underlying the business approach to higher education is that business interests—profit making, managerial control, and economic bottom lines—may not correspond with educational interests, which stress intellectual development, acquisition of knowledge and skills, and critical thinking. Although university and business interests clearly overlap, an overemphasis on profit making and managerial control sets business at odds with university interests. Wood and Valenzuela assert that therein lies the answer to the question of why California higher education can be seen as a social problem: To the extent that business interests dominate California colleges and universities, there may be negative effects on the quality of California higher education, access to it, and its affordability. The authors discuss a number of trends connected to new planning for California colleges and universities, beginning in the 1990s, which will weaken the quality, affordability, and accessibility of higher education in California, thereby significantly weakening the Master Plan. These trends are (a) increased use of distance learning to provide instruction, (b) increased tuition, (c) increased attacks on academic tenure, (d) increased loss of professors' intellectual property rights over the courses they teach to maximize profits for others, and (e) increased use of part-time instructors.

The authors then assert that higher education will be continually diminished if these trends are not soon reversed. Indeed, they see one negative dynamic already under way. Bright young scholars who had planned on academic careers are now rethinking this career choice. If those who have spent much of a decade in graduate school arduously preparing themselves to become professors believe that their only opportunities lie in nontenured, low-paid, part-time instruction, delivered by distance learning, they may decide not to bother. The authors suggest that the greatest minds of the current generation will surely be lost to higher education if these trends are not reversed.

Finally, the authors offer numerous suggestions on what can be done to reverse these trends.

NOTE

1. To pull the country out of the 1929 Depression, President Franklin Roosevelt established the Works Progress Administration, which basically paid unemployed men to work on various public construction and maintenance projects, from buildings to bridges.

REFERENCE

California Citizens Commission on Higher Education. (1998). *A state of learning: California higher education in the twenty-first century.* Los Angeles: Center for Governmental Studies.

CHAPTER 6

PROBLEMS OF FAMILIES

Robin Wolf
Robin Franck
James A. Glynn

Three important trends have altered the nature of many families in both California and the nation. These are high rates of divorce, nonmarital childbearing, and teen pregnancy and parenting. In this chapter, these will be referred to as *family trends*. Such trends are particularly strong in California. For a number of decades, California's **divorce rate** was above the national average, and it is still high. California's nonmarital **birthrate** is also high. Although California no longer leads the nation in teen pregnancy and birthrates are declining, there is still cause for concern, as all of these trends may adversely affect the well-being of children. In this chapter, we will first briefly track the history of these three family trends. Then, we will examine the current situation and compare the California experience with that of the nation. Next, we will attempt to explain these trends. In conclusion, we will speculate about the future and suggest solutions.

Complicating our discussion is the debate over whether or not the high rates of divorce, nonmarital childbearing, and teen pregnancy and childbearing are actually social problems. How you view the issues will determine whether the definition of a problem is appropriate. Two commonly used positions in this debate are the social change perspective and the social problems perspective.

The social change perspective. Some social scientists argue that the increase in divorce and nonmarital childbearing does not necessarily constitute a social problem. From this perspective, these trends simply reflect neutral social change, creating diverse family forms, such as one-parent families, that are no better or worse than two-parent families. In other words, social change is inevitable, not necessarily problematic. Moreover, divorce can represent new beginnings and renewed chances for happiness (Amato, 1987; Weiss, 1979). To put it another way, the rising divorce rate may simply mean that more people are dissolving—instead of tolerating—unhappy marriages (Zastrow, 2000, p. 264). Those who take the social change perspective view the alarm over divorce and nonmarital childbearing as a nostalgic attempt to return to the traditional two-parent family of the 1950s, demonstrating an unwillingness to recognize that social change is inevitable (Coontz, 1997). This argument has merit. Why, then, is there so much public distress over the family change that is taking place?

The social problems perspective. Glynn, Hohm, and Stewart (1996, p. 3) define *social problems* as situations, policies, or trends that are (a) distressing to large numbers of people, (b) contrary to the moral beliefs of a society, and (c) at least partly correctable. From a social problems perspective, divorce and nonmarital childbearing have weakened the institution of marriage and generated concern for the well-being of children. In *Marriage in America, A Report to the Nation,* the Council on Families in America (1995) asserts that marriage is the most effective vehicle that societies have devised for assigning adults social and financial responsibility for children. The high rates of divorce and nonmarital childbearing may indicate that Americans are withdrawing some of their support from marriage as a social institution.

This chapter takes the social problems position and examines the downside of these family trends, all of which have resulted in financial hardship for women and children. However, it is important to clarify one point before we move forward. We are not arguing that all children should live with two married parents. Instead, we maintain that children benefit when households contain more than one adult who has made a serious long-term commitment to take responsibility for the child's well-being.

THE HISTORY OF FAMILY TRENDS IN CALIFORNIA

Like the rest of the nation, Californians have witnessed numerous changes within the family. In this section, we will discuss the changing divorce rate, the dramatic increase in nonmarital childbearing, teenage childbearing, demographic variables, the **feminization of poverty,** and teen pregnancy in the Golden State.

DIVORCE

For decades, Californians have appeared to be more open to the idea of divorce than residents elsewhere in the nation. In a state that has an image of exalting personal freedom, a higher-than-national divorce rate is not surprising. California gave rise to Hollywood, with film stars who appeared to divorce casually and very publicly; to the hippie movement of San Francisco's Haight-Ashbury in the 1960s, with its emphasis on sexual freedom, which was widely publicized in media portrayals of the region; and to the human potential movement, which stressed self-development and individualism, resulting in a *California Task Force to Promote Self-Esteem and Personal and Social Responsibility* in the 1980s. However, we must look beyond the state's seemingly freewheeling lifestyle to understand California's high divorce rate.

In 1998, California had the largest immigrant population of any state (Rein, Jacobs, & Seigel, 1999). Many of these immigrants left behind their extended kin, becoming detached from family members who otherwise might have urged troubled couples to work out their problems. Moreover, in the early 1970s, California was the first state to introduce no-fault divorce, which allows either partner to seek a divorce on the grounds of irreconcilable differences, without having to prove wrongdoing on the part of the spouse. This made divorce much easier to obtain and helped keep California's divorce rate high.

In spite of liberalized access to divorce, during the 1980s, California's rising divorce rate reversed direction and started to drop, then appeared to level off in the 1990s. We do not yet know if this decline and leveling off in divorce represents a long-term trend. Even so, California's divorce rate is still high. A case in point is that Southern California is home to *Divorce Magazine,* where the "Big D Divorce Ring" is marketed to those who would wear it to spark conversations with future friends or potential lovers (Harvey, 1998).

NONMARITAL CHILDBEARING

There is little historical information on nonmarital childbearing in California, but it is informative to look at the national experience. Figure 6.1 reveals how unusual childbirth outside of marriage was in the United States as recently as 1960. In that year, only 4% of all children in one-parent families were living with a never-married mother. In contrast, today, 4 in 10 children in one-parent families live with a never-married mother. Moreover, the number of children living in one-parent families grew from 5 million in 1960 to more than 17 million in 1998. This change took place during an era when the size of the U.S. population was relatively constant. Today, the number of children living with a never-married mother exceeds the total number of children living in all one-parent families in 1960.

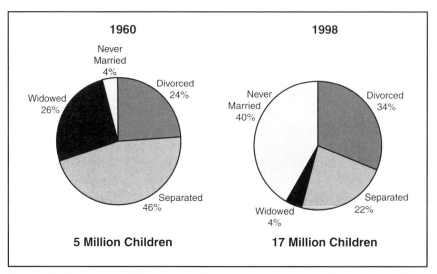

Figure 6.1. Percentage of U.S. Children Living in Mother-Only Families by Marital Status of Mother
SOURCE: U.S. Bureau of the Census, 1998a.
NOTE: Includes children under age 18.

TEENAGE CHILDBEARING

Teenage pregnancy and childbirth in California are among the highest in the nation. Over the latter part of the last decade, an overall national decline in teen birthrates was reflected in a decline in California's teen birthrates. However, California's teen birthrate ranked 14th among the states in 1996 (Meckler & Duerksen, 1999) and 11th in 1998 (Ventura, Mathews, & Curtin, 1999), indicating that other states made more progress in decreasing births to teenaged mothers than the Golden State did.

When we look at teen pregnancy, the significant development during the last 40 years is not that teens become pregnant and enter parenthood but that the vast majority of teens who give birth do so outside of marriage. This should not be interpreted incorrectly. Ventura et al. (1999, p. 1) point out that most births to unmarried women are not to teenagers. About 7 out of every 10 **out-of-wedlock births** in 1998 were to women who were at least 20 years of age.

Pregnancy among teenagers was not unusual in the 1950s, but teenagers overwhelmingly gave birth as married women, even if that required a wedding ceremony with a pregnant bride. In 1956, the **median age at first marriage** in the United States reached a low point of 20.1 years of age for females and 22.5 for males (U.S. Bureau of the Census, 1975, A/158-159, p. 19). However, teenage pregnancy and marriage did not present a serious social problem in the 1950s. This was an era when jobs for men were plentiful and wages were rising. Often, one breadwinner could

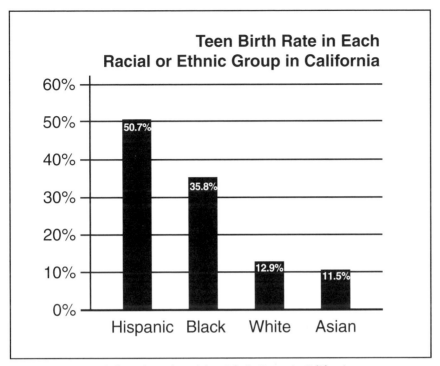

Figure 6.2. Teen Birthrate in Each Racial or Ethnic Group in California
SOURCE: California Department of Health Services, 1999.
NOTE: The teen birthrate refers to the number of births for every 1,000 females age 15 to 19. "White" refers to non-Hispanic whites. "Asian" includes Pacific Islanders and Native Americans.

support a family. Housing was more affordable, helped by low interest rates and a suburban housing boom (Newman, 1993; Skolnick, 1991). All of these factors made early marriage possible for teens, whether or not they became pregnant. Today, well-paid factory jobs are far more difficult to find, higher education is increasingly required to earn a good wage, and two incomes are typically necessary to achieve a middle-class lifestyle. By 1997, the median age at first marriage for women had risen to 25 and for men to 26.8 (U.S. Bureau of the Census, 1998a). Today, teenage women are far less likely to marry, regardless of whether or not they are pregnant.

The growth in teen births among Latinas. California's teen birthrate has declined over the last decade, but to a lesser degree for Latinas, including women of Mexican, Central American, and South American descent. Latinas accounted for 60% of California's births to teens in 1998 (Garvey, 1998). The teen birthrate among whites was one fourth the Latina rate, and African Americans accounted for a little over a third of the teen births in California (Figure 6.2).

The growing tendency for Latina teens to enter unmarried motherhood in California represents a change from the traditional family pattern among U.S.-born

Mexican Americans. In the past, when a Latina teenager became pregnant in rural areas of the American Southwest, both the girl's and boy's families joined to press the couple to marry. A folk saying revealed the attitude of the girl's family: *Ese bribon no va a burlarse de ti* (That rascal is not going to make a fool of you) (Gallegos y Chavez, 1980). Motherhood is viewed as the primary social role for women in Hispanic culture, and early marriage was acceptable, if not promoted (Rodriguez, 1988).

FAMILY TRENDS IN CALIFORNIA TODAY

Now that we have briefly examined the history of the family trends we are discussing, let us shift our focus to California at the present time.

DEMOGRAPHIC FACTORS

Recall that California's divorce rate, along with that in the rest of the nation, appears to have leveled off from the decline of the 1980s. This may be associated with a change in the demographic makeup of California's population. Because Latinos and Asians tend to have lower divorce rates along with a greater tendency to disapprove of divorce, the larger the portion of Californians who are Latino or Asian, the greater the segment of the California population that would be disinclined to divorce (Gonzales, 1992).

California's Latino and Asian populations are, of course, highly diverse. However, Latinos currently account for nearly half of all recent immigrants to the United States, and that figure is even higher for California (Rein et al., 1999). Generally speaking, the more recent a group's immigration, the more likely that family patterns will reflect origin cultures. Such cultural characteristics as the tendency to judge a people's behavior by how it reflects on their family might help account for the fact that in 1997, only 7% of adult Latinos were divorced, compared with 8% of whites and 10% of blacks (Lugaila, 1998).

THE FEMINIZATION OF POVERTY

High rates of nonmarital childbearing and divorce in California have led to a *feminization of poverty*. This term refers to the growing tendency for low-income families to be headed by single women. As Figure 6.3 shows, single women head

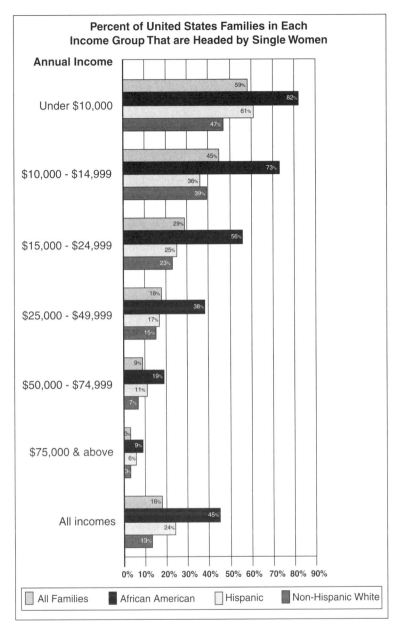

Figure 6.3. Percentage of U.S. Families in Each Income Group That Are Headed by Single Women
SOURCE: U.S. Bureau of the Census, 1998b

18% of all families, and 59% of all families with an annual income of under $10,000. The lower a family's income, the more likely that family is to be headed by a single woman. This pattern is particularly strong among African Americans and Latinos.

Nationally, about 84% of children who were living with just one parent in 1998 lived with their mother (U.S. Bureau of the Census, 1998a). Thus, the economic fate of single mothers affects their children. Divorced mothers are more likely than never-married mothers to be awarded child support. However, many women who are awarded child support never actually receive full child-support payments, and some mothers receive no child-support payments at all. Only three quarters of the children eligible for child support following divorce actually receive any, and most get paid only a partial amount. White women and better educated women have the resources to pursue the issue of child support in court and are more likely than Latina or African American women to receive child support (Ventura et al., 1999).

Marital status is becoming an important marker distinguishing between the "haves" and the "have-nots." When we compare the median annual income of U.S. families, those headed by single mothers fall at the bottom of the list, well below those headed by single fathers and by married couples (Figure 6.4).

TEEN PREGNANCY IN CALIFORNIA TODAY

Currently, in California, more than 1 in 10 births is to a teenager. Teen pregnancies resulted in 59,000 California births to women age 19 and younger in 1999, the last year for which data are currently available (California Department of Health Services, 1999). As Figure 6.2 shows, the highest rate of teen births is found among Latinas and the lowest rate among Asians.

Who is at risk for becoming an unmarried teen mother? Social class plays an important part in the risk of teen motherhood. Many teenagers who give birth in California grow up in families that live near or below the poverty line (Roan, 1995b). The pregnancy risk is highest for teenagers who grow up in a one-parent home, have poor academic performance in high school, and have a mother who has not completed high school (McLanahan & Sandefur, 1994). An important factor that insulates many teens against teen pregnancy is high educational and occupational aspirations. In contrast, teenagers who feel discouraged about their future often perceive the **opportunity costs** of pregnancy and childbirth as low (Murray, 1991). That is, adolescents who think their chances of obtaining a well-paid job and marriage are not bright often feel that they are not giving up many opportunities by becoming pregnant (Staples, 1985; Ventura, Curtin, & Mathews, 2000). This is not to say that teens plan their pregnancies; rather, it takes strong motivation and an enormous effort for a sexually active teenager to avoid pregnancy.

Why should we be concerned about teen parenthood? Parenting as a teenager is fraught with difficulties. Because teens are in the stage of identity formation, they tend to have an unstable identity. Parenthood brings a new identity with new re-

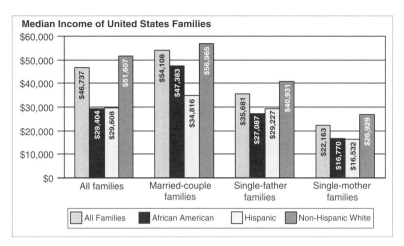

Figure 6.4. Median Income of U.S. Families, by Race/Ethnicity
SOURCE: U.S. Bureau of the Census, 1998b.

sponsibilities, but it also places the teen mother "off time" in her stage in the life course. She experiences the limits and responsibilities of parenting while many of her friends enjoy more freedom to spend time with friends, go on dates, and live a seemingly more carefree existence.

Like most teens, adolescent parents do not typically have well-developed problem-solving skills and often have an incomplete knowledge of child development. Not only are teen mothers less skilled at parenting, they are also less likely to complete high school, and they possess little economic and social support. Even so, almost half will have a second child within 3 years (Kendall, 2000, p. 324). The perception of teenage parenting as problematic remains despite the growth in support services for teen parents. Some high schools and community centers are beginning to provide support programs, as are health centers. The support programs include nutritional services, primary and preventive health care, vocational counseling, legal and educational services, and parent training courses (Lamanna & Reidmann, 1997, p. 324).

Many of the negative consequences of teen parenting for children can be mitigated by the involvement of grandparents in the parenting process. Grandparents can offer baby-sitting, help with housework, and provide advice and empathy for teen mothers (Nath, Borkowski, Whitman, & Schellenbach, 1991). However, many grandparents are reluctant to take over responsibility for their grandchildren. This is particularly the case for young grandparents, who have made sacrifices, worked hard, and hold a job that they like. Moreover, when the grandmothers take over much of the parenting role, a new conflict is often introduced into the extended family. Infants often develop an attachment to their grandmothers, and teen mothers feel left out as they compete with the grandparents for the infants' affection and with their infants for their own mothers' attention. A study of African American teen parents revealed that the grandmother's involvement in rearing her grandchild appears to work out best when the grandmother acknowledges the teen mother as the parent and models parenting skills for the teen mother to learn. Then, as the teen mother gains competence in parenting, the grandmother gradually relinquishes some of the parenting responsibility to the mother (Apfel & Seitz, 1991).

FAMILY TRENDS IN CALIFORNIA AS COMPARED TO THE UNITED STATES

Demographic trends in general tend to occur earlier and are often stronger in California than in the United States as a whole. Let us now compare the family trends in California with the rest of the nation.

COMPARING DIVORCE IN THE UNITED STATES AND CALIFORNIA

For a number of decades California's divorce rate ran above the national average. Moreover, California's "divorce revolution," a dramatic upsurge in divorce, occurred in the 1960s, a decade earlier than the national divorce revolution. The decline and leveling off of the divorce rate over the last decade could mean that California's families are stabilizing, but we should be guarded in our optimism. The drop in California's divorce rate probably occurred because people with a high risk of divorce have been removed from the pool of married people. The age at which a person first marries is the major predictor of divorce, and people who marry before they reach age 20 have the highest divorce rate of any age group. Today, this age group tends to form families through nonmarital childbearing rather than through marriage. In other words, divorce-prone families will not show up in the divorce data because they do not marry.

NONMARITAL CHILDBEARING IN CALIFORNIA AND THE UNITED STATES

Traditionally, California's nonmarital birthrate has been higher than the national average. Until the mid-1990s, 34% of all California births each year were to unmarried mothers, compared to 26% of births in the United States as a whole (U.S. Bureau of the Census, 1995). Now, about one third of all births nationwide are out of wedlock (Ventura & Bachrach, 2000). Nonmarital births are widespread and cut across all age groups and all racial and ethnic lines.

As Figure 6.5 shows, U.S.-born women in every racial and ethnic group (with the notable exception of Hawaiians) are more likely to bear children out of wedlock. However, it must be remembered that Hawaii became a state in 1959, so women under the age of 40 who were born on the islands were born in America. Consequently, the apparent anomaly presented by women of Hawaiian origin is probably not statistically significant.

Of those women who were born in the United States, the highest percentage of nonmarital births is found among non-Hispanic black women. If we define a traditional family as a man and woman who get married and then begin the process of childbearing, then more than 7 out of 10 ten black children are born into nontraditional families. Of course, the unmarried mother may get married at some point thereafter, but the point here is that the child does not begin life within a traditional setting.

In addition, more than half of Puerto Rican, Native American, and Filipino births to U.S.-born women of each category are out-of-wedlock births. Close to half of all children born to native-born mothers of Mexican, Central American, and South American descent are also nonmarital births. The lowest percentage of out-of-wedlock births is those to East Asian women, especially those not born in the United States. But notice that an American-born woman of Chinese heritage is almost twice as likely as a Chinese immigrant to bear a child out of wedlock, whereas an American-born woman of Japanese descent is three times more likely to have a nonmarital birth than a Japanese immigrant. For this second generation of women, perhaps, the traditional restraints of their mothers' country of origin do not apply as rigidly as they did for the immigrant generation.

COMPARING TEEN PREGNANCY IN THE UNITED STATES AND CALIFORNIA

When we examine teen pregnancy in the United States, there is good news and bad news. The good news is that the pregnancy rate for teenagers, girls age 15 to 19 who are sexually active, has declined over the last two decades. In 1972, one in every four teenage girls who were having sexual intercourse became pregnant that year. By 1990, only one out of five sexually active teens became pregnant. This decline in pregnancy among sexually active teens resulted primarily from greater use of contraceptives among adolescents. The birthrate for teenagers declined 3% between 1998 and 1999, to a rate of 49.6 births per 1,000 women ages 15 to 19, the lowest rate in the 60 years data on teen births have been recorded (U.S Department of Health and Human Services, 2000b).

The bad news about teen pregnancy is that the overall teen pregnancy rate is still rather high, especially when compared to the rate in other industrialized countries. U.S. teenagers have higher rates of pregnancy and parenthood than do teens in

CHALLENGES TO CALIFORNIA INSTITUTIONS

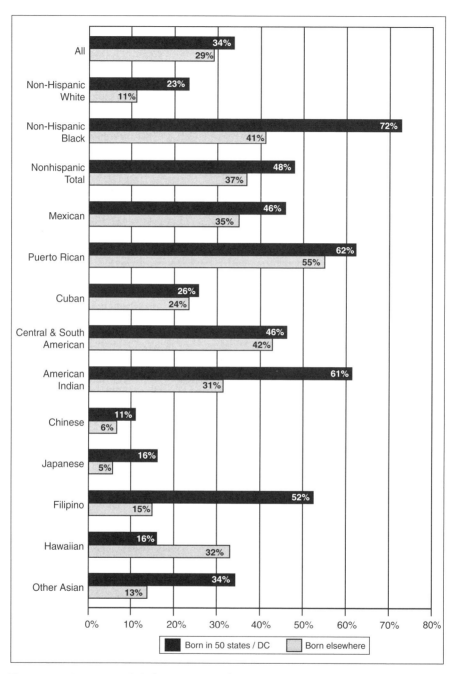

Figure 6.5. Percentage of Births to Unmarried Women by Mother's Place of Birth, Hispanic Origin, and Race or National Origin, United States, 1998
SOURCE: Adapted from Ventura & Bachrach, 2000.

other developed countries, despite the identification in the mid-1970s of a teen pregnancy epidemic (Lamanna & Reidmann, 1997, p. 322). In 1996, President Clinton introduced a National Campaign to Reduce Teenage Pregnancy. The campaign proposed actions that included taking a clear stand against teen pregnancy. A multiple component model was identified, involving one or more aspects of poverty, lack of opportunity, family dysfunction, and social disorganization in general (Dryfoos, 1998, p. 129). Much more attention is given now to looking at adolescence in multiple contexts rather than solely at families, or schools, or peers, or neighborhoods (Furstenberg, 2000).

CALIFORNIA TEEN PREGNANCY AND ABORTION

Although the other states seem to be catching up (as previously stated), California's teen pregnancy rate is still higher than the overall national rate. California's teen pregnancy rate is also higher than California's teen birthrate (California Legislature Office of Research, 1995). California ranks first in the nation in teen pregnancy and 28th in the nation in teen births as a percentage of all births in the state (*States in Profile,* 1995, Table I4). This discrepancy between teen pregnancies and teen births occurs because a substantial number of teen pregnancies in California terminate in abortion rather than in childbirth. In fact, one out of every five abortions in the United States takes place in California, although California residents make up only 12% of the U.S. population (Fay, 1993, Table 2.13.93). In both California and the nation, there appear to be two tracks for pregnant teens. Middle-class teens are more likely to use abortion, whereas low-income teens have a higher probability of giving birth. Both in the United States and in California, at least 80% of teens who give birth come from poor or low-income families (Alan Guttmacher Institute, 1994; California Legislature Office of Research, 1995).

Fathers in teen births. It is tempting to assume that teenage girls become pregnant by teenage boys of the same age. However, this does not appear to be the case. Of the California men who father babies born to teen mothers, two thirds of those for whom age is recorded are adults age 20 or older (Roan, 1995a). The California pattern, in which teen mothers typically become pregnant by older adult men, parallels the national experience. On the national level, the Alan Guttmacher Institute (1994) reports that more than half of the fathers in teen births were adults age 20 or older. This is not terribly surprising, because 60% of California teen births are to young women age 18 and 19. However, it is surprising to discover that one fifth of the fathers in U.S. births to teen mothers and 14% of fathers in California births to teen mothers are 6 or more years older than the teen mother (Alan Guttmacher Institute, 1994, Figures 41 and 42; California Legislature Office of Research, 1995).

Given their age, many of the fathers probably have jobs and are capable of paying child support. The father's age also probably indicates that younger females may find it difficult to resist the sexual advances of substantially older dating partners. In addition, there is also evidence that some teens who become pregnant look forward to moving out of their parental home and establishing a separate residence. Many of these girls may mistakenly assume that their older male partners are ready to take on the responsibilities of a family (Roan, 1995a).

There is growing evidence that the majority of teens who give birth have a personal history that includes sexual abuse (Dryfoos, 1998). The younger a sexually active teen, the more likely she is to have experienced involuntary sex (Alan Guttmacher Institute, 1994; California Legislature Office of Research, 1995). Young sexually active teens are more likely than older sexually active teens to have experienced sexual abuse at the hands of a mother's boyfriend, a stepfather, a stepmother's son, or a male family acquaintance. Adolescents who have been sexually abused often feel that they do not have control over their bodies and are more susceptible to sexual pressure in dating situations than teens who have not experienced sexual abuse (Roan, 1995a). The risk of sexual abuse for young teens is higher for teens in one-parent families than for teens in two-parent families. To some extent, the presence of a father in the home acts as a barrier making it more difficult for a male family friend to take sexual advantage of a daughter. Sexual abuse tends to lower the age of sexual initiation, and the younger the age at which the teen first becomes sexually active, the greater her pregnancy risk (Alan Guttmacher Institute, 1994). In a 3-year study of girls in poor America by Dodson (1998), girls asserted that sexual abuse was prevalent. Childhood abuse, molestation by strangers, and date or peer rape were regarded as forceful elements in the sexual landscape that a girl must manage. Sadly, the risk of sexual abuse was seen as an inherent part of a girl's sexual development.

WELFARE IN THE UNITED STATES AND IN CALIFORNIA

The welfare system is a joint federal and state program that provides cash benefits to needy children and the parents or other adults who care for them. In both the United States and California, this program has undergone substantial change. In 1996, the U.S. Congress ended Aid to Families with Dependent Children (AFDC), which allowed parents to receive welfare benefits continuously until dependent children reached age 18. At the same time, Congress created Temporary Assistance to Needy Families (TANF). The goals for the two programs differ. Whereas the old AFDC program focused on providing financial assistance to nonworking parents, the goal of TANF is to move parents from welfare to work. Thus, the emphasis in the new TANF program is on the word *temporary*. TANF places time limits on the receipt of benefits. The maximum time allowed for any one episode of continuous receipt of TANF assistance is 18 months in California and no more than 2 years in any U.S. state. The lifetime cumulative period during which a person can collect TANF benefits is 5 years in all states, including California. However, to avoid extreme hardship to families, each state may exempt 20% of its caseload each year from the 5-year lifetime limit (U.S. Department of Health and Human Services, 1998).

California has a disproportionate share of TANF recipients. Although California has 12% of the U.S. population, it is home to 22% of families receiving TANF (U.S. Department of Health and Human Services, 1998). What are the implications of these facts? Figure 6.6 provides information on nonmarital childbearing and poverty among young children in California. This figure lists selected California counties and ranks them in descending order by their poverty rate for children under the age of 6. Alpine County has the highest poverty rate (42%) for children under age 6, and Marin County has the lowest at 6%. The next bar shows the percentage of births to unmarried mothers. In Alpine County, almost three fourths of births are to unmarried mothers. Imperial, Fresno, Kern, and Humboldt Counties all show birthrates to unmarried mothers of about 40%. The lowest birthrate to unmarried mothers is in Marin County, with 5%. The third bar shows the percentage of births to mothers without a high school diploma. Again, Alpine County tops the list at 57%. Next highest is Monterey, followed by Fresno and Merced Counties. Finally, the percentage of births to teenage mothers under the age of 18 is listed. Alpine outranks all other counties, with Fresno, Merced, Imperial, and Stanislaus at about half the rate of Alpine.

What these data suggest is that a cluster of social characteristics surrounds poverty. Mothers who give birth while unmarried, who have a low level of educational attainment, or who give birth as teenagers are at risk for economic hardship. Their children are disproportionately represented in the poverty rates (Santelli, Lowry, Brener, & Robin, 2000). Today, it is difficult to achieve a middle-class lifestyle without two incomes, so it is not surprising that children born to unmarried mothers stand a greater chance of facing poverty. Likewise, when mothers have not completed high school, chances are slim that these mothers will be able to qualify for well-paying jobs and so substantially improve economic conditions for themselves and their children. There is probably a good deal of overlap between the two populations: unmarried mothers and poorly educated mothers.

The last bar of Figure 6.6 gives the percentage of births in 1998 that were to teenage mothers for each county. If these percentages seem small, it points to the fact that unmarried motherhood is no longer an experience that happens primarily to the young. Today, unmarried women who give birth may also be in their twenties or thirties (Bock, 2000). Also, some of these older unmarried mothers might have first given birth as teenagers and then had a second unmarried birth in their twenties. (For further information about the welfare system, please see Chapter 8.)

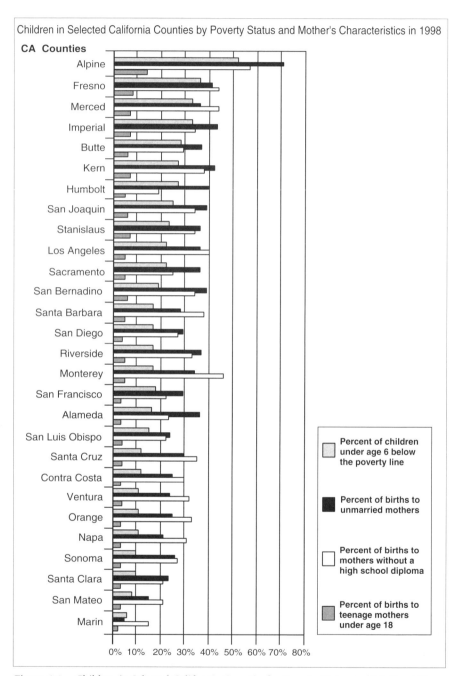

Figure 6.6. Children in Selected California Counties by Poverty Status and Mothers' Characteristics in 1998

SOURCE: California Department of Health Services, 1999.

ALTERNATIVE EXPLANATIONS FOR FAMILY TRENDS

Before we explore explanations for family trends, it is useful to look at the nature of causality. Social causation is usually multidimensional. Risks are cumulative, and rarely does any one factor cause a single outcome, such as teen pregnancy or nonmarital birth. Along this line, we will explore the high divorce rate, the **sexual revolution,** extended transitions, the rise in **cohabitation,** one-parent families, and economic restructuring.

HIGH DIVORCE RATE

Some social scientists argue that a high divorce rate weakens the institution of marriage. A society with a high divorce rate experiences a snowball effect. As the divorce rate rises, the willingness of people to commit themselves fully to marriage declines (Gill, 1993). Thus, a high divorce rate affects not only those who divorce but also single people and married couples, creating insecurity in regard to marriage.

Marriage itself has undergone a redefinition. Originally perceived as a social institution for the practical purposes of economic support and responsible child rearing, today, marriage emphasizes the relationship aspect over the institutional benefits of marriage. As a result, marriage is redefined not as necessarily permanent but as possibly nonpermanent or perhaps semipermanent (Lamanna & Reidmann, 1997, p. 472).

THE SEXUAL REVOLUTION

The sexual revolution of the 1960s and 1970s brought a more accepting attitude toward sex outside of marriage. The sexual revolution was made possible by the mass marketing of oral contraceptives in the mid-1960s and the landmark U.S. Supreme Court decision, *Roe v. Wade,* in 1973, which legalized abortion across the United States. In California, where abortion was already legal, *Roe v. Wade* loosened restrictions on abortion (D'Emillo & Freedman, 1988).

Surveys indicate that most females and males in the United States have engaged in premarital intercourse by the age of 19 and that most young women between the ages of 18 and 24 are not virgins at marriage (Andersen & Taylor, 2000, p. 178).

EXTENDED TRANSITIONS

Typically, Americans experience an extended transition between sexual initiation and marriage. The median age at first marriage for women has risen from 20.3 in 1960 to 25.5 in 1998. A similar rise has occurred among men, whose median age at marriage rose to 28.1 in 1998 (U.S. Bureau of the Census, 2000). Although the age at first intercourse leveled off or possibly even declined in the 1990s, the gap between onset of sexual activity and marriage is now huge compared to what it was in the 1950s. About half of all women today have sexual relations by age 18, more than 7 years earlier than the median age of marriage (Furstenberg, 2000).

THE RISE IN COHABITATION

There has been a significant rise in cohabitation, which refers to unmarried couples living together in a sexual relationship. More than three times as many couples lived together without being married in the 1990s as in the 1970s. In fact, sociologists contend that most people will adopt this lifestyle at some point in their lives (Andersen & Taylor, 2000, p. 388).

California has enacted a groundbreaking law permitting the creation of **domestic partners.** These partnerships enable couples to file a declaration of domestic partnership and receive rights such as:

- A domestic partner of a public employee may be covered for health benefits.

- A health facility must allow a patient's domestic partner, the children of the patient's domestic partner, and the domestic partner of the patient's parent or child to visit the patient (Rabenn, 2000).

The legal acknowledgment of domestic relationships will no doubt have an impact on cohabitation and marriage rates. Although it is too early to tell what that impact will be, it is worth noting that the California Legislature has recognized domestic partnerships as a legal status. Under the domestic partnership provision, it is also possible for same-sex couples to enjoy the benefits of health insurance coverage and hospital visitation rights. (For further discussion of domestic partners of the same sex, please see Chapter 4.)

THE IMPACT OF ONE-PARENT FAMILIES ON THE NEXT GENERATION

In their book, *Growing Up With a Single Parent*, McLanahan and Sandefur (1994) point out that children who spend some part of their childhood in a one-parent family have an increased risk of teen pregnancy and high school dropout when they reach adolescence. The authors reviewed four surveys with nationally representative samples, three of them **longitudinal studies** that tracked a group of children over time. All of these studies revealed that teenagers who spent some part of their childhood living in a one-parent family were twice as likely as those from two-parent families to bear a child as a teenager, twice as likely to drop out of high school, and one and a half times as likely to be "idle," neither working nor going to school (McLanahan & Sandefur, 1994, p. 2). Moreover, this risk remains when economic class and race are held constant. McLanahan and Sandefur calculate that half the risk of negative outcomes is attributable to family income and half to the absence of the father. A father in the home provides resources over and above his financial contribution to the family. Fathers often play an important role in reinforcing the authority structure in the family, thereby providing additional parental oversight and guidance for teenagers. Fathers in the home can help to enforce curfews, see that teens spend time on homework, and set limits on teen behavior, including sexual behavior. Thus, the father's presence in the home makes it more likely that children will achieve economic success in life. His absence increases the risk of teen pregnancy.

ECONOMIC RESTRUCTURING

We cannot understand nonmarital births by simply looking at the unmarried mothers. We also need to look at the economic position of the unmarried fathers (Wilson, 1995). Men's wages, measured in constant dollars that control for the effect of inflation, rose during the decades of the 1950s and 1960s, reaching a peak in 1973. After that, real wages of men began to decline. This loss of purchasing power created a **cost-of-living squeeze** (Harrison & Bluestone, 1988). If women had not entered the workforce, transforming the majority of married couples into two-earner families, American families would have experienced substantial downward economic mobility (Newman, 1993). Over the last three decades, competitive pressures in the global marketplace have squeezed American wages. Many U.S. factories have closed or have farmed out orders to global suppliers, cutting many higher paid jobs in American cities. Corporate downsizing has cut workforces, and automation has replaced a number of U.S. workers (Ehrenreich, 1986). Economic restructuring has been particularly hard on working-class families, especially families of African Americans, who have an unemployment rate that is double the rate for white men of the same age. African American sociologist Elijah Anderson (1989) points out that young African American men face **triple discrimination** in employment: They are stigmatized because of their age, race, and gender. The resulting economic insecurity means that fewer young adult men are willing to make a commitment to marriage, even if their girlfriend becomes pregnant.

THE FAMILY IN GLOBAL CONTEXT

In 2000, California had a population that was very close to 34 million people, making it by far the most populous of the 50 states. It was growing by a little more than 1% a year and had a median household income of more than $40,000 year. Fewer than 6 of every 1,000 live births died during the first year of life, and 28% of its adult population were college graduates (Population Reference Bureau, 2000a). By contrast, the countries of Southeast Asia were growing about one-and-a-half times as rapidly, with an annual **per capita income** of little more than $1,200 (perhaps about $5,000 for a household of four people). Moreover, some of those countries have an extremely high **infant mortality rate** (IMR). The IMR refers to the number of babies per 1,000 live births who do not live long enough to see their first birthday. In Laos, the IMR is 104, and in Cambodia, it is 143.

The countries of Central America were growing twice as fast as California, with a per capita annual income of a little more than $3,000 (perhaps between $12,000 and $13,000 for a household of four people). Among the major problems facing these countries was **doubling time,** the number of years it will take a population to double in size. No advanced industrial nation will ever double in size. However, the doubling time for Nicaragua, Belize, Honduras, and Guatemala is 26 years or less.

In middle Africa, countries are growing three times as fast as California, and per capita annual income is a scant $320 (perhaps $1,300 for a household of four, but a household that small would indeed be rare). On average, these countries double in population in about 23 years, despite an average IMR of 106. Moreover, nearly 7 out of every 10 cases of **human immunodeficiency virus** (HIV) infection occur in sub-Saharan Africa (Population Reference Bureau, 2000b).

In California, HIV/AIDS was long thought to be a gay disease, affecting mostly men who had sex with other men or who shared hypodermic needles. (For a full discussion, please see Chapter 5). In Africa, HIV/AIDS is spread mostly by heterosexual intercourse (Glynn et al., 1996). This may be an insurmountable problem because, according to local custom, a wife may not refuse sexual intimacy with her husband if he demands it. Furthermore, in many of these countries, girls and young women undergo the ritual of **clitoridectomy,** a form of female genital mutilation in which the clitoris is surgically removed. The purpose is to keep women virgins until marriage and to prevent them from straying after marriage (Mallery, 1997). The removal of the clitoris prevents women from experiencing pleasure from sexual intercourse. Consequently, women see their role in society as baby makers. So, in war-torn countries such as Mozambique, where many men have died, women often share a man for the purpose of getting pregnant (Bryjak & Soroka, 2001). If that man is HIV-positive, he may infect many women, who may then pass the disease to their children during pregnancy, childbirth, or breast-feeding.

The scope of the problem in this part of the world challenges the imagination. In 16 sub-Saharan countries, more than one tenth of the population between the ages of 15 and 49 is HIV-positive. In 5 of those countries, at least one fifth of the population is affected. In Botswana, more than one in three adults carries the disease. The effect on the family structure and the economies of such countries is devastating. Orphaned children are not adopted by other clan members, as was the traditional custom, because relatives fear that contact with surviving children will present danger to their own families.

Orphans drop out of school to find work, thus lessening their chance to improve their lives. The economy suffers, as businesses go bankrupt because of the lack of skilled, educated staff members. This, in turn, lowers the gross national product (the sum value of all goods and services produced), and that is reflected in even drearier lives for already impoverished people. Medicine is too expensive for most families, and it is also scarce. By contrast, AIDS patients in California are living longer because anti-AIDS drugs have reduced the number of deaths due to the disease by half and led to huge savings on hospital bills (Nullis, 2000). So, although families in sub-Saharan Africa are ravaged by this problem, things seem to be improving in California and the United States, as a whole.

Elsewhere in the world, women are having fewer children than their mothers did, but world population continues to grow. And, as people worldwide live longer and fewer children are born, the population ages, putting a greater burden on the working-age adults. This is now causing problems in Scandinavia, Western Europe, and Japan. Within the next couple of decades, it will also be a problem for the United States (*The World's Women*, 2000).

Births to unmarried women have increased dramatically in most economically developed areas. Figure 6.7 shows that in a number of economically developed countries, the percentages of births to unmarried women exceed the rate in the United States. Nearly two thirds of all children in Iceland are born out of wedlock. About half of the children of Sweden and Norway are born to unmarried mothers. Even Denmark, France, the United Kingdom, and Finland surpass the United States in this category. As we might suspect from our previous examination of out-of-wedlock births to selected racial and ethnic groups in California, Japan has the lowest rate (Ventura & Bachrach, 2000).

Because the United States often follows certain social trends set in Western Europe, we might anticipate further increases in such social indicators as age of entry into first marriage, number of people living alone, number of unmarried couple households, and percentage of children born out of wedlock. In the next section, we will take a look at some of the possibilities.

170 ■ CHALLENGES TO CALIFORNIA INSTITUTIONS

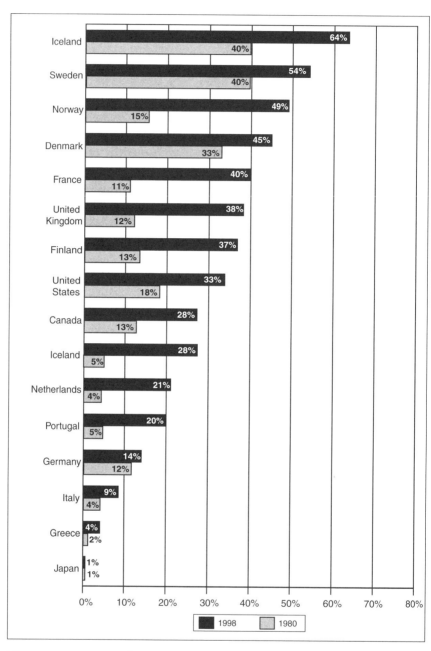

Figure 6.7. Percentage of Births to Unmarried Women, Selected Countries, 1980 and 1990
SOURCE: Adapted from Ventura & Bachrach, 2000.

FUTURE PROSPECTS

Let us turn now to the future of family trends in the United States and California. When we observe the recent divorce data for both California and the United States, it is tempting to think that marriages are stabilizing. However, we should be guarded in our assessment. What we are observing might be a shift away from marriage among young Americans toward unmarried childbearing. That is, the groups with the highest risk of divorce, namely those who marry before age 20, have been largely removed from the population of married people.

Strong social forces tend to keep the rate of nonmarital childbearing high. The idea of committing to marriage often raises anxiety in a society that features a high divorce rate. Economic restructuring and a global economy continue to keep wages of unskilled employees low, making it financially difficult to commit to marriage. The sexual revolution, which encourages an early age of sexual initiation, combined with a trend toward marrying at a later age, will probably continue to keep the risk of pregnancy high for youth and young adults.

SUGGESTIONS FOR THE FUTURE

After several decades, in which the focus has been on mothers in one-parent families, it is time to shift attention to fathers. There is some evidence that employed fathers are more likely to marry the mothers of their children than unemployed fathers are (Testa, Astone, Krogh, & Neckerman, 1989). William Julius Wilson (1995) has advocated a program modeled on the **Works Progress Administration** (WPA) agenda developed by Franklin Roosevelt during the **Great Depression** of the 1930s, a program that provided jobs for low-income breadwinners. In addition, government agencies could strengthen programs to locate fathers who are delinquent in child support payments. Recently Los Angeles County installed sophisticated computers to locate fathers who owe child support. Moreover, adult men should be discouraged from beginning sexual relationships with young teens. Some California legislators have advocated a public education campaign to emphasize that men who father children outside of marriage are taking on an 18-year financial responsibility, just as married fathers do (Roan, 1995a).

Pregnancy prevention. Among high school students, some programs have been particularly effective in reducing pregnancy. These use a combined approach emphasizing assertive communication in sexual situations, helping teens foresee a more positive future, informing youths about the risk of sexually transmitted diseases, and also providing contraceptive information (Garvey, 1998). Some social analysts have suggested that raising the driving age in California to 17 and eventually to 18 could help reduce the teen pregnancy rate by making the private back seat of the car less accessible. In addition, more teens who have been sexually abused need to come to the attention of Child Protective Services, and more of the adults who sexually abuse them should come into the criminal justice system.

Encouragement for three-generation families. Requiring teenage TANF recipients to live with responsible adults in a family setting could benefit children. Pregnant teens in abusive families would need to be placed in foster care where a responsible adult could oversee the teen mother's parenting efforts.

Skills and social support. Classes in high school and college on problem solving and communication skills could perhaps help to reduce the divorce rate. Parenting skills and life coping skills, if they were made more available to teens, might help manage some of the stress of teen parenting.

Although these suggestions are not likely to eliminate teen pregnancy, nonmarital childbearing, and divorce, they might mitigate some of the detrimental effects and improve the future outlook for California's families and children.

REFERENCES

Alan Guttmacher Institute. (1994). *Sex and America's teenagers.* New York: Author.
Amato, P. (1987). Family processes in one-parent, stepparent, and intact families: The child's point of view. *Journal of Marriage and the Family, 49,* 327-337.
Andersen, M., & Taylor, H. (2000). *Sociology: Understanding a diverse society.* Belmont, CA: Wadsworth.
Anderson, E. (1989). Sex codes and family life among poor inner-city youths. *Annals of the American Academy of Political and Social Science, 501,* 59-78.
Apfel, N. H., & Seitz, V. (1991). Four models of adolescent mother-grandmother relationships in black inner-city families. *Family Relations, 40,* 421-429.
Bock, J. D. (2000). Doing the right thing? Single mothers by choice and the struggle for legitimacy. *Gender & Society, 14*(1), 62-86.
Bryjak, G. J., & Soroka, M. P. (2001). *Sociology: Cultural diversity in a changing world* (3rd ed.). Boston: Allyn & Bacon.
California Department of Health Services. (1999). *Live births by age of mother* [On-line]. Available: http://www.dhs.ca.gov/hisp/applications/vsq/screen.age
California Legislature Office of Research. (1995). *Teen pregnancy and parenting in California.* Sacramento: Author.

Coontz, S. (1997). *The way we really are: Coming to terms with America's changing families.* New York: Basic Books.

Council on Families in America. (1995). *Marriage in America, a report to the nation.* New York: Institute for American Family Values.

D'Emillo, E. F., & Freedman, C. D. (1988). *Intimate matters: A history of sexuality in America.* New York: Harper & Row.

Dodson, L. (1998). *Don't call us out of names: The untold lives of women and girls in poor America.* Boston: Beacon.

Dryfoos, J. (1998). *Safe passage: Making it through adolescence in a risky society.* New York: Oxford University Press.

Ehrenreich, B. (1986, September 7). Is the middle class doomed? *The New York Times,* p. 44.

Fay, J. S. (1993). *California almanac* (6th ed.). Santa Barbara, CA: Pacific Data Resources.

Furstenberg, F. (2000). The sociology of adolescence and youth in the 1990s: A critical commentary. *Journal of Marriage and the Family, 62,* 896-910.

Gallegos y Chavez, E. (1980). Northern New Mexico women: A changing silhouette. In A. D. Trejo (Ed.), *The Chicanos* (pp. 67-80). Tucson: University of Arizona Press.

Garvey, M. (1998, June 1). A bundle of responsibility: Middle-school program seeks to lower pregnancy rate for teenagers. *Los Angeles Times,* p. 1.

Gill, R. T. (1993). Family breakdown as family policy. *Public Interest, 110,* 84- 91.

Glynn, J. A., Hohm, C. F., & Stewart, E. W. (1996). *Global social problems.* New York: HarperCollins.

Gonzales, J., Jr. (1992). *Racial and ethnic families in America.* Dubuque, IA: Kendall Hunt.

Harrison, B., & Bluestone, B. (1988). *The great u-turn: Corporate restructuring and the polarizing of America.* New York: Basic Books.

Harvey, S. (1998, April 30). Only in L.A. *Los Angeles Times,* p. 3.

Kendall, D. (2000). *Sociology in our times* (2nd ed.). Belmont, CA: Wadsworth.

Lamanna, M. A., & Reidmann, A. (1997). *Marriages and families.* Belmont, CA: Wadsworth.

Lugaila, T. A. (1998). *Marital status and living arrangements: March 1997 (update) and summary tables* (Current Population Reports: Population Characteristics, p 20-506). Washington, DC: Government Printing Office.

Mallery, S. (1997). *Zar possession as psychiatric diagnosis: Problems and possibilities.* Unpublished doctoral dissertaion, Fuller Theological Seminary, Pasadena, CA.

McLanahan, S., & Sandefur, G. (1994). *Growing up with a single parent.* Cambridge, MA: Harvard University Press.

Meckler, L., & Duerksen, S. (1999, April 29). Drop seen in rate of teen births. *San Diego Union-Tribune,* p. A-1.

Murray, V. M. (1991). Socio-historical study of black female sexuality: Transition to first coitus. In R. Staples (Ed.), *The black family: Essays and studies* (pp. 73-87). Belmont, CA: Wadsworth.

Nath, P. S., Borkowski, J. G., Whitman, T. L., & Schellenbach, C. J. (1991). Understanding adolescent parenting: The dimensions and functions of social support. *Family Relations, 40,* 411-420.

Newman, K. S. (1993). *Declining fortunes.* New York: Basic Books.

Nullis, C. (2000, June 28). 50% of teens to die of AIDS in parts of Africa. *Fresno Bee,* pp. A1, A14.

Population Reference Bureau. (2000a). *United States population data sheet 2000.* Washington, DC: Author.

Population Reference Bureau. (2000b). *World population data sheet 2000.* Washington, DC: Author.

Rabenn, G. (2000, December 6). *California miscellaneous-summary of statutes, domestic partner registration* [On-line]. Available: http: www.divorcesource.com/Ca/Articles

Rein, M. L., Jacobs, N., & Seigel, M. (1999). *Immigration and illegal aliens-burden or blessing?* Wylie, TX: Information Plus.

Roan, S. (1995a, July 10). The invisible men. *Los Angeles Times,* pp. E1, E4.

Roan, S. (1995b, July 9). A sign of the times. *Los Angeles Times,* pp. E1, E3.

Rodriguez, J. (1988). Labor migration and familial responsibilities: Experiences of Mexican women. In M. B. Melville (Ed.), *Mexicanas at work in the United States* (pp. 47-63). Houston, TX: University of Houston, Mexican American Studies.

Santelli, J. S., Lowry, R., Brener, N. D., & Robin, L. (2000). The association of sexual behaviors with socioeconomic status, family structure, and race/ethnicity among U.S. adolescents. *American Journal of Public Health, 90*(10), 1582-1588.

Skolnick, A. (1991). *Embattled paradise: The American family in an age of uncertainty.* New York: HarperCollins.

Staples, R. (1985). Changes in black family structure: The conflict between family ideology and structural conditions. *Journal of Marriage and the Family, 47,* 1005-1013.

States in Profile: The State Policy Reference Book. (1995). Birmingham, AL: U.S. Data on Demand, Inc. and State Policy Research, Inc.

Testa, M., Astone, N., Krogh, M., & Neckerman, K. M. (1989). Employment and marriage among inner-city fathers. *Annals of the American Academy of Political and Social Science, 501,* 79-91.

U.S. Bureau of the Census. (1975). *Historical statistics of the United States, colonial times to 1970.* Washington, DC: Government Printing Office.

U.S. Bureau of the Census. (1995). *American women: A profile* (Statistical brief 95-19). Washington, DC: Government Printing Office.

U.S. Bureau of the Census. (1998a). *Marital status and living arrangements.* Washington, DC: Government Printing Office.

U.S. Bureau of the Census. (1998b). *Selected characteristics of families by total money income in 1998* (Current population surveys, March supplement, FINC-01) [On-line]. Available: http://ferret.bls.census.gov/macro/031999/faminc/new01_003.htm

U.S. Bureau of the Census. (2000). *Statistical abstracts of the United States.* Washington, DC: U.S. Department of Commerce.

U.S. Department of Health and Human Services, Administration for Children and Families, Office of Planning, Research, and Evaluation. (1998). *Characteristics and financial circumstances of TANF recipients, fiscal year 1998.* Washington, DC: Author.

U.S. Department of Health and Human Services. (2000a). *National vital statistics report* (Vol. 48, No. 14). Washington, DC: Author.

U.S. Department of Health and Human Services, National Center for Health Statistics. (2000b, August 8). *New CDC birth report shows teen birth rates continue to drop* [On- line]. Available: http://www.cdc.gov/nchs/releases/00nes/00news/newbirth.htm

Ventura, S. J., & Bachrach, C. A. (2000). Nonmarital childbearing in the United States, 1940-99. In *National vital statistics reports* (Vol. 48, No. 16). Washington, DC: U.S. Department of Health and Human Services.

Ventura, S. J., Curtin, S., & Mathews, T. J. (2000). Variations in teen birth rates, 1991-1998: National and state trends. In *National vital statistics report* (Vol. 48, No. 6). Washington, DC: U.S. Department of Health and Human Services.

Ventura, S. J., Mathews, T. J., & Curtin, S. C. (1999). Declines in teenage birth rates, 1991-1998: Update of national and state trends. In *National vital statistics reports* (Vol. 47, No. 26). Washington, DC: U.S. Department of Health and Human Services.

Weiss, R. S. (1979). *The family life and the social situation of the single parent.* New York: Basic Books.

Wilson, W. J. (1995, August 20). *Reviving the environmental perspective in the public policy debate.* Address at the annual meetings of the American Sociological Association in Washington, DC.

The world's women 2000: Trends and statistics. (2000). New York: United Nations Publishing.

Zastrow, C. (2000). *Social problems: Issues and solutions* (5th ed.). Belmont, CA: Wadsworth.

CHAPTER 7

INEQUALITY AND POVERTY

Robert Enoch Buck

If California were an independent nation, it would have the fifth-richest economy in the world. It would also be, by far, the most unequal of all advanced industrial societies. It is ironic that such a rich state has a rate of inequality that is more similar to Malaysia, Kenya, and Zambia than to developed countries such as Japan, Italy, and France (Office of Economic Research [OER], 2000a; World Bank, 2000). Why is California so unequal? Is inequality like this really a social problem? If so, is anybody doing anything about it? Why hasn't a state as wealthy as California overcome this problem?

Structural social inequality is the outcome of the **stratification system** of a society, a social institution that consists of the institutionalized processes through which things are distributed and hierarchies of inequality are produced. The hierarchy of wealth is divided into social classes, each of which has a different pattern of **income determinants,** factors that determine the incomes received by members of that class. This means that factors increasing the incomes of one class, such as a booming stock market, will not necessarily have the same effect on members of other classes. For example, the fundamental source of income for the upper class is returns on wealth invested in financial markets. For the middle class, which depends on occupational earnings, the way the labor market matches individuals with jobs

---

Author's Note: I would like to thank Anna C. Walden and the editors for comments and suggestions on previous drafts of both this and the following chapter.

and determines what they will be paid is of paramount importance. The labor market consists of a large number of **job queues** within which this matching process takes place. It is easiest to understand a job queue as consisting of people in line waiting to be matched with all of the jobs of a given type in a given area. Those in the line are matched to jobs on the basis of their qualifications, often referred to as *human capital*, that is, the amount of education, training, and experience that they have. This process is mediated by employer preferences, including their prejudices regarding race, sex, ethnicity, religion, personal appearance, sexual orientation, and the like. The labor market is also the primary income determinant for the poor: The matching process either places them in a job that does not supply them with secure employment and a decent standard of living or it does not provide them with any employment. For many of the poor, public assistance provided by the government has been the basic source of their material quality of life.

Two additional characteristics of labor markets are important in the production of inequality. First, the qualifications that are required to get a specific job can change over time depending on the availability of workers with various levels of human capital. Second, the amount that a job pays can also vary depending on the availability of labor and on power relations within the labor market. If the supply of qualified workers in a job queue increases relative to the number of jobs available, wages will go down, and employers may raise the level of qualifications they require. Whether or not workers are unionized and how powerful unions are relative to management also affect wages.

The government also influences inequality in a couple of important ways. First, it redistributes wealth directly by collecting taxes and other revenues and then spending this money in a pattern that is different from the way in which it was originally distributed. Every penny the government spends redistributes wealth. Every time the government buys products such as computers or airplanes, much of what it spends is redistributed to the owners of the corporation. Although a lot of people own stock in corporations, those who benefit most from this are, by far, members of the upper class. Every time the government creates a payroll, either through hiring people or buying products that lead vendors to hire more workers, it is redistributing wealth to these middle-class employees. Finally, the government redistributes wealth to the poor, and especially to poor children, in the form of income assistance, food stamps, school lunches, and the like. However, recent changes associated with **welfare reform** promise to reduce such assistance drastically.

In addition to the government and the economy, two other social factors also have a major effect on inequality. The first is the manner in which education is distributed. This is becoming increasingly important because the value of a college education relative to that of a high school diploma has been increasing dramatically over the past two decades. The final major determinant of inequality is culture: the manner in which stereotypes lead to discrimination against various types of individuals, ideas about why people are poor and how the system of distribution and redistribution works, and cultural values that confer prestige on individuals based on their wealth.

But isn't inequality inevitable? How can it be a social problem? Does an analysis of it really belong in this book? Although some inequality is inevitable in complex societies, there are four conditions under which it constitutes a serious social problem:

1. where wealth confers privilege and esteem, and poverty confers low respect
2. where the poor are numerous, and their material quality of life sinks below societal standards for health, nutrition, housing, medical care, and personal safety
3. where employment security erodes to the point that the possibility of experiencing poverty over the life course becomes widespread
4. where government efforts to prevent or mitigate inequality through the provision of income supports and social services are unsuccessful in overcoming its human costs

All four of these conditions exist both in California and the United States as a whole; they have been generally worsening since the 1970s.

HOW UNEQUAL IS IT?

There are two important aspects of inequality: the distribution of income and the concentration of wealth. Income refers to all the money an individual or household receives from all sources in a given year. Wealth is accumulated over time and is usually measured as net worth, the value of all of a person's assets minus all debts. Wealth is difficult to measure, and data are gathered on it relatively infrequently. Income, on the other hand, is measured annually by the U.S. Bureau of the Census. As a result, most research on inequality examines the distribution of income. Because the primary problem associated with inequality is the poverty faced by those at the bottom, I will use the ratio of the size distribution of income (or inequality ratio, for short) as my measure of income inequality. This is calculated by dividing the percentage of all personal income that is received by the richest fifth of the population (also known as the richest income quintile) by the percentage received by the poorest fifth (or bottom income quintile). I will also take a look at trends regarding changes in real income for the poorest income quintile, the richest 5% of the population, and, just to get a quick look at how the middle class is doing, the middle income quintile as well.

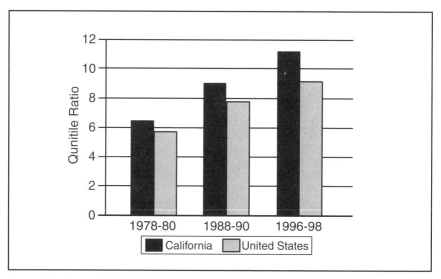

Figure 7.1. Inequality Ratios for California and the United States, 1978 to 1998
SOURCE: Adapted from Bernstein, McNichol, Mishel, & Zahradnik, 2000.

INEQUALITY OF INCOME

Income inequality in the both California and the United States decreased considerably between the beginning of World War II and the mid-1970s, and it has increased considerably, especially since 1980, and even more so in California than in the United States. I will look at data for three comparable periods, 1978 to 1980, 1988 to 1990, and 1996 to 1998, all rather high points economically. In 1978 to 1980, the income share of the poorest income quintile in the United States, including welfare and before taxes, was 6.5%, just slightly more than for California at 6.3%. At the same time, the income share for the richest fifth of Californians was 39.9%, which was slightly higher than for the United States as a whole. The inequality ratio at that time was 6.33 for California, which was slightly higher than that of the United States at 5.91 (See Figure 7.1).

By 1988 to 1990, the income share of the poorest income quintile in the United States had dropped to 5.3% of total personal income. In California, it had declined even more rapidly, reaching 4.9%. This means that during this decade, the income share of the poor declined by 19% in the United States, whereas the share for the poor in California, which was smaller to begin with, decreased by 22%. To put this in perspective, think about having your salary cut by nearly a fourth, then think about it happening when you already didn't have enough to live on in the first place. During the same period of time, the income share of the richest fifth of Californians rose quite rapidly to 44.5%, whereas that in the United States rose to 43.1%. As a result,

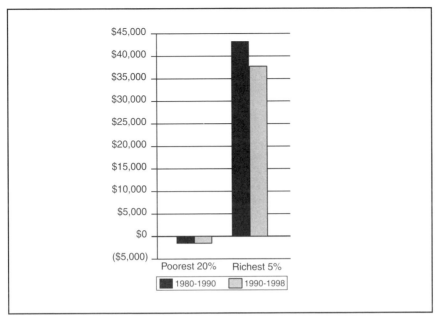

Figure 7.2. Changes in Income of Poorest and Wealthiest Californians, 1980 to 1998
SOURCE: Adapted from Bernstein, McNichol, Mishel, & Zahradnik, 2000.

the inequality ratio for the United States climbed to 7.94%, whereas in California, it rose more steeply to 9.08%, an increase of more than 40% in just one decade (Figure 7.1).

By 1996 to 1998, the income share of the poorest fifth in the United States had dropped to 4.9%, where it had been in California almost a decade earlier, whereas California's had dropped to 4.2%. This constitutes an additional decline of 8% in the income share for the poor in the entire United States and a decline of more than 14% for the poor in California. Once again, the income share for the richest income quintiles in both California and the United States grew as well, resulting in an inequality ratio of 9.27 for the United States and 11.57 for California. This means that the rate of inequality in California increased by 83% in only 18 years.

Another important aspect of income distribution is the actual increases in the amount of income for the poor and the wealthy. The last two decades of the 20th century were very good for wealthy Californians. During that time, the average income of the richest 5% of the population, correcting for inflation, increased by nearly 50%. During the same time period, the average income in constant dollars for the poorest fifth of the population in California declined by nearly $3,000.

As Figure 7.2 shows, the income declines for the poor between 1980 and 1990 and between 1990 and 1998 were extremely similar, whereas the gains of the rich decreased slightly. Figure 7.3, which shows the income trajectories of both groups, re-

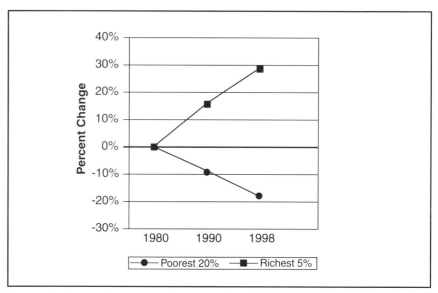

Figure 7.3. Income Trajectories of Poorest and Wealthiest Californians, 1980 to 1998
SOURCE: Adapted from Bernstein, McNichol, Mishel, & Zahradnik, 2000.

veals both the great difference between them and the steadiness of the trends for both.

From the time that the government started calculating the poverty rate in the 1960s until 1980, California had a significantly lower family poverty rate than the United States as a whole. The rates converged rapidly over the next 3 years, then remained about the same between 1983 and 1990. Over the next 5 years, the family poverty rate in California increased to 25% greater than the national rate, the level where it has remained since then. Most of this change was due to an increase in the poverty rate for single-parent families in California. About 36.2% of such families were poor in 1979, increasing to 45.2% by 1997–1998, during a period in which the national poverty rate for such families was declining (Johnson & Tafoya, 1999; U.S. Bureau of the Census 1983, Table 72).

It is not just the poor who are losing ground in California. The average income of the middle fifth of California's families increased only 1.5% between 1980 and 1990, losing ground to the more rapid income growth of the richest quintile. During the 1990s, however, the middle fifth not only fell further behind the wealthiest group, its average family income in constant dollars declined by $2,252, despite the rapid economic growth of the period. Finally, the income share of the fourth income quintile, that is, the second-richest fifth of households, has also been declining since the early 1980s. Only the richest 29% of Californians have registered any real income gains since 1979, with the richest 9% showing the only appreciable increases (U.S. Bureau of the Census, 1999c).

INEQUALITY OF WEALTH

Since 1980, concentration of wealth among the rich has been growing at an even more rapid rate than inequality of income. Even if we converted their wealth into 1999 dollars, only three men whose economic careers ended prior to 1980 were billionaires (Kafka & Pappas, 1999). By 1983, there were 15 billionaires, 2 of them Californians, including Gordon Peter Getty, the richest of them all with $2.2 billion, all derived from inheritance (Gissen with Behar, 1983). At that time, it required $125 billion to qualify for the richest 400 people in the United States, and one out of eight of them were from California. By 1999, the number of billionaires had increased astronomically to 267, of whom 68, or about one in four, were Californians (Forbes, 1999). The wealth of this group had also grown astronomically as well, with Bill Gates's $85 billion being nearly four times as much as the total wealth of all of the billionaires of 1983 combined. Getty, with roughly the same amount of wealth as when he was the richest in 1983, was now tied for 131st-richest. In 1976, the richest 1% of the population held 19.9% of the total wealth of the nation. By 1997, their share of the wealth had doubled to 40.1% (Collins, Hartman, & Sklar, 1999).

SOCIAL CLASSES AND INEQUALITY

What has caused the surge in inequality in California? Primarily, the stratification determinants for different social classes have been pushing in different directions. Two primary determinants account for this. Growth in financial markets has greatly increased wealth opportunities for the wealthy, whereas changes in the labor market have been the basis for both the decline of much of the middle class and the increase in poverty. The government has abetted all of these trends through changes in taxation, the regulation of business, and provision of social services, especially to the poor. I examine these in relation to the changing fortunes of social classes.

THE UPPER CLASS

The past 20 years have been very good to the rich in California. People get rich quickly because of market disequilibria, and there have been more of them than usual lately, with surges both in older markets such as real estate, entertainment, and finance, and in newer ones such as computers, cable television, and wireless communication. Governmental policy has also played a key role in the rapid surge in wealth concentration since 1980. One way in which this has been done is taxation. A series of tax reforms during the 1980s, including much lower marginal tax rates on high incomes and a tax break on income from capital gains, that is, income deriving from investment rather than from paid employment, greatly reduced the federal income tax rate for the richest Americans. It also increased payroll taxes, such as social security, which effectively shifted more of the tax burden to the middle class. At the same time, many former federal responsibilities were passed to the states. Like other states, California has tended to rely on taxes that hit the poor and the middle class the hardest.

The government also deregulated business practices in a number of industries in a way that, among other things, allowed a continuing wave of mergers that facilitated much higher levels of economic concentration. Finally, the Reagan administration, especially through the president's firing of striking air traffic controllers, greatly reduced enforcement of the National Labor Relations Act, the law protecting the rights of workers to organize, to bargain collectively for better pay, and to use the threat of strikes as a power resource. This greatly increased the power of corporations relative to that of workers in the labor market, contributing to lower wages and higher profits.

THE MIDDLE CLASS

Since the 1970s, the middle class has been losing ground overall, while dividing into three segments. The decline began as Japan and other industrial societies that had been damaged by World War II finished paying off their war debts and began investing more in their industrial base, greatly increasing competition for U.S. firms in the world economy. This led to cutbacks in industrial production and the movement of labor-intensive jobs to Third World countries with lower labor costs, both of which led to job reductions. This was followed by the wave of mergers noted above, which, in addition to concentrating wealth in the upper class, also led to large-scale layoffs of redundant workers, costing thousands of Americans their jobs.

Another cost-cutting effort has been the use of more part-time and temporary workers. California now has more nonvoluntary part-time workers than any other state, and, indeed, it has as many as the next three states, New York, Texas, and Pennsylvania, combined (U.S. Bureau of Labor Statistics, 1999a). This is significant because part-time workers not only earn less because they work fewer hours, but they are paid far less per hour than full-time workers in the same occupational categories. For example, the average hourly wage for full-time blue-collar workers in Los Angeles is $13.60, more than 50% more than the $8.99 wage for part-time workers in the same category (U.S. Bureau of Labor Statistics, 2000). Finally, part-time and temporary workers very seldom get fringe benefits such as employer-based medical insurance, paid sick leave, or savings and pension plans.

The impact of declining conditions in the labor market on middle-class family incomes was offset to some degree by an increase in the number of working wives.

Furthermore, during the 1970s and 1980s, California had an unusually large number of growing industries with high-wage jobs, such as computers, communications, entertainment, defense, and aerospace (Dertouzos & Dardia, 1993). However, the end of the Cold War and concomitant government cutbacks in military and aerospace spending in the late 1980s caused employment in those industries to decline by half between then and the mid-1990s. This resulted not only in the loss of about 150,000 jobs in these industries but also in significant declines in manufacturing jobs among suppliers to those industries as well (California Department of Finance, 1999; Office of Economic Research, 2000a). Manufacturing is extremely important in determining inequality for several reasons. First, it is the primary way in which individuals with less than a college education can make a decent living. Second, in manufacturing, a much higher percentage of workers are generally members of unions than in other industries. This is important because unions increase the power of workers relative to the power of employers in the labor market, which results in higher wages and better fringe benefits and working conditions. Finally, most manufacturing is labor intensive and, thus, creates large numbers of jobs for blue-collar workers.

Industrial growth since the decline of the aerospace sector has failed to replace either the number of lost jobs or their incomes. Much of the growth has been in electronic equipment and computers, which have added about 45,000 jobs between them, but these sectors are top heavy with professional employees and require relatively few production workers. Apparel manufacturing has also added about 20,000 jobs, but it pays very low wages (California Department of Finance, 1999; Foster, 2000; Office of Economic Research, 2000a, 2000b; U.S. Bureau of Labor Statistics, 1999a, 1999c). In 1998, the average weekly earnings for production workers in aerospace were more than 60% greater than the average for all other manufacturing workers (calculated from California Department of Finance data). For example, the weekly earnings for the few remaining mechanical workers in aircraft in Los Angeles in 1999 was $861 per week, as opposed to $395 a week for electronic assemblers and $321 per week for textile sewing machine operators (U.S. Bureau of Labor Statistics, 2000).

The most important impact of restructuring on inequality has come through changes in the occupation structure, both overall and within industries. Table 7.1 shows the overall shift in the occupation structure of California between 1980 and 1999.

Managerial occupations, which constituted 12.0% of all jobs in 1980, grew to 15.8% in 1999, an increase of 31.7%. The share of all jobs for professional, sales, and unskilled labor occupations also increased by more than 10% during this time period. Furthermore, skilled and semiskilled labor occupations, technical occupations, and administrative support and clerical occupations all experienced more than a 10% decrease in their share of all occupations. Although these data attest to the magnitude of change, they are misleading in that some of the occupational cate-

TABLE 7.1   Changes in the Distribution of Jobs Across Occupation, California, 1980 to 1999

Occupation	1980	1999	Change
Managerial	12.0	15.8	3.8
Professional	13.1	15.2	2.1
Technical	3.3	2.8	−0.5
Sales	10.8	12.4	1.6
Clerical	18.5	14.4	−4.1
Protective services	1.5	1.6	0.1
Personal services	3.6	3.8	0.2
Skilled labor	12.3	10.2	−2.1
Semi-skilled labor	7.1	5.0	−2.1
Transportation	3.6	3.2	−0.4
Unskilled labor	3.8	4.2	0.4
Low pay services	7.5	8.0	0.5
All non-urban	2.7	3.6	0.9

gories with high rates of change are too small to have much of an effect on employment opportunities, and, hence, inequality of income. We can get a better feel for the impact of these changes by looking at an index of job change for occupational categories (See Figure 7.4).

Clearly, the changes that have had the greatest effect have occurred in six categories. There have been major increases in managerial, professional, and sales occupations and declines in clerical, skilled, and semiskilled labor. All of these changes have contributed considerably to inequality. Managerial and professional occupations are the two highest-paying occupational categories, increasing concentration at the top of the income hierarchy. Some of the growth in sales has been in very high-paying jobs that require considerable human capital, such as sales representatives in industry and finance, whereas much of it has been in retailing, where there is significant discrimination against African Americans (Browne, 2000). The decline in clerical jobs is crucial because it is the primary occupational category providing a decent income to women without a college degree. Finally, skilled labor is the main occupational category that pays males without a college education a middle-class wage, with wages for many of the lost semiskilled jobs in defense and aerospace also having paid at this level.

In addition to the labor market, several of the governmental changes discussed with regard to the upper class also had an important negative effect on the middle class. The changes in taxes, for example, shifted a major portion of the tax burden from the rich to this group. Likewise, weakened enforcement of labor laws led to a decline by half in the percentage of workers represented by unions. This was a major factor in the deterioration of wages. Unions are the primary source of labor market power for workers, with the result that union workers are paid more than nonunion workers. In addition, the union-nonunion wage gap is greater in California than for the United States. For example, among blue-collar workers in Los Angeles, California's leading industrial city, the average earnings of union workers are 57% greater than those of nonunion workers, as opposed to only 38% for the nation as a whole (Foster, 2000; U.S. Bureau of Labor Statistics, 1999b).

As a result of these changes, the middle class began to fragment into three parts. At the top is the stable middle class. It is made up primarily of two-parent families, with at least one college-educated adult who holds a stable, white-collar job and many with two such earners. This group has been able to maintain the middle-class lifestyle that emerged in the period immediately following World War II.

The second group is the declining middle class. It is made up of families that have found it increasingly difficult to maintain a middle-class lifestyle. For many, this is due to a slowing of growth of family income relative to inflation. This has happened to a diverse group, ranging from blue-collar families facing wage erosion due to de-unionization to public employees facing declining governmental pay scales. Another segment of the declining middle class is made up of families whose breadwinner was a victim of a layoff due to corporate restructuring, an event that generally results in an extended reduction in lifetime earnings (Schoeni & Dardia, 1996). Still another segment of this group was created by the rapid increase in single-parent families, coupled with the lower income opportunities faced by female workers due to labor market discrimination.

The third segment of the new middle class is the marginal middle class. This group consists of families that are currently not poor but, because of the changing composition of the labor market, do not have a family member who holds a stable full-time, full-year job that pays above the poverty level. At best, families in this category maintain a somewhat diminished middle-class lifestyle, with some key elements, such as home ownership and the ability to put children through college far less likely. These families maintain a barely self-sufficient family income through having multiple job holders, often with some holding two or more jobs at the same time. The lower boundary for this group is really not suffering from want of the basic necessities. The percentage of single-parent families is higher in this category than for the rest of the middle class, although the ability of families to remain in this category and not fall into poverty is much greater for those families with two adult earners. However, even these families fall into poverty fairly frequently due to the instability of the kinds of jobs that their members hold. Thus, the loss of a part-time job by a child could plunge such a family into poverty, whereas another member

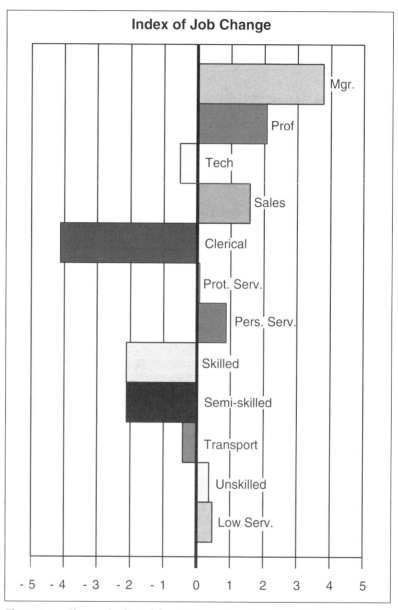

Figure 7.4. Changes in the California Occupation Structure, 1980 to 1999

picking up such a job could extricate them from it. For this reason, quite a few families cycle between this group and poverty on a more or less regular basis, and most experience poverty at some point over an extended period of time.

THE POOR

The primary cause of the increase in inequality in California, and the reason it is a major social problem, is the increase in poverty. During the 1990s, although the state was in the midst of a great economic recovery, the poverty rate in California increased by 30%. This is especially problematic because it occurred at a time when the poverty rate nationwide was decreasing by 6% (U.S. Bureau of the Census, 1993a, 1993b, 1999d). As a result, in a period of just 8 years, the percentage of families that were poor in California rose rather alarmingly from 7% below the national rate to 23% above it (U.S. Bureau of the Census, 1993a, 1993b, 1999d).

California has pockets of grinding ethnic poverty, with rates of 48% for those of Mexican origin in Isla Vista, 49% for Native Americans in Eureka, 53% for Koreans in Berkeley, 69% for Laotians in Modesto, 75% for Vietnamese in San Francisco, 76% for Cambodians in Fresno, and 79% for Hmong in Visalia (U.S. Bureau of the Census, 1993b). In addition, although only about 16% of Californians are poor at any given point in time, the percentage of the population that experiences difficulties maintaining a decent quality of life at some time or other over an extended period of time is much greater. There are two reasons for this. The first has to do with the way in which the official poverty rate is calculated, and the second has to do with the dynamic nature of poverty. The latter is extremely important, because many people interpret the poverty rate as supporting the stereotype that the poor are a small minority who do not want to work because of defects in their character.

Measuring Poverty

The official federal poverty line ($14,962 for a single mother with two children in 2000) was established by the government in the early 1960s as a means of measuring the progress of policies and programs created as part of the War on Poverty, a government effort during the 1960s that created or expanded a number of anti-poverty programs. It was based on the U.S. Department of Agriculture's minimum emergency diet, the least costly diet that could maintain minimal nutritional standards without being hazardous to health over a short period of time. Because the average household at that time spent about a third of its income on food, the official poverty rate was pegged at three times cost of the Department of Agriculture's minimum diet (Schwartz & Volgy, 1992). Although the government has regularly adjusted the measure for inflation, it has not taken into account the cost of unmeasured items such as housing and medical care, which have grown faster than the cost of food. As a result, the official poverty measure has increasingly underestimated the cost of living, and, as a result, the amount of poverty. Scholars who have attempted to recalculate the government's standard to take these changes into account suggest that the true poverty line was actually 50% higher than the official mark as early as 1982 and was 67% higher by 1990 (Schwartz & Volgy, 1992). An alternative measure, the California Self-Sufficiency Standard, takes into account a detailed list of costs including food, housing, medical care, clothing, taxes, utilities, child care, and transportation, as well as income enhancements such as the earned income tax credit and differences in the cost of living in different parts of the state. It suggested that the minimum earnings needed to achieve minimal self-sufficiency for a single mother with two children in 1995 ranged between $17,000 and $26,000, depending on the county in which the family resides and the ages of the children (Equal Rights Advocates, 1996, 2000). Thus, a large number of families are not under the official poverty level but are nevertheless unable to meet all of their basic needs on a regular basis.

Who Are the Poor?

Many of those who are at some distance from the poor in the socioeconomic system have the misconception that only a few people are poor, that the majority of them are more or less permanently unemployed, and that they are a major burden on governmental budgets. The poor are also commonly stereotyped as being welfare dependent, being addicted to drugs or alcohol, and lacking motivation and moral character. Although some fit one or more of these descriptors, the poor in this society are actually quite diverse. Although the official poverty rate at any given point in time ranges between about 12% and 15%, less than 2% of the population are permanently poor, but about 40% experience poverty at least once over a period of two decades (Devine & Wright, 1993). Prior to welfare reform, only 36% of poor families in California received any public assistance income (U.S. Bureau of the Census, 1993b). Among those who did, the median duration of a spell of participation in major means-tested programs was only 7 months, with only 43% of participants in **Aid to Families with Dependent Children** (AFDC) receiving assistance for 24 consecutive months (U.S. Bureau of the Census, 1995b). Finally, prior to welfare reform, the tax burden created by the welfare they received constituted a fraction more than 1% of the federal budget (Buck, 1998).

There are really three different types of poor people in the United States, each with its own different profile and problems. These are the episodically poor, the cyclically poor, and the chronically poor.

The *episodically poor* consists of individuals who experience at least one significant episode of poverty within two decades but are poor less than 15% of the time overall. This group constitutes roughly 13% of the total population and nearly half of those who are ever poor, although it constitutes only about 1% to 2% of those who are poor at any given point in time (see Table 7.2).

Episodes of poverty for this group result from an economic crisis, most often loss of income by the main breadwinner, either through layoff, death, divorce, or illness. When such families experience an episode of poverty, they usually right themselves in a matter of months.

TABLE 7.2   Distribution of Poor Households by Poverty Category

	Percentage of Population	Percentage of Current Poor
Episodically Poor (poor less than 15% of the time) employment: very low	13%	1%-2%
Cyclically Poor (poor 15%-80% of the time) employment: more than 60 percent	18%	55%-65%
Chronically Poor (poor more than 80% of the time) employment: sporadic to none	12%	33%-45%
5-year poverty episode	8%	21%-28%
Poor all 20 years	1.3%	8%

SOURCE: Devine & Wright 1993.

Some families in this category have no one who holds a full-time, full-year job that pays above the poverty level. They maintain a position above the poverty level most of the time by piecing together a family occupational mosaic consisting of several earners, some of whom hold more than one job. Because so many of the jobs of this quality are temporary or unstable, family occupational mosaics often change frequently, such as when a child loses an afterschool job, or the father loses one of his two jobs, then someone finds something else, and so on. Most families with occupational mosaics in this category have two working parents, yet many still live so close to the margin that they are liable to experience more than one episode of poverty over two decades and to face situations during which they cannot meet all of their needs even more frequently.

What differentiates episodically poor families from others who face economic crises is the lack of sufficient assets to sustain them until their income flow is restored to its regular level. Thus, the greatest need for such families is enough aid to weather the current storm, although greater employment stability would certainly fill a major need as well. For family occupational mosaic households in this category, educational programs to upgrade the skills of adult members so that they can qualify for more stable jobs could also be of great benefit.

The *cyclically poor* are so named because they cycle in and out of poverty on a regular basis. They are poor more than 15% of the time but are also not poor at least 20% of the time over an extended period (see Table 7.2). About 18% of the overall population is in this category, and they make up about one third of those who are ever poor. Because of the frequency of their episodes of poverty, they constitute between one half and nearly two thirds of the poor at any given point in time. Although this category is somewhat diverse, its major component is single-parent families, many of which have an occupational mosaic. It also includes some two-parent families with less employment stability than similar families among the episodically poor. This category contains the majority of families receiving assistance through **Temporary Assistance to Needy Families** (TANF, which replaced AFDC as a result of welfare reform) and, thus, most of those who are directly affected by welfare reform. Employment among the cyclically poor is surprisingly high, although it is frequently interrupted due to the prevalence of unstable jobs.

The *chronically poor* are poor more than 80% of the time. About 12% of the population is in this category, and two thirds of them experience episodes that last 5 years or longer. At the bottom of this category are those who are more or less permanently poor, a group that constitutes slightly more than 1% of the overall population (see Table 7.2). The chronically poor constitute about 15% of those who are ever poor and between one third and a little less than one half of those who are poor at any given point in time, depending on shortages of jobs with decent rates of pay and relatively low education requirements, which push some of the cyclically poor down into this category. Many in this group are lone individuals. Barriers to entry into the labor market, such as mental health, family violence, and drug or alcohol problems, are fairly common. However, many other types of people are in this group as well. Indeed, 18% of female-headed households in the United States are chronically poor (U.S. Bureau of the Census, 1995a).

Chronic poverty has increased significantly since the mid-1970s, an occurrence that appears to be correlated with the deterioration of the lower end of the labor market. Whereas less than 8% of the population experienced a 5-year period of chronic poverty during the 1970s, about 12% did in the 1990s.

A few members of this category experience much longer episodes of poverty. Those who experience 5-year periods of unbroken poverty constitute 8% of the overall population and between one fifth and slightly more than one fourth of those who are poor at any given point in time. Finally, only 1.3% of the population is poor virtually all of the time (see Table 7.2).

THE CAUSES OF POVERTY

Why are people poor in the richest economy in the world, especially during its longest period of economic growth since World War II? In the previous section, I examined some personal characteristics and experiences that make some individuals more likely to be poor than others, including experiencing a change in the family structure or economic situation or having multiple barriers to employment, such as drug addiction, poor mental health, or a history of family violence and abuse. Having been raised in a family in which no one was in the labor force would also fit into this category. Family background and educational opportunities, especially with regard to the acquisition of human capital, are even more important as personal determinants of poverty, as is discrimination based on skin color, ethnic origin, gender, age, sexual orientation, religion, and physical appearance (Catanzarite, 2000; Darity & Mason, 1998; Darity & Myers, 1998; Holzer & Stoll, 2000; Roscigno, 2000).

However, large-scale social and economic factors, not personal characteristics, are the fundamental cause of poverty and inequality. Poor people are poor because of the kinds of jobs they get (or do not get), how much they get paid for them, how much things like food and housing cost, how much the government takes in taxes, and how much it provides to them in services. In the rest of this chapter, I will look at factors associated with jobs and incomes. In the next chapter, I turn to the impact of governmental policies and programs on the poor.

HUMAN CAPITAL-LABOR MARKET MISMATCH

The fundamental cause of poverty in the labor market is the mismatch between the human capital of the poor and the availability of jobs that require such skills. The structure of the labor market consists of a large number of job queues, each of which is made up of all of the jobs within a specific occupational category within a specific region. As previously noted, job queues match potential workers with jobs in that occupation based on the characteristics of applicants and employer preferences. It is best to think of a job queue as a group of individuals with various qualifications standing in line for jobs within a specific occupation. Whether they get a job, what job they get, and how much they get paid for it are all a result of the matching process of that job queue. Thus, even though there is a very low unemployment rate, and even labor shortages in some job queues, such as those for electronic engineers and financial sales representatives, there is still crowding caused by an oversupply of workers for the number of jobs available in job queues that require low levels of education (Bernstein, 1997). This helped drive down the average wages of low-wage workers in California, which, in constant dollars, declined from $8.13 per hour in 1979 to a low of $6.51 per hour in 1997 (Economic Policy Institute, 1999). Finally, hiring decisions within job queues are ultimately made on the basis of employer preferences, which have to do not only with human capital but also with personal likes and dislikes that result in discrimination.

THE POOR, HUMAN CAPITAL, AND JOBS

The labor market mismatch faced by the poor is a major problem for efforts to reduce poverty. Nearly half of low-income Californians do not finished high school,

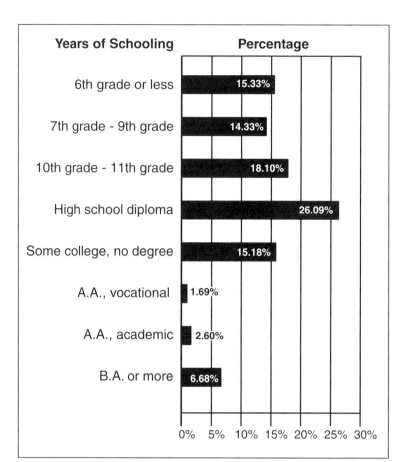

Figure 7.5. Educational Distribution of the Low-Income Population of California, 1998 to 1999
SOURCE: U.S. Bureau of the Census, 1999a, 2000.

whereas more than one fourth have a high school diploma and nothing more. Another 15% have attended college but received no degree, with 4% having completed community college programs and not quite 7% holding bachelor's degrees (see Figure 7.5).

This means that about half of low-income Californians qualify for jobs in the three lowest occupational categories: farm labor, unskilled urban labor, and lower-level service occupations. These are occupations in which hourly wages are low, and full-time, full-year jobs are rare. Even more important, these positions constitute only a small percentage of all jobs to begin with, and they are shrinking in this regard relative to other occupations (U.S. Bureau of the Census, 1995b, Table 2).

TABLE 7.3 Educational Profile of Recent Hires for Selected Occupational Categories

	Less Than High School	High School Only	Some College	Percent of U.S. Jobs	Percent of California Jobs
Clerical	5.7	48.7	45.6	17.11	16.57
Skilled labor	18.4	49.0	32.6	13.51	11.06
Semi-skilled labor	28.4	51.7	19.9	14.64	9.13
Unskilled labor	31.4	47.8	20.8	4.53	3.72
Service	20.7	44.1	35.2	11.99	12.39
Lower-level service					7.28
Household service	36.3	41.2	22.5	5.28	

SOURCE: U.S. Bureau of the Census 1993a, 1993b.

Because employers prefer individuals with the highest qualifications available for even lower-level jobs, the real supply of jobs available to people with a low level of education depends less on the minimal standard for an occupation than on the competition they face in the labor market. For that reason, it is best to infer the educational requirements for an occupation from the education profile of recent hires (see Table 7.3).

The urban occupation category with both the lowest educational distribution and income is lower-level service occupations (e.g., maids, janitors, busboys), in which 36.3% of recent hires have less than a high school education and 41.2% have a high school education but no college. Even at this level, the percentage of non-high school graduates among jobholders is less than the percentage of such individuals among those receiving income assistance. Furthermore, this is not a large category, constituting just 7.3% of all jobs in California (see Table 7.3).

The next lowest category with regard to human-capital requirements is unskilled labor occupations such as handlers, equipment cleaners, helpers, and laborers, a category in which 31.4% of recent hires have less than a high school education and 47.8% have a high school education and nothing more. Wages in this category are still extremely low. Furthermore, this sector, too, cannot accommodate many welfare recipients, especially single mothers, because it constitutes only 3.7% of all jobs in California and has a predominantly male labor force (see Table 7.3).

The other occupational category with low human-capital requirements is semi-skilled labor. Only 28.4% of recent hires in this category have less than a high school education, 51.7% have a high school diploma and nothing more, and 19.9% have at least some college. It includes jobs such as machine operators and assemblers. Many jobs in this category, especially those exhibiting the fastest rates of growth, have become increasingly technical, thus requiring higher levels of education. It is useful to separate these semitech jobs from the rest of semiskilled jobs because of these requirements. Although the educational data presented here do not include job training programs, it is certain that many workers in this category have had such training and that the percentage of workers in this category with college-level training will continue to grow. Semiskilled jobs as a whole constitute only 9.1% of all jobs in California, far less than the national figure of 14.6%, thus exacerbating the misfit problem in the state.

Precision production, craft, and repair jobs, the so-called skilled-labor occupations, constitute 11.06% of jobs in California and pay rather well, but only 18.4% of recent hires in this category have less than a high school diploma, whereas 32.6% have some college, making the job category a bad fit for the education profile of welfare-to-work clients. Like semiskilled occupations, jobs of this type are more prevalent in states with a relatively high concentration of heavy industry, and they constitute just 11.1% of jobs in California, as opposed to 13.5% for the United States as a whole (see Table 7.3).

Because the distribution of California jobs is skewed toward white-collar occupations and the great majority of welfare-to-work participants are women with less than a college education, their best opportunity for self-sufficient employment appears to be the lowest white-collar category, which consists of administrative-support and clerical occupations. This category not only constitutes 16.6% of jobs in California, but 76.9% of those holding such jobs are female. Furthermore, this category has a relatively low educational profile for a white-collar occupation. Although only 5.7% of recent hires have less than a high school education, the majority have no education at the college level (see Table 7.3).

Are there any rapidly growing jobs that might accommodate welfare-to-work participants? There are 18 occupations that are currently producing a net increase of 3,000 or more jobs per year in California (see Figure 7.6).

Six of these jobs—general managers, public school teachers, registered nurses, electronic data processing systems analysts, computer engineers, and support specialists—pay quite well but have human-capital requirements far beyond the level of the great majority of the poor. Although some among the pool of welfare recipients could achieve the academic qualifications for such jobs, given the educational opportunity, it is unlikely that this will occur often enough for such jobs to provide a significant contribution to employing the poor. Six more fast-growing jobs—cashiers, watchmen, food service, hand packers, groundskeepers, and janitors and maids—pay far below a self-sufficiency wage. Retail sales and waiting tables have highly variable rates of pay depending on the place of employment, although the pay for even the better jobs is mitigated by the fact that so many jobs in these categories are part-time. Three more fast-growing occupations—secretaries, medical assistants, and office clerks—have a high percentage of female employees, are gener-

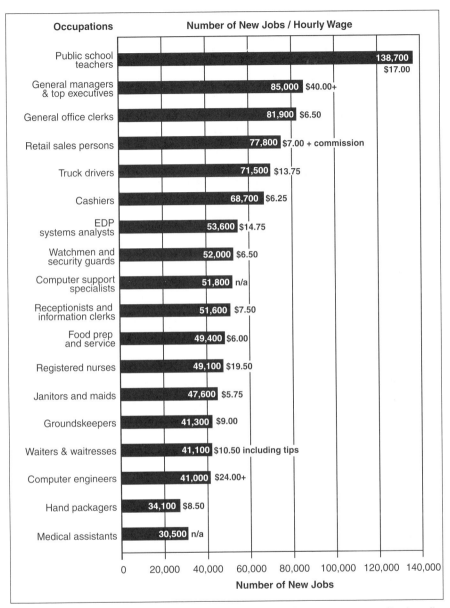

Figure 7.6. Occupations With the Greatest Job Growth in California, 1998 to 2008 (Projected)
SOURCE: California Employment Development Department, 2000a; U.S. Bureau of Labor Statistics, 1999c.

ally full-time, full-year, and generally pay at or above the self-sufficiency income for a family of three. However, once again, recent hires in such jobs suggest that some training beyond the high school level is necessary for employment.

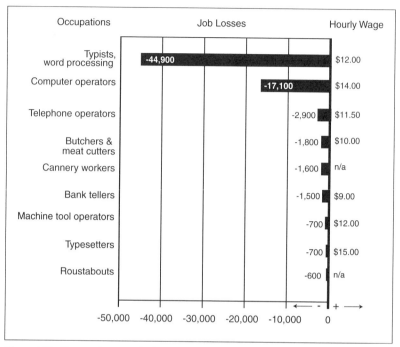

Figure 7.7. Occupations With the Greatest Loss of Jobs in California, 1998 to 2008 (Projected)
SOURCE: California Employment Development Department, 2000b; U.S. Bureau of Labor Statistics, 1999c.

Finally, among the few occupations that are a good fit for welfare-to-work participants in terms of paying a self-sufficient wage and having lower human-capital requirements, many are experiencing a declining number of jobs, even during the economic recovery of the late 1990s (see Figure 7.7).

Even more important, five of them—typists, computer operators, telephone operators, typesetters, and bank tellers—are jobs in which there is a relatively high concentration of women, making them a good match for single female heads of households. Furthermore, they require training that easily fits under the time limits of **CalWORKs,** California's welfare-to-work program.

CONCLUSIONS

The rate of inequality in California is growing rapidly, and has been doing so for nearly two decades. It is as if society is dividing, with economic growth lifting one group and the rest falling further and further behind. It is significant that slightly less than 10% of the population of California experienced major income gains during the economic boom that started in the late 1990s, with about 70% losing ground during that period of time (Daly & Royer, 2000).

The primary cause for this increase of inequality has been the deterioration of the bottom end of the labor market. Californians are aware of the role in this played by changes in the occupation structure due to the rapid decline of the aerospace industry during the late 1980s. However, this deterioration has been going on since the mid-1970s, partly as a result of increased competition in the world economy and partly due to changing business practices with regard to both increasing the percentage of part-time and temporary jobs and exporting labor-intensive jobs to foreign countries with lower wage rates. The deterioration of opportunities for workers with low human capital is continuing as a result of ongoing patterns of occupational growth and shrinkage. The problem is further exacerbated by increasing economic returns to education, which are widening the pay gap between those with a college education and those without one. The government has also played a role in labor market deterioration, both by choosing not to enforce the National Labor Relations Act rigorously and by business deregulation, which allowed a massive wave of mergers that led to large-scale layoffs of workers. People falling through the cracks of the economy is not a new phenomenon. It is as old as industrial society, when people first entered the cities and became dependent on the labor market for their living. Industrial societies have recognized this for nearly a century and have, in varying degree, created systems of social services to ameliorate the misery it causes. Thus, cutbacks in the provision of such services since the early 1980s, and especially welfare reform, which began in 1997, have also contributed to increases in inequality and poverty. It is to these matters that I will turn in the next chapter.

REFERENCES

Bernstein, J. (1997). *Low wage labor market indicators by city and state: The constraints facing welfare reform.* Washington, DC: Economic Policy Institute.

Bernstein, J., McNichol, E. C., Mishel, L., & Zahradnik, R. (2000). *Pulling apart: A state-by-state analysis of income trends.* Washington, DC: Center on Budget and Policy Priorities.

Browne, I. (2000). Opportunities lost? Race, industrial restructuring, and employment among young women heading households. *Social Forces, 78,* 907-929.

Buck, R. E. (1998). *Wealth, status, and power.* San Diego: KB Books.

California Department of Finance. (1999). *California statistical abstract 1999.* Sacramento: Author.

California Employment Development Department. (2000a). *Occupations with the greatest growth, 1998-2008.* Sacramento: Author.

California Employment Development Department. (2000b). *Occupations with projected declines: California, 1998-2008.* Sacramento: Author.

Cantanzarite, L. (2000). Brown-collar jobs: Occupational segregation and earnings of recent immigrant Latinos. *Sociological Perspectives, 43,* 45-75.

Collins, C., Hartman, C., & Sklar, H. (1999). *Divided decade: Economic disparity at the century's turn.* Boston: United for a Fair Economy.

Daly, M. C., & Royer, H. N. (2000). Cyclical and demographic influences on the distribution of income in California. *The Federal Reserve Bank of San Francisco Economic Review, 43,* 1-13.

Darity, W. A., Jr., & Mason, P. L. (1998). Evidence on discrimination: Codes of color, codes of gender. *Journal of Economic Perspectives, 12,* 63-90.

Darity, W. A., Jr., & Myers, S. L., Jr. (1998). *Persistent disparity: Race and economic inequality in the United States since 1945.* Cheltenham, UK: Elgar.

Dertouzos, J., & Dardia, M. (1993). *Defense spending, aerospace, and the California economy.* Santa Monica, CA: Rand Corp.

Devine, J. A., & Wright, J. D. (1993). *The greatest of evils: Urban poverty and the American underclass.* New York: Aldine de Gruyter.

Economic Policy Institute. (1999). *State hourly wage rates for low-wage (20th. Percentile) workers, 1979-1998.* Washington, DC: Author.

Equal Rights Advocates. (1996). *California self-sufficiency standard.* San Francisco: Author.

Equal Rights Advocates. (2000). *California self-sufficiency standard* (2nd. ed.). San Francisco: Author.

Forbes Magazine. (1999). *The 400 richest people in America* [On-line]. Available: http://www.forbes.com

Foster, A. C. (2000). *Union-nonunion wage differences, 1997.* Washington, DC: Government Printing Office.

Gissen, J., with Behar, R. (1983). The Forbes four hundred. *Forbes, 132*(9), 71-192.

Holzer, H. J., & Stoll, M. A. (2000). *Employer demand for welfare recipients by race.* Evanston IL: Joint Center for Poverty Research.

Johnson, H. P., & Tafoya, S. (1999). *The basic skills of welfare recipients: Implications for welfare reform.* San Francisco: Public Policy Institute of California.

Kafka, P., & Pappas, B. (1999). Decades of dough. *Forbes* [On-line]. Available: http://www.forbes.com

Office of Economic Research. (2000a). *California: An economic profile.* Sacramento: California Trade and Commerce Agency.

Office of Economic Research. (2000b). *California economic review* (Second Quarter, 2000). Sacramento: California Trade and Commerce Agency.

Roscigno, V. J. (2000). Family/school inequality and African-American/Hispanic achievement. *Social Problems, 47,* 266-290.

Schoeni, R. F., & Dardia, M. (1996). *Earnings losses of displaced workers in the 1990s.* Santa Monica, CA: Rand Corp.

Schwartz, J. M., & Volgy, T. J. (1992). *The forgotten Americans.* New York: Norton.

U.S. Bureau of the Census. (1983). *1980 census of population: Vol. 2, Part 6. Social and economic characteristics: California.* Washington, DC: Government Printing Office.

U.S. Bureau of the Census. (1993a). *1990 census of population, Vol. 2, Part 1. Social and economic characteristics: United States.* Washington, DC: Government Printing Office.

U.S. Bureau of the Census. (1993b). *1990 census of population, Vol. 2, Part 6. Social and economic characteristics: California.* Washington, DC: Government Printing Office.

U.S. Bureau of the Census. (1995a). *Dynamics of well-being: Poverty* (CPR P70-42). Washington, DC: Government Printing Office.

U.S. Bureau of the Census. (1995b). *Dynamics of well-being: Program participation* (CPR P70-41, SIPP Data Series). Washington, DC: Government Printing Office.

U.S. Bureau of the Census. (1999a). *Current population survey, March, 1998: Supplement data set.* Washington, DC: U.S. Government FERRET Data System.

U.S. Bureau of the Census. (1999b). *Health insurance coverage: 1998* [On-line]. Available: http://www.census.gov

U.S. Bureau of the Census. (1999c). *Money income in the United States, 1998.* Washington, DC: Government Printing Office.

U.S. Bureau of the Census. (1999d). *Poverty in the United States, 1998.* Washington, DC: Government Printing Office.

U.S. Bureau of the Census. (2000). *Current population survey, March 1999: Supplement data set.* Washington, DC: U.S. Government FERRET Data System.

U.S. Bureau of Labor Statistics. (1999a). *Geographic profile of employment and unemployment, 1998.* Washington, DC: Government Printing Office.

U.S. Bureau of Labor Statistics. (1999b). *Los Angeles-Riverside-Orange County, CA national compensation survey, August 1998.* Washington, DC: Government Printing Office.

U.S. Bureau of Labor Statistics. (1999c). *National compensation survey: Occupational wages in the United States, 1998.* Washington, DC: Government Printing Office.

U.S. Bureau of Labor Statistics. (2000). *Los Angeles-Riverside-Orange County, CA National Compensation Survey, July 1999.* Washington, DC: Government Printing Office.

World Bank. (2000). *World development indicators 2000.* New York: Author.

CHAPTER 8

POVERTY AND WELFARE REFORM

Robert Enoch Buck

Despite its great wealth, California has a poverty rate of 16.3%, a figure that is 23% higher than the national average (U.S. Bureau of the Census, 1999c). Indeed, during the long economic boom beginning in the early 1990s, the poverty rate in California increased by 30% at the same time that the U.S. poverty rate was declining (U.S. Bureau of the Census 1993a, 1993b, 1999c). In the last chapter, I examined how changes in the labor market have made it difficult for increasing numbers of people, and especially those with a high school education or less, to achieve economic self-sufficiency. Historically, such individuals were able to rely on assistance from the federal government to help them through periods when the labor market failed to provide them with the means to maintain a decent quality of life. The dismantling of this support since the early 1980s, and especially since the advent of **welfare reform** in 1997, has greatly changed the probability of people being poor, as well as how poor they will be. This is especially true for single-parent families in which there is only one adult earner, families that include a disproportionately large number of California's children. How has the **welfare state,** as this system of governmental support is known, changed? How are single parents and their children doing as a result of the combination of these changes in the system of social support and the continuing deterioration of the bottom part of the labor market?

THE WELFARE STATE

Welfare states consisting of broad-based governmental social service programs began to emerge in response to the social dislocations resulting from the rise of urban industrial labor markets. The first were in Europe, with U.S. programs starting during the mid-1930s, at the low point of the Great Depression. There are two basic kinds of social programs run by welfare states: social insurance programs in which all can participate, such as social security, and means-tested programs that provide services to individuals on the basis of their need, such as welfare. Unlike Europe, the United States has shown a greater preference for social insurance programs and, as a result, has erected a far less complete social safety net for the poor.

During the 1960s, the War on Poverty and Great Society programs of the Kennedy and Johnson presidencies expanded the safety net and created education, training, and job programs to help the poor move into the labor market. From the inception of the welfare state until welfare reform began in 1997, people were assumed to have a right to welfare, although not necessarily to assistance sufficient to lift them above the poverty level.

The Humphrey-Hawkins Bill of 1976 took a very different approach to the problem of poverty, declaring that people had a right to a job. Had Congress passed it, it would have guaranteed jobs for all, with the government acting to generate jobs and serving as employer of last resort by creating public jobs at the prevailing wage (Weir, 1988).

Throughout the War on Poverty era, opposition grew against what some termed "black programs," so labeled because a large percentage of those eligible for assistance were African American victims of segregation and discrimination. This opposition grew into a war against welfare, which, especially during the 1980s, produced legislation that repealed or scaled back many poverty programs. A media campaign generated public support for this by selling the idea of the "undeserving poor" through the use of stereotypes about laziness and welfare queens driving to pick up their monthly checks in Cadillacs. As a result, welfare programs increasingly became seen by the public as unfair and expensive ways of redistributing wealth from the rich and the middle class to the unworthy (Katz, 1989).

Welfare reform was the culmination of this movement. The name of the act that created it reflects its ideological heritage: the Personal Responsibility and Work Opportunity Reconciliation Act of 1996 (PRWORA). This law ended the right to welfare, replacing **Aid to Families with Dependent Children** (AFDC) with **Temporary Assistance for Needy Families** (TANF), which require 32 hours of work activity per week from those who received assistance and established a 60-month lifetime limit on aid. TANF funds are distributed through large block grants to states, allowing the states to establish and administer programs based on their own rules and requirements for eligibility, so long as they meet broad federal guidelines. Previously, states received matching funds from the federal government, getting one dollar of federal money to spend on assistance to the poor for every dollar of their own that they spent for such purposes. Under the new law, if a state cuts its welfare expenditures, it gets to keep the rest of the money. In return for this largesse, the federal government requires states to spend a specified amount on what it calls maintenance of effort (MOE). These MOE funds can be used for a variety of support, training, and educational programs to assist those moving from welfare to work. The amount that a state must commit to MOE funds depends on the size of its federal block grant and the degree to which it meets certain federal incentives, such as getting a certain number of participants into jobs or reducing the number of children being born out of wedlock. The federal government also created additional funds administered by the Department of Labor to be used for job creation and other forms of support for those with significant personal barriers to labor market entry (Blank, 1997; Mermin & Steuerle, 1997).

CALWORKS

California was one of the last states to enact a welfare-to-work program when the California Work Opportunity and Responsibility to Kids Act, or **CalWORKs,** was signed into law in August 1997. The California law added several requirements beyond those established by the federal government, most notably shorter time limits for welfare-to-work participants, 24 months for those on welfare when the program began and 18 months for those entering after that time. Under this law, counties are allowed to provide a variety of programs to help prepare individuals to work, to do a job search, and to get a job. Participants who do not hold unsubsidized jobs after their welfare-to-work time limit expires must do 32 hours of community service per week to continue to receive TANF until their federal 60-month limit is reached.[1]

CalWORKs has a strong work-first emphasis, meaning that participants should get a job as quickly as possible, as opposed to first getting such things as more education or job training. It also has a state-enforced program for recovering child support payments from noncustodial parents. Participants can lose their grant if they fail to cooperate with child support-enforcement officials, to put in their 32 hours per week, to obtain immunizations for their children and make sure the latter do not have too many absences from school, and to complete a host of procedural requirements successfully.

One way in which state programs differ from one another is the kind of work incentives they offer. The primary type of incentive is the income disregard, which allows participants who get jobs to continue to receive a portion of their benefits in excess of what their grant would be if they did not work.[2] California has an income disregard of $245 plus 50%; that is, participants who work are allowed to keep the first $245 they earn, with half of the rest of their earnings subtracted from their monthly grant, a fairly generous incentive compared to those of other states.

CalWORKs gives counties a strong fiscal incentive to cut costs, allowing them to keep 75% of all savings from their grant from the state. It also allows counties considerable latitude in designing their own programs. Most counties, however, modeled their programs on one of several welfare reform demonstration projects run in the years prior to PRWORA. A wide variety of programs was tried in these projects, including guaranteed annual income programs, which provided poor people with an income with no strings attached; second-generation programs, which focused on the needs of children in an attempt to break family poverty cycles; workfare programs, which required community service in return for assistance; work-first programs, which required participants to enter the labor market and provided both pressure and assistance to see that this was done; and human capital development models, which focused on providing parents with education and training needed to achieve economic self-sufficiency. Federal and state regulations both have work-first elements, such as their time limits and required weekly hours put into work-related activities. Nevertheless, both federal and state laws left room for a wide variety of other elements, including allowing educational programs as one of the ways in which participants could do their required weekly hours of work-related activities and using TANF funds and state MOE money for second-generation intervention programs, housing assistance, and the like. However, with minor variations, most California counties adopted some kind of work-first model.

How does CalWORKs work? The official state plan runs something like this. When people apply for TANF, an eligibility clerk decides if they are eligible. The clerk also decides if a person is a candidate for diversion, a one-time lump sum given in lieu of TANF. An oft-used example is the case in which someone who needs his or her truck for work is given the money needed to repair it rather than being put on assistance. Diversion was designed for those who can be put back into the labor market immediately if a simple need such as this is met. Such individuals never enter the welfare-to-work program. Those deemed eligible enter CalWORKs and begin receiving TANF, at which time their 5-year federal clock begins. They then should receive an appraisal to determine what they need. At this stage, some may be classified as work exempt for things such as being disabled or being the primary caregiver for someone else who is disabled. Those not ruled work exempt are sent to 4 weeks of job club. There, they are taught how to do a resume and a job search, and then are sent out to look for a job. Program designers envisioned a number of possible employment outcomes at this point, with some failing to get jobs and others allowed to continue postsecondary education programs started prior to seeking assistance (called self-initiated programs or SIPs). Those failing to get a full-time job are supposed to get an assessment of their needs and meet with their case manager to jointly reach a welfare-to-work agreement about how they would spend the next portion of their welfare-to-work time. This agreement is a binding decision regarding what participants will do for 32 hours a week in exchange for aid. Participants have a legal right to appeal this agreement and to petition to alter it if they so desire.

The state law allows for a wide variety of possible assignments at this point, although counties have generally not been allowed to provide all of the possible opportunities. Among other things, these include finishing a high school diploma, receiving job training, going to community college, continuing a SIP, getting microenterprise training to prepare participants for starting their own business, getting a low-wage job, and doing what is called work experience, that is, working for your TANF payment at a site supervised by a contractor with the county. Those completing their welfare-to-work agreement with time still remaining on their CalWORKs clock do a revised welfare-to-work agreement regarding how they will put in the hours needed for the time they have left. It was expected that most would be employed by this time. Those still requiring aid after their CalWORKs time elapses were to be assigned to community service until their federal clock ran out, at which time they would not be eligible for assistance for the rest of their lives.

The practice of many counties has differed considerably from the state model. For example, whereas the model has several points at which an individual may be evaluated and a different course of action taken, many participants have only a single meeting for signing their welfare-to-work agreement. There is often little real assessment, with some case managers assigning what clients will do rather than allowing them to participate in reaching the agreement. Part of the problem may be that case managers in many counties are carrying a large overload of cases, forcing them to dispose of clients as quickly as possible for as long as possible, for example, assigning clients to work experience for their entire welfare-to-work time period. Although some counties allow women with newborn babies a 12-month exemption from welfare-to-work activities, some have lowered that to 12 weeks. Counties also differ in the degree to which they allow program participants to pursue educational activities, the key characteristic of human capital development programs (Select Committee on Welfare Reform Implementation, 1998; also, see Figure 8.5). San Francisco developed a successful employer-led job development and placement program, as well as a transitional job program leading from workfare in the public sector to employment in permanent civil service jobs (Bliss, 2000; Roos, 1999). Also, Alameda and Sacramento Counties have microenterprise programs. These are especially important among immigrant communities, which often have a number of individuals who owned small businesses in their country of origin but lack the capital and knowledge of U.S. business practices to do that here.

THE OUTCOMES OF WELFARE REFORM

Welfare reform has been widely heralded as a success in the political arena as a result of the decrease in caseloads since the program began. Critics, however, have questioned whether welfare reform has done much to alleviate poverty or has just gotten poor people off of the welfare rolls. Furthermore, some economists suggest that individuals leaving welfare are creating crowding in low-income job queues, which will force other low income workers out of jobs and suppress wages. Another question should be asked as well: How have the rolls been reduced?

REDUCING THE ROLLS

Most who see welfare reform as a success are assuming that the reduction in the welfare rolls is due to program participants getting jobs that pay so well they no longer need assistance. However, caseloads can also decline if there is a reduction in recidivism (the number of individuals returning to welfare after having left it) or if the duration of episodes of receiving welfare is shortened, or if barriers are created that prevent those who become poor from getting on the rolls (Gittelman, 1998).

Are welfare rolls decreasing at an unprecedented rate because people going through the program are getting good jobs and getting off the rolls? No. Welfare rolls were declining before welfare reform was in place. They began to decline during fiscal year 1996-1997, coincident with the takeoff of California's economic recovery (California Department of Finance, 1999; MaCurdy, Mancuso, & O'Brien-Strain, 2000). The number of AFDC cases in California declined by 1,517 a month between 1994-1995 and 1995-1996, before the recovery took off, then increased to 4,648 cases per month the following year and to 9,063 the year after that. During the first 2 years of welfare reform, the rate at which rolls declined fell to 5,956 per month and to 4,743 the year after that (see Figure 8.1).

This suggests that declining rolls are primarily due to economic factors, not welfare reform. Indeed, research on one of the pre-PRWORA pilot programs that greatly influenced the development of welfare reform policy in California found that it had at best a modest impact on caseload reduction (Albert & King, 1999).

But what about people leaving the rolls? The rate at which families leave welfare is far greater than the rate at which the number of individuals who are on the rolls is reduced. This is because the pool of individuals who receive assistance is quite fluid. In the last chapter, I noted that the majority of the poor are members of the episodically poor or the cyclically poor, categories of families who enter and leave poverty relatively quickly. Indeed, the median duration of a period on income support before welfare reform was only about 6 months (U.S. Bureau of the Census, 1995a, 1995b). So, in any given month, a relatively large number of families are actually leaving the welfare rolls, and also, a relatively large number of new families are entering the welfare rolls. Thus, welfare rolls shrink during a given period of time when the number leaving the rolls is significantly greater than the number entering the rolls. During the year before welfare reform began, nearly 50,000 families per month left the welfare rolls. During the next period for which data are available, Oc-

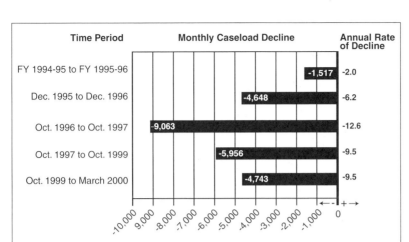

Figure 8.1. Monthly Decline in Aid to Families With Dependent Children (AFDC) and Temporary Assistance to Needy Families (TANF) Caseloads in California, 1994 to 2000

SOURCE: California Department of Social Services, 1994, 1995, 1996, 1997a, 1997b, 1997c, 1997d, 1997e, 1999a, 1999b, 2000a, 2000b, 2000c, 2000d.

tober 1999 through May 2000, at the peak of welfare reform, only about 40,000 left the rolls per month (California Department of Social Services, 1997a, 1997c, 2000a, 2000b, 2000c).

If people are leaving welfare at a slower rate now than before welfare reform began, why are caseloads declining by almost 10% per month? The major source of caseload reduction under welfare reform in California has been the smaller number of individuals entering the program. There has been a moderate decline in the number of applicants, from about 33,000 per month just before welfare reform began to about 25,000 in 2000, with a disproportionate percentage of the decline occurring among Asian and immigrant groups. Among the possible reasons for this is the well-publicized political proposals to cut off aid to immigrants. Also a factor are difficulties with the application process, especially due to a lack of forms in many of the languages that recipients use, along with few, if any, eligibility workers who are fluent in these languages (California Department of Social Services, 1997a, 1997b, 1997c, 2000a, 2000b, 2000c; Zimmermann & Fix, 1998).

Even more important, however, has been a decline in the number of applications approved. In the winter of 1997, just before welfare reform began, 84.7% of applications were approved. By March 2000, the approval rate had declined by nearly half, to 48.8% (California Department of Social Services, 1997a, 1997b, 1997c; see Figure 8.2).

There are two reasons for this decrease: an increase in the number of applications that are denied and the striking number of applications that are withdrawn. As shown in Figure 8.2, about one third of all applications statewide are denied, and

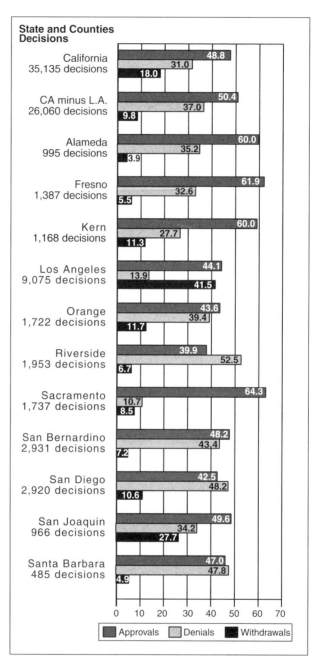

Figure 8.2. Disposition of Applications for Temporary Assistance to Needy Families (TANF) in California, All Cases, State and Selected Counties, March 2000, in percentages

SOURCE: California Department of Social Services, 2000d, 2000e

there is considerable variation from county to county, ranging from only 10.7% denied in Sacramento County to 52.5% in Riverside County. Such a wide variation suggests that different standards are applied in different places. Even more important, more than half of the denials are on procedural grounds rather than the applicants' failure to meet eligibility requirements (California Department of Social Services, 2000d, 2000e).

Even more surprising is the large number of applications withdrawn before a decision has been reached. It is difficult to understand why so many would withdraw their applications for assistance once they initiated the process. Nevertheless, this is the outcome for 18% of applications in California. Once again, there is considerable variation from county to county, resulting in a pattern that suggests three different categories of outcomes. A number of counties, and especially the smaller ones, have a withdrawal rate of less than 6%, which may indicate that little in the nature of their application procedures induces withdrawals. A second group of counties have withdrawal rates that are about twice the rate for the first group, including San Diego, Orange, and Kern Counties, suggesting that certain program characteristics in those places may encourage withdrawals. Finally, a few counties have much higher rates, most notably Los Angeles County, with a withdrawal rate of 41.5%, suggesting a rather serious problem or set of problems causing such a large number of cases of this type (see Figure 8.2). One apparent possibility for an increased rate of withdrawals in San Diego, a county in the second category, is the requirement that applicants must consent to unannounced inspections of their homes by peace officers. The purpose of this is to look for evidence that people other than official family members live in the household, for example, a pair of men's shoes or too many toothbrushes. There are many reasons why someone might not wish to be subjected to such treatment other than being guilty of welfare fraud. Anticipating that many applicants may not be willing to be subjected to such a search, officers carry printed forms for them to fill out to terminate their applications.

These barriers to entry into the welfare-to-work program have also contributed to the much lower rate of recidivism since the inception of welfare reform. Only 7% of those leaving TANF under welfare reform returned to it by 1999, considerably lower than might be expected (California Department of Social Services, 2000g).

HOW ARE THOSE LEAVING WELFARE DOING?

One of the fundamental goals of welfare reform is to get program participants out of poverty by moving them from welfare to self-sufficient employment. Although many have left the welfare rolls, are they really doing any better financially? What has happened to those who have left welfare?

Data on those who have left CalWORKs are not good, but leaver studies from other states paint a fairly consistent picture. Nationwide, on average, slightly more

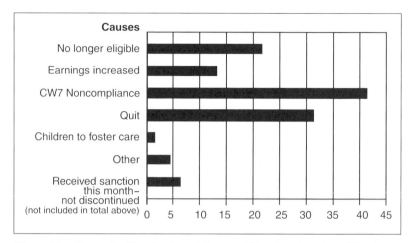

Figure 8.3. Causes for Termination of Cases and New Sanctions in California, Spring 2000
SOURCE: California Department of Social Services, 2000f.

than half of leavers appear to be working, although Connecticut has a rate of 83% and several other states are hovering around 35%. Many former welfare recipients work part-time, with the rate of full-time work being higher in states with larger work incentives. Most are working at low-wage jobs that pay between $5.50 and $7 an hour, and they have experienced modest increases in earnings since leaving welfare. The overall incomes of such families, however, are generally lower than when they were still receiving cash assistance, with most remaining below the poverty level. About 40% of leavers nationwide are not working, with a significant percentage of these having no income at all. Finally, about one fourth of leavers return to welfare within a few months, although there is considerable variation between states on this point as well, with California's recidivism rate among the lowest (Cancian, Haveman, Meyer, & Wolfe, 2000; Hunter-Manns & Bloom, 1999; Loprest, 1999; Loprest & Zedlewski, 1999; Tweedie, Reichert, & O'Connor, 1999).

Perhaps the best information on California leavers comes from official data on the reasons for case terminations, which show that only 8.8% of those leaving assistance did so because their earnings exceeded eligibility limits. Another 13% of them exceeded family income or wealth limits, usually because another adult earner was added to the household (and sometimes additional children as well) or because the family moved in with relatives or an unrelated family (see Figure 8.3).

Nearly a third of leavers simply quit the program. Evidence regarding the income distribution among low-income, single-parent families suggests that few of these families have become economically self-sufficient and that nearly half do not have an employed head of household.

Individuals decide to leave the program for several reasons, including not liking or not being able to meet program requirements, having difficulty with the procedural red tape associated with the program, or disliking the treatment they have received from program staff. The primary reason for participants leaving CalWORKs, however, is difficulty associated with the CW7 form, which accounts for about 40% of all leavers (California Department of Social Services, 2000f). This is a form that program participants must file each month that details any changes in their family situation during the previous month, such as anyone moving in or out of the household or anything received during the month, whether cash, goods, services, or gifts. Participants and their advocates note that the form is complicated, confusing, and, apparently examined extremely closely, with many being rejected for minor errors in filling it out. In a number of cases, CW7s apparently were lost in processing before being officially recorded. This has happened often enough that the Los Angeles Legal Aid Foundation (1998) has suggested that participants hand-deliver the forms and ask for a receipt if at all possible. Furthermore, a recent law requiring case managers to make three contacts with participants before terminating them for CW7 noncompliance is apparently not followed. Because many participants do not know their rights, they interpret their first notice as meaning that their assistance has been cut off. The timing associated with the process also appears to cause problems, with participants sometimes not receiving their third notice until it is too late in the month to do anything about it before they are terminated. Finally, the form is often used in an almost punitive fashion to cut monthly grants. In one instance, after a student won a prize in an essay contest, her family's TANF grant was cut almost to zero due to this new "income" (Caring Council, 2000). Even sadder is the fact that, because the award was not classified as earned income, the family's monthly grant was reduced by the entire amount of the prize, although it would have been reduced only half as much if the mother had received the same amount in additional income, as a result of the state's earned income disregard incentive.

Creating leavers by making participants uncomfortable enough to quit appears to be an important component of work-first programs. In assessing welfare-to-work programs, researchers at the Manpower Demonstration Research Corporation, a major policy research firm, note that "the message" sent by case managers is a significant part of such programs. In assessing the welfare-to-work program in Los Angeles, the researchers noted the importance of a tough, enforcement-oriented case management, with a relatively large percentage of participants receiving sanctions that cut their monthly payment. They also noted that this approach was successful in getting participants to leave the program more rapidly (Freedman, Knab, Gennetian, & Navarro, 2000)

Interestingly, it was expected that work-first policies would quickly lead those among the poor with the highest levels of human capital to get jobs quickly and exit programs, leaving a client population made up primarily of individuals with low levels of education and little, if any, work experience. In fact, the opposite appears to be the case. In one California county's study of characteristics of program participants at the outset of welfare reform and again 2 years later, when rolls had been reduced by nearly 50% and few new participants had been added, it was revealed that those with the least education were most likely to leave the program. The percentage of those remaining in the program who had a high school education had increased by 25% during that period, with all of the corresponding decline coming among those with less than a high school education (San Diego Health and Human Services Agency, 1997, 2000a). This supports the notion that the procedural complexity of welfare reform is so difficult that only the best and the brightest of participants can navigate it. It also fits with the results of a national study, which discovered that many leavers had mental health problems, learning disabilities, or a low IQ. It is important to note that many of these were eligible for a waiver allowing them to receive assistance without time limits or work requirements and to receive Supplemental Security Income for the disabled as well. Nevertheless, many of them have left TANF, and half of those who left are not receiving SSI either (Sweeney, 2000).

Data from a small survey of California leavers suggest that half or more have had to cut back from what their lifestyle was as poor welfare recipients, with many having to forego some basic necessities. About 76% have experienced difficulty in paying their bills, whereas 15% had to give up their housing and either find something even cheaper or move in with another family. In addition, 6% reported giving up their children to foster care because they could not afford to support them (California Department of Social Services, 2000g).

These data also showed that many leaving TANF have lost other forms of assistance to which they are entitled. Only 19% of California leavers continue to receive food stamps.[3] This is 60% lower than the national average and slightly more than one fourth of the rate for Texas (see Table 8.1).

About 57% of leavers in the California survey received **Medicaid,** slightly better than the national average. Nevertheless, California has the third-highest percentage of the population without health coverage of any kind (U.S. Bureau of the Census, 1999b). There has been a dramatic increase in the cost of housing in California as well, especially in the larger urban areas where the majority of the poor live. Most housing programs are federally funded, but some states have programs as well. Only 10% of California's leavers receive such aid. In other states for which data are available, 14% (Massachusetts) to 52% (Connecticut) get housing assistance (see Table 8.1).

The greatest expense that single mothers with children must face when entering the labor force is child care. Child care assistance is probably the most crucial factor in determining whether or not a family's quality of life improves or worsens when they leave welfare. A female head of household with two children who worked and received child care assistance while on welfare would need to earn an additional $1,150 a month to maintain the same quality of life if she left TANF and had to pay for child care (Equal Rights Advocates, 1996). Only 11% of California leavers re-

TABLE 8.1   Welfare Leavers' Enrollment in Assistance Programs

	Food Stamps	Medicaid	Housing	Child Care
United States	47%	55%	—	19%
California	19%	57%	10%	11%
Arizona	44%	48%	—	14%
Connecticut	54%	61%	52%	—
Florida	57%	68%	—	18%
Illinois	20%	43%	—	—
Iowa	64%	66%	—	—
Massachusetts	18%	—	14%	—
Mississippi	58%	44%	34%	10%
Missouri	47%	39%	—	—
New York	29%	48%	—	—
South Carolina	62%	77%	—	13%
Texas	68%	74%	20%	—
Wisconsin	49%	71%	—	17%

SOURCES: U.S.: Adapted from Loprest 1999; California: CDSS 2000; Connecticut: Hunter-Manns & Bloom 1999; Illinois: Julnes & Halter 2000; all others: NCSL 2000.

ceive such assistance, slightly more than half of the national average (California Department of Social Services, 2000g).

The primary reason why so many leavers lose their food, health, and child care assistance is that applications for such programs were part of the process of applying for welfare. When AFDC was replaced by TANF, many states and counties failed to sever eligibility data for these other aid programs in their administrative records, even though federal law required them to do so (Schott, 2000). As a result, as families left cash assistance, they were often automatically dropped from the rolls of the other forms of assistance as well. Because they were generally not adequately informed that they were still eligible for such programs, most apparently assumed that they were not (Burt, Pindus, & Capizzano, 2000; Community Service Society of New York, 1999). This problem also extended to new applicants, for in many areas, the applications of those who qualified for food stamps and Medicaid but not TANF were never processed for these other types of aid.[4]

OUTCOMES FOR PROGRAM PARTICIPANTS

How are those who are participating in CalWORKs doing? It depends largely on the content of their welfare-to-work agreement and how well they do in the job search.

In a typical month, 13% of California enrollees have been given exemptions due to being disabled or caring for a person with a disability. Another 12% are being sanctioned, cutting their monthly grant. About 2% leave the program because of employment, and less than 1% exit because they have reached the time limit for receiving assistance. About a third of enrollees are in unsubsidized jobs, the majority working part-time for subpoverty wages and owing the county 10 or more hours of additional work per week to meet their 32-hour requirement. About 12% are in various types of job training programs. About 2% are in adult basic education programs for high school dropouts, and an additional 1% are young mothers in a high school completion program. About 3% are in mental health, substance abuse, or domestic violence treatment programs. Finally, 6% are in Self-Initiated Programs in higher education (California Department of Social Services, 2000i).

CalWORKs participants are somewhat better off than leavers in that they all get child care assistance and the state's rather liberal earned income disregard. Is this combination working to move families in the program out of poverty? No. If anything, families at the bottom end of the income distribution are losing ground. The poverty rate among participant families has remained steady, with four out of five having incomes (including TANF, the value of food stamps, school lunches, etc.) that remain below the poverty level. However, the percentage of participant families with incomes that are below 50% of the poverty level has actually increased slightly during the first 3 years of welfare reform, in spite of an economic boom in the state (Primus, Rawlings, Larin, & Porter, 1999; U.S. Bureau of the Census, 1998, 1999a, 2000).

CONTINUING PROBLEMS AND POLICY ISSUES

The primary problems associated with California's welfare reform policies have to do with their focus on reducing the number of people on welfare and their tendency to treat problems faced by participants after the fact rather than addressing the causes of poverty and the means required to lift the poor to economic self-sufficiency. Four areas of policy are critical for California. First is the state's effort in terms of funding and program adequacy, the degree to which programs address the causes of the problems in question (focus), and the degree to which they reach those in need (targeting). The second area of concern is how the state addresses the fundamental issue of poverty: improving income and employment to the level of self-sufficiency. The third issue is housing, especially given the shortage and rising costs in California at the beginning of the 21st century. Finally, there are questions concerning the welfare of children, the intended prime beneficiaries of TANF.

WELFARE EFFORT

California's annual federal block grant is $3.73 billion. Like most states, California allocates all of that money to a variety of purposes but does not spend all that is allocated. By the end of the 1999 fiscal year, it had not spent $1.61 billion, 43% of that year's funding. In addition, it had still not spent $7.9 million from the previous year. States generally hold back obligated grant money to create a "rainy day fund" to be used in case of a recession. Nevertheless, the amount of unspent funds could easily address most of the problems facing California's poor. The state also receives large amounts of additional federal funds for ancillary services and job creation. For example, the state was allocated $83.7 million in matching funds for programs to ensure that parents and children did not lose MediCal coverage in the changeover from AFDC to TANF. During the first 2 years of CalWORKs, when so many families lost their MediCal coverage due to the problems this fund was created to prevent, the state spent only 7% of this allocation (Ross & Guyer, 1999). California also spent $2.91 billion of its own money as MOE funds in fiscal 1999, the minimum amount required to get its block grant. Although many other states also spend the minimum, some spend up to 25% more than what is required (Lazere, 2000).

Data such as the increase in the percentage of families with incomes less than half of the poverty level and the decline in the percentage of the poor receiving food stamps, MediCal, and child care assistance suggest that California is not doing a good job in targeting many of its programs. People who are eligible for programs are not participating because they do not know the programs exist or they do not know they are eligible for them or there are barriers that prevent them from signing up. For example, only 29% of Asian Americans in California are enrolled in programs for which they are eligible, due both to a lack of information, a lack of forms in their own languages, and the scarcity of eligibility workers fluent in them (U.S. Bureau of the Census, 2000). Outreach programs to such groups could be established by identifying key leaders within such communities and using them to spread the word and identify bilingual community members for hiring as outreach workers and ombudsmen for grievance cases.

IMPROVING INCOME AND EMPLOYMENT

In the last chapter, I said that the primary cause of poverty is the mismatch between the level of human capital of most program participants and what is required by available jobs that pay above the poverty level. Although no one thing will provide a solution to this problem, a number of things, in combination, could improve the situation greatly.

Increasing Human Capital

The primary reason that only a third of CalWORKs participants are employed in unsubsidized jobs, the great majority of them low wage and part-time, is that most participants lack the human capital (education, training, and experience) to qualify for higher paying jobs. The poverty rate among households headed by a woman with less than a high school education is more than three times the national family poverty rate. Among those with a high school diploma only, the poverty rate is slightly more than double the national rate. Among those who have attended college but did not complete a degree program, however, the poverty rate declines by 31.3% in comparison to those with just a high school diploma. Furthermore, completion of a community college vocational program reduces the poverty rate by 43.8% compared to those with just a high school diploma, whereas completion of an academic 2-year Associate of Arts degree lowers it by 50% (U.S. Bureau of the Census, 2000).

Addressing the primary cause of poverty directly requires programs to increase the human capital of as many program participants as possible to prepare them for occupations that will provide them with a self-sufficient income. Although many CalWORKs participants have low levels of schooling, nearly half of female heads of poor households in California meet community college admission requirements (California Department of Social Services, 1999b; Johnson & Tafoya, 1999; San Diego Community College District, 2000; San Diego Health and Human Services Agency, 2000b; U.S. Bureau of the Census, 1999a). However, only about 6% of current participants have been allowed to include postsecondary education as a part of their welfare-to-work program, almost all of them in SIPs, programs that were initiated before they applied for CalWORKs (see Table 8.2).

Furthermore, CalWORKs case managers disallowed the postsecondary programs that a number of applicants brought with them when they entered CalWORKs, ruling in some cases that a college degree would not improve their employability. Another frequent problem with postsecondary education programs is that it is often impossible to finish within the CalWORKs time limits. For example, the 18-month limit for new participants would disallow any 2-year community college academic programs, especially including the time elapsed between the beginning of assistance and the start of the next semester.

One proposal for overcoming these problems, called *educational triage*, would divide participants into three groups based on educational achievement, test scores, and motivation. The top group would be given 2 years to complete a postsecondary program, the second group would enter a high school-equivalency program, and the third would go directly into welfare-to-work activities (Buck, 1999b; Fein, Beecroft, Long, & Catalfamo, 2000).[5] Those assigned to education would be rigorously monitored with regard to effort and performance. Those lagging would be shifted to welfare-to-work activities, and those most successful in the high school-equiva-

TABLE 8.2 Assignment of CalWORKs Enrollees in Key Types of Educational Programs, by County, 2000

	Pre-CalWORKs Self-Initated Postsecondary Education	Employment-Related Education	Adult Basic Education
California	5.6	1.3	2.5
Alameda	6.6	0.0	2.2
Butte	9.9	8.6	3.2
Fresno	2.8	0.04	2.5
Humboldt	9.7	8.3	1.3
Imperial	4.0	1.4	9.7
Kern	5.6	4.9	6.6
Los Angeles	7.8	0.0	2.6
Mendocino	7.4	6.1	8.1
Orange	7.7	7.7	1.2
Riverside	3.1	0.0	4.0
Sacramento	3.6	0.3	0.9
San Bernardino	3.4	1.2	0.7
San Diego	5.5	1.0	3.3
San Francisco	n/a	1.3	3.3
San Joaquin	2.1	1.3	1.2
Santa Barbara	7.5	8.3	2.8
Santa Clara	7.9	0.0	5.1
Santa Cruz	15.1	0.0	3.1
Sonoma	9.0	0.003	3.6
Ventura	6.6	4.8	1.4

SOURCE: CDSS 2000h.

lency program would be allowed to enter a postsecondary education program. California has earned the right to waive work requirements for as many participants as the above program proposes by virtue of its large-scale reduction in enrollees. The only change that would be required would be to allow a waiver of the state's 18-month clock for participants. Finally, the state could fund the program either out of its huge cache of unspent federal block grant funds or through the use of state MOE money. The latter would allow the state to exempt those in the programs from having to do federally required work activities in addition to their studies (U.S. Health and Human Services Administration, 1999). As welfare reform has progressed, an increasing number of states are providing postsecondary education programs using either state funds or federal Workforce Investment Act funds. These are sometimes done outside of welfare-to-work boundaries so that those in school face no work requirements.

Increasing Real Income

Whereas human capital development programs attack poverty by improving people's position in the labor market, other methods increase the incomes of poor people directly. A number of scholars have suggested a significant increase in the minimum wage (Danziger, 2000; Jencks, 1997). This is problematic in that it would do nothing for the majority of the working poor, who earn more than minimum wage. Furthermore, to raise it enough to move a significant number of families above the poverty level, serious political opposition would have to be overcome; almost certainly, employers would eliminate some jobs. There is also clear evidence that increases in the minimum wage are offset to some degree by a reduced likelihood of getting employer-based health insurance and pension plans and a greater likelihood of a reduction in sick leave. All of these are extremely important for the working poor, who are the least able to pay for medical care, to cover themselves during a period of lost wages, or to save for their old age (Royalty, 2000).

A more fruitful approach is a refundable earned income tax credit (EITC).[6] EITCs allow low-wage workers to keep more of their earnings by paying less in taxes. Furthermore, for those whose incomes are so low that they owe little or no taxes, a refundable credit would give them the difference between their tax credit and whatever taxes they owe. The current federal refundable EITC provides a single mother with two children, working for minimum wage, nearly $2 more for each hour she works, a substantial boost to real family income. A state EITC pegged to the number of children in the family could have a major impact on poverty.[7] Currently, 12 states have such programs, the largest of them in states with very successful welfare-to-work programs: Minnesota, Wisconsin, Massachusetts, and Vermont (Children Now, 2000).

However, support targeted at earned income would only help the one third of CalWORKs participants with unsubsidized employment.[8] Another one third of participants face multiple barriers to employment, such as low education, low skills, little if any work experience, childhood in a home in which there were no labor force participants, physical or mental health problems, alcohol or drug dependence, and experience of spousal or child abuse (Danziger et al., 2000; Johnson & Tafoya, 1999). The requirements needed to prepare such individuals for the labor market are large in number. The U.S. Department of Labor is providing funds for such purposes through local Private Industry Councils, with programs including an intensive orientation to the workplace, basic job training, training in how to look for and interview for a job, participation in various sorts of treatment programs, and training in following orders, taking criticism, and maintaining control of one's emotions. Early fragmentary results suggest these programs are having a rough time, with few job placements relative to the amount of funds allocated. In addition, many placements in such programs are in temporary subsidized jobs, that is, jobs in which wages are paid out of government funds rather than by the employer, with some lasting for a limited period of time. Furthermore, most participants are trained for unstable, low-wage jobs with incomes far below the poverty level. Finally, many employers do not want to hire individuals coming out of such programs because of concerns that they have poor work habits and are members of racial or ethnic minorities (Lundgren & Cohen, 1999).

Given the cost of such programs and the low level of outcomes, some policy experts have questioned the efficacy of trying to arrange a way for all such individuals to get their incomes through the labor market. Christopher Jencks (1997), for example, has suggested that it is not economically sound to force a mother to work at a minimum wage job to put her children in child care that costs more per hour than she is earning, noting that it makes far more sense to allow such women to stay at home and care for the children themselves. This, however, flies in the face of common stereotypes about the character of the poor, individualistic notions of social justice, and popular ideas about limiting the role of the government in the social arena.

A very different approach, and one that is more in keeping with the notion that it is unfair for individuals to receive support from the government and not work, is to guarantee everyone the right to a job. The backbone of proposals of this nature is a job-stimulation and employment program created by the government, which also assumes the role of employer of last resort for those otherwise unable to find work. There have been a number of attempts at programs of this type, from the Works Progress Administration during the Great Depression through one proposed as part of the original welfare-reform proposals during the Clinton administration. In this latter version, the government would have become the employer of last resort for welfare mothers unable to find jobs during their first 2 years in the welfare-to-work program. However, Congress did not include this measure in the final version of PRWORA.

HOUSING

Housing has become an increasingly difficult problem for the poor in California. The typical CalWORKs family is a single working mother with two children and a monthly income from all sources, including TANF, of about $1,000. The median rent for a two-bedroom apartment in California is about $900 (Breslin & Carroll, 2000; Riches & Ross, 2000). With the maximum monthly TANF cash grant at $626, many families expend their entire monthly stipend for rent on the day they receive it.[9]

The most damaging of California's housing problems is the rapid increase in the number of homeless families since 1997. It is significant to note that, although changes in the housing market developed throughout the 1990s, the number of homeless families surged rather suddenly, a surge that coincided with a rapid reduction in the number of families receiving TANF support. Shelters for homeless families in San Diego County, for example, always had room to accommodate those needing assistance from spring through fall but began experiencing a surge in applicants, which led to increasing numbers of poor, single-parent families being turned away beginning in the late spring of 1999 (Caring Council, 1999). In addition to the growth in the number of homeless families, some areas have also experienced a large increase in "hidden homelessness," for example, two families with children living in a single two-bedroom apartment (Caring Council, 1999).

California's current housing problem is a result not only of rising housing prices but also of the fact that only 10% of California TANF participants receive housing subsidies of any kind (California Department of Social Services, 1999; Riches & Ross, 2000). There are three main forms of housing assistance for the poor: government-owned housing (e.g., housing projects), rent vouchers (primarily through the federal Section Eight program), and apartments built by owners who got low interest government loans in return for maintaining a certain percentage of low-cost units for a given period of time. Rent vouchers offer the greatest assistance in getting people out of poverty because they are targeted specifically to the poor. Also, vouchers are portable, allowing people to escape poor neighborhoods. However, Section Eight assistance is woefully inadequate, and the state has not seen fit to create a voucher program of its own to supplement it. The result has been that, in some cities, qualified families have been on waiting lists for as long as 10 years before receiving assistance. Furthermore, those receiving vouchers have a limited amount of time to find housing that meets Section Eight standards regarding quality and price, with many now failing to do so because of the shortage of units and high prices in the current housing market.

A significant amount of current affordable housing was created through government low-cost loan programs that required builders to maintain a certain number of low-cost units for an extended period of time, often 20 years. Many of these units are reaching their time limit, allowing the owners to raise rents to market levels, with few, if any new units being built to replace them. This has been especially true in Los Angeles and San Diego Counties, where there are large poor populations. Overall, nearly 200,000 units will become eligible for conversion between 1998 and 2003 (California Department of Housing and Community Development, 1998; California Housing Partnership Corporation, 1999, 2000).

The Importance of Housing for Welfare Reform Efforts

Stable, affordable housing is extremely important for families facing welfare-reform time limits because of the crucial role it plays in moving toward economic self-sufficiency. In a study of four California counties, it was found that rent subsidies had a major impact on the number of hours worked per month (Sard & Lubell, 2000). Indeed, one study of long-time welfare recipients found that employment gains for those with rent subsidies were double those of individuals without them (Knox, Miller, & Gennetian, 2000).[10]

Rent vouchers affect such outcomes in several ways. First, they allow poor families in inner-city neighborhoods that are a great distance from available jobs to move to areas with greater job opportunities. Second, by lowering housing costs, they free up more household income for work-related child care and transportation costs. Third, they stabilize family life in a manner that makes job seeking and job keeping less difficult by reducing the frequency of having to move and the need to live in overcrowded arrangements (Center on Budget and Policy Priorities, 2000). The impact of this on children is extremely important. They are uprooted less frequently with regard to housing and school, and their families are able to move to less poor neighborhoods with a better peer environment, especially with regard to attitudes toward school. Rent vouchers can also have an important effect on health, improving nutrition by allowing a greater percentage of family income to be spent on food and reducing the spread of communicable diseases by putting families in less crowded conditions. Children in families not receiving housing subsidies are more likely to suffer from stunted growth than those in families receiving such subsidies (Meyers, 1995). A study in Boston showed that moving to neighborhoods with lower rates of poverty also reduces the severity of chronic diseases, such as asthma, among children, as well as the risk of being a victim of violent crime (Katz, Kling, & Liebman, 2000).[11]

Although federal housing assistance in California is quite inadequate, the state can use both TANF and MOE funds for such purposes. Some of the most successful welfare-to-work programs in other states have a strong housing component (Sard & Lubell 2000). The share of the California budget going to housing has declined by nearly 70% since 1990, making it one of the least active states with regard to housing (Riches & Ross, 2000). California also spent only 56.6% of its TANF block grant during the 1999 fiscal year, leaving a pool of $1.6 billion, some of which could be allocated to state rent vouchers (Lazere, 2000). Some California cities are developing affordable housing programs, but cities generally lack the resources to solve the problem, and their programs often target the middle class rather than the poor (Breslin & Carroll, 2000).

CHILDREN AND CHILD CARE

How are poor children in California faring under welfare reform? I have already noted that the poorest of them are now poorer, that their mothers are at home less, and that many of them have lost food stamps and medical coverage. Even more problematic, many are not receiving adequate child care.

For a number of reasons, child care could be the most crucial element in the process of welfare reform in California. First, it is extremely expensive, so much so that forcing a single mother to work at a minimum-wage job is not cost-effective. Second, the supply of child care is inadequate to meet the needs of all who are eligible. Third, the administration of child care assistance in California is divided among a number of different jurisdictions at both the state and local level, with the result that many families eligible for assistance fall through the administrative cracks. Finally, and perhaps most important, child care is crucial in preventing the transmission of poverty to the next generation.

The average monthly child care cost for a single parent with two children in California is more than $700, with care for just one infant costing about $600 and care for toddlers costing more. After-hours care, needed by those working night shifts, is not only much more expensive than regular child care but also far more difficult to find (California Child Care Resource and Referral Network, 1999; Equal Rights Advocates, 1996). One third of families need infant care, with 43% needing care for a preschooler, which is twice as expensive as care for a school-age child. In addition, 9% of all families require after-hours care, with women in low-wage jobs far more likely than average to need it (California Child Care Resource and Referral Network, 1999). For a poor single mother with two children, child care is likely to cost more than rent, and each of these costs more than either her monthly TANF payment or her take-home pay if she worked full time at minimum wage.

Because of this, subsidized child care for poor working mothers is extremely important. One study found that, although welfare reform resulted in a net decrease in income of 6% for low-income families, this was more than offset by subsidized child care (Witte, Queralt, Chipty, & Griesinger, 1999). Indeed, among CalWORKs leavers, those losing both TANF and subsidized child care creates a far greater risk of becoming homeless.

How is California doing in providing child care subsidies for the poor? California is one of only eight states in which 6% or less of children who are eligible for child care support actually receive it, slightly more than half of the average for all states (Child Care Bureau, 2000). The cause of this appears to be a combination of supply, funding, policy priorities, and organization.

Just as the federal government consolidated welfare funds into TANF block grants to states, it combined a number of federal funding streams for children into another block grant, the Child Care and Development Fund. Like TANF, this block grant allows states to design their own programs and requires them to spend a certain amount of MOE funds in this area. Unlike TANF, the child care program allows states money in addition to their basic grant on a matching-fund basis if they spend more money on child care and development than they did before welfare reform. States can also use a portion of their TANF grant for child care, and they can use state money to fund child care vouchers or a state child-care tax credit (Long, Kirby, Kurka, & Waters, 1998). California has committed enough MOE funds to draw all of its federal grant in this area, but it has neither transferred TANF money to child care nor using state funds to augment Head Start, both of which have been done in states with the most successful welfare-to-work programs (Long et al., 1998).

How adequate is the supply of child care in California? There are licensed child care slots for only 21% of California children who need it. Two thirds of those slots are in child care centers, and the rest in family child-care homes (California Child Care Resource and Referral Network, 1999). The majority of these are located in nonpoor areas, with most of the slots going to children of two-earner, middle-class families (Fellmeth & Weichel, 2000). One result of this shortage is the number of children who receive child care from more than one provider during a week. This is especially a problem for children under 5 years old, for whom not having one regular child-care provider can be disruptive. In California, 36% of children of low-income mothers have two or more child-care arrangements during a week, while 6% have three or more (Capizzano & Adams, 2000).

One of the hopes for welfare reform was that states would replace the fragmented, complex, multi-agency federal child-care assistance program with a seamless system of support under the aegis of a single agency. Although many states have done this, California has not (Long et al., 1998). Instead, it has divided its funds among a number of programs that operate within a complex three-stage system (Fellmeth & Weichel, 2000). Stage 1 child care is for families new to CalWORKs and whose head of household has not yet established a stable welfare-to-work placement. It is run by the State Department of Social Services, which has delegated program operations to the counties. They, in turn, often delegate parts of their programs to other local agencies. Counties were totally unprepared to meet the increased need for child care that this caused, resulting in a decision that licensed child care would not be required and essentially shifting the task of finding child care to the families themselves. Although this increased flexibility relieved one problem, it also relegated poor children to the increased risk associated with a variety of informal child-care arrangements, many with no track record and beyond the protection of licensing regulations regarding background checks and health and safety inspections. The low value of vouchers has exacerbated the problem, with critics noting that those who are willing to accept $7,500 per year to care for two toddlers 8 to 10 hours a day, 5 days a week, may not be the best people to care for those children (Fellmeth & Weichel, 2000).

Stage 2 is run by the state Department of Education. It provides up to 24 months of child care for families whose head of household's work situation is deemed to be stable by their CalWORKs employment case manager. It includes a number of programs, including child development programs. Like Stage 1, it is mandated to serve all program participants by the federal government. Like all of the stages, it is a capped entitlement, meaning that those who are eligible receive it only to the extent that funds are available. Until late in 2000, Stage 2 funds were grossly inadequate. Furthermore, the federal government mandates a preference for at-risk children suffering from abuse or neglect, with the rest going on a waiting list prioritized on the basis of family income and, to some extent, program participation. By the summer of 2000, the number of families waiting to be transferred into Stage 2 care was four times the number that were actually transferred (Adams & Schulman, 1998; California Department of Social Services, 2000j; Fellmeth & Weichel, 2000).

Stage 3 is an extension of Stage 2, an extra year for those who use up their aid under CalWORKs. There is also a greater Stage 3, which provides assistance for families that are not in CalWORKs but whose household income is less than 75% of the state median family income. Like the others, this is a capped entitlement; until late 2000, it went unfunded. At that time, the state used a portion of the budget surplus to fund Stage 2 and Stage 3 for CalWORKs participants for that fiscal year. It is not known if this funding will continue after that time. However, issues still remain regarding the funding and delivery of greater Stage 3 for low income families that are not in CalWORKs. This is of critical importance, for as the state has drastically reduced its welfare rolls, it has also reduced the percentage of poor families that are eligible for the funded portions of the child care assistance program. Thus, although funding has increased, the percentage of poor families receiving assistance has decreased considerably.

What can be done about California's child care problem? There are several possibilities. One would be simply to increase funding by using some of the state's TANF surpluses augmented with additional state funds to fund the CalWORKs-related stages and shift the budget surplus money to fund greater Stage 3. Unfortunately, this would be a very unpopular move politically. A second would be to create a refundable state child-care tax credit (Cherry & Sawicky, 2000; Ross, 2000). This could allow low income families a portion of their child care costs on a sliding scale. The federal government has such a program, but it does not have a refundable credit. Because this means that individuals can only subtract the amount of the credit from the income tax that they owe, the poorest families, with incomes so low that they have to pay very little or no taxes, receive little or no benefit, whereas those with higher incomes and, hence, more taxes to pay, benefit much more. A state refundable tax credit that would pay poor families the difference if their credit were larger than the amount of taxes owed would be of greatest benefit to poor families in California.

A third possibility would be to allow women with the lowest probability of getting jobs with decent incomes to stay home and care for their children. From the beginning, the state had planned to allow those who were unsuccessful in finding jobs to become child care providers for those who were. This would merely expand that program. States are allowed to exempt 20% of program participants from work requirements, with the amount increasing as they reduce their welfare rolls. At this point, California could exempt a major portion of participants without penalty. It certainly is not cost-effective to force a woman to work at a minimum wage job for a given number of hours and pay a child-care provider more than that to look after her children. Unless those required to work are guaranteed a job that pays a decent wage, there will be a percentage for whom welfare-to-work is not an answer.

CONCLUSIONS

Creating programs to care for the poor is the art of the possible practiced within the bounds established by politics and prejudice. George Bernard Shaw once called the presence of poverty amid the affluence of modern industrial society the greatest of evils. If this is so, then that evil lies as much within the lack of political and public will to address the problem as it does in the economic factors that engender it in the first place. Real solutions to the problem are readily available. But until ideological concerns can be replaced by a public honesty about the nature of poverty, little will be done about it. Although many factors influence poverty and a host of personal characteristics affect the likelihood of specific individuals being poor, the fundamental cause of poverty is the labor market. A lasting solution to poverty must start from this fact. Although there may be people who do not want to work and others whom employers might not want to hire, the bottom line is that there are not enough jobs available that would provide the poor with a self-sufficient income. Once we are past assigning blame, the question becomes one of simple mathematics: How many do we try to provide with human capital sufficient for labor market success? How many do we try to save by providing assistance in overcoming personal barriers to entering the labor market? How many do we provide with a guaranteed job at a self-sufficient income? And how many do we provide with a decent standard of living while allowing them to stay home and raise their children? Finally, although in this last sentence and most of the public discourse on this issue, the *them* in poverty invariably refers to adults, we must remember that the great majority of poor people are children, and it is they who suffer or survive as a result of the kinds of programs that we provide.

NOTES

1. This is equal to about $4.50 per hour, well below minimum wage.
2. A state without such an incentive would reduce the monthly grant of those who work by the amount that they earn. Thus, at the end of the month, participants who worked would have no more to show for their efforts than if they had not worked.
3. This fits a national pattern in which more than half of those who were eligible lost this benefit, resulting in a 23% increase in the number of families facing food insecurity or hunger (Center for Law and Social Policy, 1999; Tweedie et al., 1999).

4. Most, if not all states have attempted to contact those who were terminated by mail, with many notices returned because the addressee had moved, a frequent phenomenon among the poor (Health Care Financing Administration, 1999; Schott, 2000). Perhaps the best reinstatement plan was that of Pennsylvania, which automatically gave a 90-day reinstatement to all who were terminated, improper or not, then went through all of them on a case-by-case basis to determine eligibility. The state is also reimbursing medical expenses for those who were wrongfully terminated (Health Care Financing Administration, 1999; Schott, 2000).

5. The principle underlying selection would allow all people with promise the opportunity to prove themselves, with progress being closely monitored, and those not taking the program seriously immediately being shifted into Group 3.

6. See Neumark and Wascher (1999) for a comparison.

7. Furthermore, a refundable EITC for California would be relatively inexpensive, somewhere on the order of $636 million per year for one pegged at 20% of the federal credit. See Johnson (2000).

8. This number could probably be raised considerably through the introduction of a broader human-capital development program.

9. The great majority of poor families in California are spending far more than the recommended 30% of family income on housing, with many spending more than twice that much. Although apartments with lower rents are available, especially in less populated counties, it is important to note that one in eight housing units in California is substandard, virtually all of them at the low end of the market and occupied by poor people (Riches & Ross, 2000).

10. Indeed, almost all of the gains made by participants in the successful Minnesota Family Investment Project were made by those with rent subsidies. This was true for getting and maintaining jobs, increasing the number of hours worked, and increasing income (Center on Budget and Policy Priorities, 2000; Ong, 1998).

11. It is interesting that this latter program provided only housing vouchers, allowing individuals to move to a less poor neighborhood, unlike the Minnesota study noted above, which combined rent subsidies with a work-incentive program. Whereas the Minnesota program resulted in economic as well as health and well-being gains, the Boston program failed to produce significant employment or income gains for participants.

REFERENCES

Adams, G., & Schulman, K. (1998). *California: Child care challenges.* Washington, DC: Children's Defense Fund.

Albert, V. N., & King, W. C. (1999). The impact of a mandatory employment program on welfare terminations: Implications for welfare reform. *Journal of Social Sciences Research, 25,* 125-150.

Blank, R. (1997). *Policy watch: The 1996 welfare reform.* Evanston, IL: Institute for Poverty Research.

Bliss, S. (2000). *San Francisco works: Toward an employer-led approach to welfare reform and workforce development.* New York: Manpower Demonstration Research Corporation.

Breslin, K., & Carroll, D. (2000). *Housing matters for CalWORKs families.* Sacramento: California Budget Project.

Buck, R. E. (1999b). *The poor, welfare to work, and the San Diego labor market: Some neglected considerations* (Policy Paper No. 1). San Diego, CA: Caring Council.

Burt, M. R., Pindus, N., & Capizzano, J. (2000). *The social safety net at the beginning of welfare reform: Organization of and access to social services for low-income families* (Working paper). Washington, DC: Urban Institute.

California Budget Project. (2000). *Health coverage programs available to low-income Californians.* Sacramento: Author.

California Child Care Resource and Referral Network. (1999). *The California child care portfolio 1999.* Sacramento: Author.

California Department of Finance. (1999). *California statistical abstract 1999.* Sacramento: Author.

California Department of Housing and Community Development. (1998). *California's housing markets 1990-97: Statewide housing plan update phase II.* Sacramento: Author.

California Department of Social Services. (1994). *Public welfare in California annual report, 1993-1994.* Sacramento: Author.

California Department of Social Services. (1995). *Public welfare in California annual report, 1994-1995.* Sacramento: Author.

California Department of Social Services. (1996). *Public welfare in California annual report, 1995-1996.* Sacramento: Author.

California Department of Social Services. (1997a). *Public welfare in California, December 1996.* Sacramento: Author.

California Department of Social Services. (1997b). *Public welfare in California, February 1997.* Sacramento: Author.

California Department of Social Services. (1997c). *Public welfare in California, January 1997.* Sacramento: Author.

California Department of Social Services. (1997d.) *Public welfare in California, March 1997.* Sacramento: Author.

California Department of Social Services. (1997e). *Public welfare in California, October 1997.* Sacramento: Author.

California Department of Social Services. (1999a). *California Work Opportunity and Responsibility to Kids (CalWORKs) cash grant caseload movement and expenditures report, October 1999.* Sacramento: Author.

California Department of Social Services. (1999b). *Temporary Assistance to Needy Families characteristics survey, federal fiscal year 1998.* Sacramento: Author.

California Department of Social Services. (2000a). *CalWORKs cash grant caseload movement and expenditures report, December 1999.* Sacramento: Author.

California Department of Social Services. (2000b). *CalWORKs cash grant caseload movement and expenditures report, February 2000.* Sacramento: Author.

California Department of Social Services. (2000c). *CalWORKs cash grant caseload movement and expenditures report, January 2000.* Sacramento: Author.

California Department of Social Services. (2000d). *CalWORKs cash grant caseload movement and expenditures report, March 2000.* Sacramento: Author.

California Department of Social Services. (2000e). *CalWORKs report on reasons for denials and other non-approvals of applications for cash grant, March 2000.* Sacramento: Author.

California Department of Social Services. (2000f). *CalWORKs report on reasons for discontinuances of cash grant, March 2000.* Sacramento: Author.

California Department of Social Services. (2000g). *CalWORKs leavers survey.* Sacramento: Author.

California Department of Social Services. (2000h). *CalWORKs welfare-to-work monthly activity report all (other) families, April 2000.* Sacramento: Author.

California Department of Social Services. (2000i). *CalWORKs welfare-to-work monthly activity report all (other) families, March 2000.* Sacramento: Author.

California Department of Social Services. (2000j). *Child care monthly report—CalWORKs families, May 2000.* Sacramento: Author.

California Housing Partnership Corporation. (1999). *Preserving California's housing stock: A risk assessment of the potential loss of HUD-assisted multifamily housing in California.* Sacramento: Author.

California Housing Partnership Corporation. (2000). *Federally assisted multifamily housing prepayments completed and Section 8 terminated, State of California.* Sacramento: Author.

Cancian, M., Haveman, R., Meyer, D. R., & Wolfe, B. (2000). *Before and after TANF: The economic well-being of women leaving welfare.* Madison: University of Wisconsin-Madison, Institute for Research on Poverty.

Capizzano, J., & Adams, G. (2000). *The number of child care arrangements used by children under five: Variations across states.* Washington, DC: Urban Institute.

Caring Council. (1999, November 17). *Report on homeless families and shelter overcrowding.* Paper presented at Welfare Policy Forum, San Diego, CA.

Caring Council. (2000, September 11). *Report on cuts in cash grants.* Paper presented at meeting, San Diego, CA.

Center for Law and Social Policy. (1999). *Hunger and food stamp participation.* Washington, DC: Author.

Center on Budget and Policy Priorities. (2000). *Research evidence suggests that housing subsidies can help long-term welfare recipients find and retain jobs.* Washington, DC: Author.

Cherry, R., & Sawicky, M. B. (2000). *Giving credit where credit is due: A "universal unified child credit" that expands the EITC and cuts taxes for working families.* Washington, DC: Economic Policy Institute.

Child Care Bureau. (2000). *Access to child care for low-income working families.* Washington, DC: Administration for Children and Families.

Children Now. (2000). *Where credit's due: What a state earned income credit means for California's children.* Oakland, CA: Author.

Community Service Society of New York. (1999). *The unfinished business of welfare reform: Fixing government policies that exclude the working poor from benefits.* New York: Author.

Danziger, S. (2000). *Approaching the limit: Early lessons from welfare reform.* Chicago: Joint Center for Poverty Research.

Danziger, S., Corcoran, M., Danziger, S., Heflin, C., Kalil, A., Levine, J., Rosen, D., Seefeldt, K., Seifert, K., & Tolman, R. (2000). *Barriers to the employment of welfare recipients.* Ann Arbor: University of Michigan, Poverty Research and Training Center.

Equal Rights Advocates. (1996). *California self-sufficiency standard.* San Francisco: Author.

Fein, D. J., Beecroft, E., Long, D. A., & Catalfamo, A. R. (2000). *College as a job advancement strategy: An early report on the new visions self-sufficiency and lifelong learning project.* Bethesda, MD: Abt Associates.

Fellmeth, R. C., & Weichel, E. D. (2000). *California children's budget, 2000-2001.* San Diego, CA: University of San Diego, School of Law, Children's Advocacy Institute.

Freedman, S., Knab, J. T., Gennetian, L. A., & Navarro, D. (2000). *The Los Angeles Jobs-First GAIN evaluation: Final report on a work first program in a major urban center.* New York: Manpower Demonstration Research Corporation.

Gittelman, M. (1998). *Declining caseloads: What do the dynamics of welfare paricipation reveal?* Washington, DC: National Technical Assistance Center on Welfare Reform.

Health Care Financing Administration. (1999). *HHS approves California expansion of children's health insurance plan* [On-line]. Available: http://www.hcfa.gov

Hunter-Manns, J. A., & Bloom, D. (1999). *Connecticut post-time limit tracking study: Six-month survey results.* New York: Manpower Demonstration Research Corporation.

Jencks, C. (1997). The hidden paradox of welfare reform. *The American Prospect,* No. 32.

Johnson, H. P., & Tafoya, S. (1999). *The basic skills of welfare recipients: Implications for welfare reform.* San Francisco: Public Policy Institute of California.

Johnson, N. (2000). *Estimating the cost of a state earned income tax credit.* Washington, DC: Center on Budget and Policy Priorities.

Julnes, G., & Halter, A. (2000). *When families leave welfare behind: Illinois families in transition.* Springfield: University of Illinois at Springfield, Institute for Public Affairs.

Katz, L. F., Kling, J. R., & Liebman, J. B. (2000). *Moving to opportunity in Boston: Early results of a randomized mobility experiment* (Industrial Relations Section Working Paper No. 441). Princeton, NJ: Princeton University.

Katz, M. (1989). *The undeserving poor: From the war on poverty to the war on welfare.* New York: Random House.

Knox, V., Miller, C., & Gennetian, L. A. (2000). *Reforming welfare and rewarding work: A summary of the final report on the Minnesota Family Investment Program.* New York: Manpower Demonstration Research Corporation.

Lazere, E. (2000). *Welfare balances after three years of TANF block grants: Unspent TANF funds at the end of fiscal year 1999.* Washington, DC: Center on Budget and Policy Priorities.

Long, S. K., Kirby, G. G., Kurka, R., & Waters, S. (1998). *Child care assistance under welfare reform: Early responses by the states.* Washington, DC: Urban Institute.

Loprest, P. (1999). *Families who left welfare: Who are they and how are they doing?* Washington, DC: Urban Institute.

Loprest, P., & Zedlewski, S. R. (1999). *Current and former welfare recipients: How do they differ?* Washington, DC: Urban Institute.

Los Angeles Legal Aid Foundation. (1998). *Defending your dreams: How to participate in your CalWORKs welfare to work plan.* Los Angeles: Author.

Lundgren, L., & Cohen, I. (1999). The new skills mismatch? An examination of urban employers' perceptions about public job training participants as prospective employees. *Journal of Social Services Research, 25,* 109-124.

MaCurdy, T., Mancuso, D., & O'Brien-Strain, M. (2000). *The rise and fall of California's welfare caseload: Types and regions, 1980-1999.* San Francisco: Public Policy Institute of California.

Mermin, G., & Steuerle, C. E. (1997). *The impact of TANF on state budgets.* Washington, DC: Urban Institute.

Meyers, A. (1995). Housing subsidies and pediatric undernutrition. *Archives of Pediatric Adolescent Medicine, 149,* 1079-1084.

National Conference of State Legislatures. (2000). *Tracking recipients after they leave welfare data sets* [On-line]. Available: http://www.ncsl.org

Neumark, D., & Wascher, W. (1999). *Using the EITC to increase family earnings: New evidence and a comparison with the minimum wage.* Chicago: Joint Center on Poverty Research.

Ong, P. (1998). Subsidized housing and work among welfare recipients. *Housing Policy Debate, 9,* 775.

Primus, W., Rawlings, L., Larin, K., & Porter, K. (1999). *The initial impacts of welfare reform on the economic well-being of single mother families.* Washington, DC: Center on Budget and Policy Priorities.

Riches, E., & Ross, J. (2000). *Locked out: California's affordable housing crisis.* Sacramento: California Budget Project.

Roos, C. (1999). *Legislative analysis: Employment opportunities for workfare workers under the city's civil service system.* San Francisco: Board of Supervisors Legislative Analyst's Office.

Ross, D. C., & Guyer, J. (1999). *Congress lifts the sunset on the "$500 million fund" extends opportunities for states to ensure parents and children do not lose health coverage.* Washington, DC: Center on Budget and Policy Priorities.

Ross, J. (2000). *Strategies to reward work: How can a state earned income tax credit assist California's working poor?* San Francisco: California Budget Project.

Royalty, A. B. (2000). *Do minimum wage increases lower the probability that low-skilled workers will receive fringe benefits?* Chicago: Joint Center for Poverty Research.

San Diego Community College District. (2000). *Admission requirements* [On-line]. Available: http://www.sdccd.cc.ca.us

San Diego Health and Human Services Agency. (1997). *CalWORKs baseline characteristics survey.* San Diego, CA: Author.

San Diego Health and Human Services Agency. (2000a). *CalWORKs characteristics survey 1999.* San Diego, CA: Author

San Diego Health and Human Services Agency. (2000b). *CalWORKs data.* San Diego, CA: Author.

Sard, B., & Lubell, J. (2000). *The value of housing subsidies on welfare reform efforts.* Washington, DC: Center on Budget and Policy Priorities.

Schott, L. (2000). *Issues for consideration as states reinstate families that were improperly terminated from Medicaid under welfare reform.* Washington, DC: Center on Budget and Policy Priorities.

Select Committee on Welfare Reform Implementation. (1998). *Analysis of CalWORKs county plans: Sample of eleven counties.* Sacramento: California State Assembly.

Sweeney, E. P. (2000). *Recent studies indicate that many parents who are current or former welfare recipients have disabilities and other medical conditions.* Washington, DC: Center on Budget and Policy Priorities.

Tweedie, J., Reichert, D., & O'Connor, M. (1999). *Tracking recipients after they leave welfare* [On-line]. Available: Available: http://www.ncsl.org

U.S. Bureau of the Census. (1993a). *1990 census of population: Vol. 2, Part 1. Social and economic characteristics: United States.* Washington, DC: Government Printing Office.

U.S. Bureau of the Census. (1993b). *1990 census of population: Vol. 2, Part 6. Social and economic characteristics: California.* Washington, DC: Government Printing Office.

U.S. Bureau of the Census. (1995a). *Dynamics of well-being: Poverty* (CPR P70-42). Washington, DC: Government Printing Office.

U.S. Bureau of the Census. (1995b). *Dynamics of well-being: Program participation* (CPR P70-41, SIPP Data Series). Washington, DC: Government Printing Office.

U.S. Bureau of the Census. (1998). *Current population survey, March 1997, supplement data set.* Washington, DC: U.S. Government FERRET Data System.

U.S. Bureau of the Census. (1999a). *Current population survey, March 1998, supplement data set.* Washington, DC: U.S. Government FERRET Data System.

U.S. Bureau of the Census. (1999b). *Health insurance coverage: 1998* [On-line]. Available: http://www.census.gov

U.S. Bureau of the Census. (1999c). *Poverty in the United States, 1998.* Washington, DC: Government Printing Office.

U.S. Bureau of the Census. (2000). *Current population survey, March 1999, supplement data set.* Washington, DC: U.S. Government FERRET Data System.

U.S. Health and Human Services Administration. (1999). *Temporary Assistance for Needy Families final rule.* Washington, DC: Author.

Weir, M. (1988). The federal government and unemployment: The frustration of policy innovation from the New Deal to the Great Society. In M. Weir, A. S. Orloff, & T. Skocpol (Eds.), *The politics of social policy in the United States* (pp. 149-190). Princeton, NJ: Princeton University Press.

Weir, M. (1992). *Politics and jobs: The boundaries of employment policy in the United States.* Princeton, NJ: Princeton University Press.

Witte, A. D., Queralt, M., Chipty, T., & Griesinger, H. (1999). *Unintended consequences: Welfare reform and the earnings of low-icome women.* Washington, DC: U.S. Department of Health and Human Services.

Zimmermann, W., & Fix, M. (1998). *Declining immigrant applications for Medi-Cal and welfare benefits in Los Angeles County.* Washington, DC: Urban Institute.

CHAPTER 9

THE CONTINUING CRISIS IN HIGHER EDUCATION
Threats to California's Master Plan

James L. Wood
Lorena T. Valenzuela

We began our chapter in the first edition of this book by saying,

> "California Higher Education as a Social Problem" might appear to be an oxymoron, as it seems to embody contradictory notions. The State of California has been a pioneer in providing higher education for all its citizens. The California Master Plan for Higher Education, created in 1960, was truly visionary in providing higher educational opportunity for all Californians through the California State University (CSU) system of 22 campuses, the nine-campus University of California (UC) system, and an extensive statewide network of community colleges. Millions of postsecondary students have received an outstanding higher education in the state, and many more hope to receive the same opportunities. (Wood & Valenzuela, 1997, p. 81)

This statement may seem even more puzzling in 2000, because various higher education budget problems have been addressed in recent years. So why does California **higher education** remain a social problem?

In the early 1990s, the state of California reduced its financial commitment to quality education for all, beginning with the lower grades of kindergarten through high school (K-12) and continuing into higher education, a situation that created many difficulties throughout California. This decreasing commitment to education corresponded with the serious economic recession in the early 1990s. However, its roots went back to Proposition 13 in 1978, which curtailed property tax funding of local schools, and to the conservative economic policies of President Ronald Reagan (Reaganomics) in the 1980s, which were aimed at getting "government off your backs." In practice, this meant lower taxes for those with high incomes, reduced regulation over key industries, and fewer public services such as education for the rest of the population (Greider, 1981, p. 46; Lo, 1990; Sherman & Wood, 1989, pp. 41, 258-259).

Specifically, California public higher education in the early 1990s suffered the structural legislative problem of a steadily declining state budget in terms of the proportionate share allotted to higher education. For example, in California's General Fund Budget, the allotment for CSU and UC was reduced from 9.62% ($3.84 billion) in 1990-1991 to 8.27% ($3.27 billion) in 1993-1994. This meant a loss of more than half a billion dollars for those comparison years, with other estimates closer to $1 billion (Baker, 1995; CSU Chancellor's Office, 1995; Ristine, 1993).

But the situation in California changed from 1995 to 2000, when a series of compacts (non-legally binding agreements) were worked out between Governors Pete Wilson and Gray Davis, on the one hand, and the heads of the three branches of California public higher education on the other, heads of the University of California (UC) system, the California State University (CSU) system, and the California Community College (CCC) system. Drawing on a recovered—and once again strong—California economy, the compacts called for a series of 4% yearly increases in the budgets of the three systems of California higher education, an infusion of funding that significantly improved the operation of each of the branches of higher education. This funding permitted a series of student fee *decreases* of 5% per year for 4 years, which helped reverse a series of earlier fee increases and enhanced the ability to hire more faculty, buy new technology such as modern computers, and support departments in their teaching, research, and service functions.

Beginning the new millennium in the year 2000, the California Legislature passed and Governor Gray Davis signed Cal Grant legislation, which pays full tuition at the state's public colleges and universities and up to nearly $10,000 toward tuition at the state's private universities for good students and those from less financially well-off backgrounds. In addition, students receive grants for going into teaching and scoring well on a standardized high school test, greatly expanding access to higher education for California students. At a cost of more than $1 billion a year, this financial aid plan for students has been called the most generous plan in the country ("Governor Increases," 2000; National Education Association, 2000).

The recent 4% increases in higher education budgets have taken pressure off California students to pay increasingly higher proportions of their education costs in public colleges and universities. Prior to these budget increases, a one third/two thirds financial policy had been adopted by the CSU trustees, supposedly to sustain the quality of higher education by increasing student fees until they were one third of the cost of their education. The UC system developed a parallel but misnamed, Affordability Model, which was based on assumptions concerning various kinds of expenses for students and their families and ultimately aimed at increasing student fees. UC President Emeritus Clark Kerr and former Governor Edmund G. "Pat" Brown, architects of the California Master Plan for Higher Education, designed it to provide quality college and university educational opportunity for all California students (Douglass, 2000). The budget increases, along with the recent Cal Grant program, have brought their vision closer to reality for many California students.

Of course, if education were even more affordable and accessible, students would not need to work simultaneously. Some would have only one job (versus two or three), and class schedules could accommodate them. This is often *not* the experience of college students in California and elsewhere in the United States. Nearly four of five students in the CSU system work at least part-time (79%), for example, and more than a third (36%) work full-time, with many students needing flexible class schedules (CSU Chancellor's Office, 2000).

Course offerings are related to access. According to a mid-1990s SNAPS survey of students in the CSU system, access in terms of course offerings was not being adequately provided by the university, and students ranked this as their greatest need. Recent investigations have indicated that less than the full amount of money budgeted for faculty has actually gone to faculty positions, which may partly explain why students have had trouble getting their classes ("California Audit," 2000).[1] Nonetheless, between the late 1990s and the present, there has been a considerable improvement over the situation college students faced in the early 1990s.

Indeed, it has been argued that more stable funding for higher education in general should be written into legislation, or even into the California constitution, to preserve the quality of and access to California higher education even in periods of economic recessions. Throughout the 1990s, this kind of legislation actually was formulated and passed by the legislature—for example, Assembly Bill 1415 and Assembly Constitutional Amendment 25—but it has yet to be sent to the voters in a statewide election or signed by the governor. The lack of such legislation leaves California higher education quite vulnerable in an economic recession, because higher education is part of the 15% to 25% of California's general fund budget that is unprotected and can be drastically cut if there is a significant reduction in the overall California budget. Responding to this major problem, Master Plan creator Clark Kerr (1993) called for a Resource Master Plan to fund California higher education properly. In contrast to the funding situation of the CSU and UC systems, Grades K to 14 are guaranteed more than 40% of California's yearly budget as a result of Proposition 98. In 2000, this was enhanced by dedicating additional amounts of lottery money to K to 14 education ("Region Update," 2000). Thus, in any upcoming recession, the UC and CSU systems will likely be damaged once again as they were in the early 1990s, unless legislation like Proposition 98 is passed for the UC and CSU systems. But from 1995 to 2000, the series of 4% increase compacts, based on California's strong economy, provided improved economic support for all of higher education.

THE PROBLEM IN CALIFORNIA TODAY

In spite of the improved funding, serious problems remain for California higher education, problems that are often found nationally and even internationally, as well. Beginning in the early 1990s, a series of **strategic plans** were presented to—and at times imposed on—the faculty, students, and staff by university administrations. These plans often entail very different views of governing the university, as well as similarly different views of what constitutes quality higher education. *Strategic plan,* a term used by university administrators but borrowed from business, refers to broad-scale projects for organizations, often entailing anticipated large changes, parallel to the major downsizing of corporations in the 1990s. These strategic plans have at times involved serious attempts to erode the shared governance of colleges and universities by concentrating academic decision making more and more in central university administrations and away from the faculty.

One consequence of this centralized decision-making trend has been the rather uncritical acceptance by many academic administrations of **distance learning** as a general mode of college and university instruction, moving away from the classroom as the usual center of academic instruction. It will be shown that powerful outside interests, often connected to corporations involved with selling modern computer technology, are attempting to influence academic decisions by helping devise these new strategic plans for colleges and universities. Their influence can directly affect the quality of higher education provided to California students.

A central problem underlying this business approach to higher education is that **business interests,** which stress profit making, managerial control, and economic bottom lines, may not correspond with educational interests, which stress intellectual development, acquisition of knowledge and skills, and critical thinking. Although university and business interests clearly overlap, with modern business being dependent on a well-educated labor force, an overemphasis on profit making and managerial control sets university interests at odds with those of business. And therein lies the answer to the question of why California higher education can be seen as a social problem: To the extent that business interests dominate California colleges and universities, there can be negative effects on the quality of California higher education, access to it, and its affordability. In other words, business domination of California higher education poses a threat to the three world-famous tenets of California's original (1960) Master Plan for Higher Education: affordable, accessible, quality higher education for all Californians.

As will be shown, several trends connected to new planning for colleges and universities between the 1990s and the present—such as distance learning, increased tuition, attacks on **tenure,** and increased hiring of part-time faculty—weaken the quality, affordability, and accessibility of higher education in California (and in America generally), thereby significantly weakening the Master Plan.

This chapter will focus on several strategic plans that have been influenced by corporations, often with high technology connections. We will argue that these plans diminish the three tenets of the original Master Plan for Higher Education, devised by then-UC President Clark Kerr and signed into law by then-California Governor Pat Brown via the Donohue Act. The Master Plan has been in effect for 40 years. Since the 1990s, it has faced serious challenges, some of which reflect broader issues, as different conceptions of higher education are strenuously contested. We will focus at first on distance learning, that is, using computers and the Internet to teach college and university courses. We will then show how distance learning fits into an overall business approach to higher education, which we feel is a threat to the Master Plan (Shipps, 2000). This threat to the Master Plan for Higher Education makes higher education a continuing social problem in California.

DISTANCE LEARNING AS A STRATEGIC PLAN

Distance learning uses modern computer technology to transmit higher education from a central location to many students, who sit in front of computer or television screens at various physical distances from the academic point of origin. A recent conference in Sonoma, California, keynoted by an administrator from Britain's Open University, purported to show participants how to teach 5,000 students per class section.

The following institutions have initiated distance education programs and exemplify the increasing use of distance learning in higher education:

- Open University in Britain
- UCLA, with its OnLineLearning.net (previously called The Higher Education Network, or THEN)
- The California Education Technology Initiative (CETI) of the CSU system, which recently collapsed
- The California Virtual University (CVU), which includes universities and colleges throughout California and which almost collapsed recently

- The Western Governors University (WGU), which includes several entire university systems throughout the western United States except California

- Michael Milken's Knowledge Universe

- Probably the best known of them all, the University of Phoenix

This is only a partial list and does not include the American corporations that are initiating distance learning programs leading to degrees and certificates. However, it does reflect a new direction of higher education that requires serious evaluation (Marchese, 1998).

One of the authors first heard of distance learning as a proposed solution to presumed systemwide higher educational problems, a solution that arose in 1990 during discussions by the CSU Chancellor's Office about the projected future lack of Ph.D.s (holders of doctoral degrees) to teach students in the CSU system. Given the large numbers of Ph.D.s in the United States and in California alone, this purported scarcity seemed odd to those in attendance, who were unaware of any such looming crisis. Other agendas seemed to be involved when administrators argued about the need to install widespread and costly technology throughout the 20-plus campus system to teach students at a physical distance from college classrooms. In particular, it appeared that corporations selling this expensive hardware and software would be the ones to gain if CSU and other universities instituted distance learning on a wide scale, not the students to whom such an education was aimed. Finally, the latest rationale for widespread use of distance learning has focused on Tidal Wave II, the projected addition of about 714,000 college and university students in California over the next decade (California Postsecondary Education Commission, 2000). This represents the next large generation of college students coming after the initial expansion of college enrollments in the 1960s (Tidal Wave I) (Kerr, 1964).

Over the years, suspicions about distance learning have been realized. Recently, four major corporations—Microsoft, GTE, Fujitsu, and Hughes Electronics—joined the CSU system in a multiyear, multimillion-dollar scheme involving computer hardware, software, and the marketing of distance learning courses. Due to many financing problems and potential legal monopoly problems, CETI collapsed after 2 years of backstage negotiations of which few faculty or students were even aware. Although many faculty and students—and even campus administrators—breathed a sigh of relief when this unwieldy plan folded, they were soon to learn about an attempt to revive such a plan with most of the previous drawbacks but still without consultation of the groups most affected, namely students and faculty. This new scheme was labeled "Son of CETI," with all the pejorative implications of that phrase, but it, too, has apparently been deemphasized. Yet, there are real differences between faculty and students, on the one hand, and administrators, on the other, regarding these issues. Indeed, one administrator (Ingram, 1999) even said faculty angst regarding distance learning was unjustified. Many faculty strongly disagree, as they feel justified in their concerns about this method of instruction.

What is wrong with distance learning as a general approach to providing quality higher education? Professors and students have commented on the key differences between distance learning and in-class instruction, including an eloquent discussion by Jerry Farber (1998), who shows the many benefits of classroom interchanges and stimulation, growth in intellectual breadth and flexibility, and the advantage of personal inspiration that contact with professors can provide. These benefits are impossible to obtain in a similar fashion through computer screens. Other commentators have focused on the high dropout rate of distance learners as compared to students in classrooms, the inability of professors to know if those taking distance education examinations are the students who are receiving grades and diplomas, and the tendency to make universities merely *digital diploma mills,* David Noble's (1998a, 1998b) colorful term for distance learning. Supporters of distance learning argue that it increases access for students. But when large numbers of students are taught by distance learning (e.g., the 5,000 students per section noted above and even less extreme numbers of students), the quality of higher education is diminished.

Distance learning as a general way to provide higher education is problematical because—in addition to the students' loss of intellectual stimulation and inspiration from classroom interactions—it is part of a much larger scheme to transform and, in our view, diminish higher education. This is essentially unknown to the public in California, or for that matter in the United States and internationally. Because decisions now being quietly made will significantly affect the general public, these decisions require considerable evaluation and debate, the type of debate recently sponsored by the Faculty Coalition for Public Higher Education (1999).

RELATED STRATEGIC PLANS

In the monograph, *C. P. Snow Revisited: The Two Cultures of Faculty and Administration,* Wood (1999) showed that a series of strategic plans for higher education created fundamental differences between the faculty and administration, differences that have emerged to the detriment of higher education and the public interest. Responses by faculty and students to the problems arising from these differences were similarly discussed (Ristine, 1993; Snow, 1959; Wood, 1999; see also Wood, 1998a).

Distance learning is one component of this series of strategic plans developed by university administrators, who are attempting to refocus higher education on the following issues:

1. *Learning instead of teaching* is emphasized when learning from a physical distance is the model, and this view often comes from the same administrators who call for better teaching.

2. *Accountability measures of learning* are used, particularly standardized, multiple-choice testing procedures devised by off-campus commercial agencies quite removed from college classrooms, agencies similar to the organizations that produce tests such as the Scholastic Aptitude Test (SAT) and the Graduate Record Examination (GRE). Their tests would be available for mass distribution to distance learning (or in-class) students. This plan often comes from the same administrators who call for better writing from students. The CSU statement of accountability is found in its *Cornerstones* policy document—and related accountability documents—whereas *Partnership for Excellence* is the CCC's accountability document (which includes "pay for performance" elements such as financially rewarding or punishing community colleges for the number of students they transfer to 4-year universities), with UC not having a parallel document to date.

3. *The commercialization of the university* gives businesses a much larger and more direct role in universities than ever before, while those same businesses make large profits from universities, students, professors, and even alumni (this was the aim of the CETI from a business standpoint).

4. *The related commodification of higher education* means that products (courses) are purchased (through distance learning) for consumption (obtaining a digital diploma mill degree).

5. *The loss of professors' intellectual property rights* over the courses they teach will maximize profits for others.

6. *The increasing reliance on underpaid part-time instructors* instead of full-time, tenured faculty is well under way (part-time instruction has increased to more than 40% for many American universities, with California public higher education even worse: Part-time faculty constitute 52% of the CSU system, 67% of the CCC system, and 66% of the world-famous UC system, which will not likely maintain this high status with such a high proportion of part-time faculty (CPEC, 1999). This is a great disservice to students, who need fully professional and committed faculty who will be around later to write them letters of recommendation for jobs and graduate school.

7. *Recent attacks on tenure* could greatly undermine the academic freedom necessary for universities to continue producing great innovations in science, the economy, and technology itself (Finkin, 1996).

8. The increasing control over universities by administrators, who favor an imperial approach to higher education, enables the above-connected tendencies.

9. The lessened control of departments results when faculty hiring is based on administrative plans instead of departmental curricular needs.

10. The gradual elimination, or at least significant weakening, of targeted departments is fostered by the administrative refusal to hire tenure track faculty for any purpose.

Higher education will be continually diminished if these trends are not soon reversed. Indeed, a negative dynamic is already under way. Bright younger scholars who had planned on having an academic career are now rethinking this career choice. If those who have spent much of a decade in graduate school arduously preparing themselves to become professors believe that their only opportunities lie in nontenured, low-paid, part-time instruction, delivered by distance learning, they may decide not to bother. Several of the authors' university colleagues have indicated they no longer plan to encourage their children to become professors because of the rapidly declining state of American higher education. In his jarring poem of the 1950s, *Howl*, Allen Ginsberg indicated that the greatest minds of his generation were being lost to madness. The greatest minds of the current generation may not be lost to madness, but they surely will be lost to higher education if these trends are not reversed.

Ever since World War II, American higher education has been at the center as the United States attained its powerful position in the world. This has certainly been the case in California with its many universities and colleges, including those that are world class. Without the crucial university atomic research in World War II—carried out, for example, at the University of California, Berkeley—the Allies might have been defeated.

More generally, as will be discussed below, colleges and universities contribute directly and indirectly to the regions in which they reside, as well as to the state and the nation. The University of California, San Diego, and San Diego State University (SDSU) combined, for example, contribute more than $2.2 billion a year to the San Diego region (SDSU, 1997; "UCSD Building," 2000). This is in addition to the many impacts of these two institutions alone have on the region, state, and nation in terms of providing highly educated and technically trained employees, discovering life-preserving medical treatments, and providing multiple cultural benefits from stage plays to football games to lecture series.

In 1983, the widely read report on American education, *A Nation at Risk*, warned that public high school education was being organized as if by a foreign enemy.[2] Since the 1990s, the same is true of public higher education in California and America. It might well be organized by the same foreign enemy, out to destroy this essential institution that makes strong all other American institutions, from the economy to the arts and sciences to the government, and that makes economic mobility and improvement in lifestyles possible for millions of Americans. The current, unrelenting attacks on California and American higher education, including the imposition of distance learning schemes, will significantly weaken this pivotal institution if not prevented by those in power.

ORIGIN OF THE PLANS: CALIFORNIA AND THE UNITED STATES

What is the origin of all these strategic plans? Are they randomly conceived plans to deal with specific issues, or are they part of a larger scheme to transform higher education? The strategic plans may seem to have been introduced independently of one another, ostensibly addressing different issues such as use of technology in higher education, demands for student access to higher education, and the need to coordinate academic programs. However, where did all of these strategic plans *actually* come from? The answer, as will become clear, is national in scope. It applies both to California, with its elaborate higher educational institutions, and to the rest of American higher education.

It is argued that these strategic plans come from the business model that has been increasingly applied to higher education, that is, from the use of corporate principles that stress profit making, "bottom line" decision making, and top-down administrative control. Two aspects of the business model are particularly relevant: (a) the business community's desire to control higher education to make colleges and universities more receptive to immediate business interests, with teaching, research, and service to be of greater direct utility to business and (b) the high technology sphere of business's undue influence in encouraging adoption of various strategic plans involving the purchase of expensive computer hardware and software.

Examination of a recent book by Diana Oblinger and Anne-Lee Verville (1998), *What Business Wants From Higher Education,* provides many insights into the business origin of these academic strategic plans. Interestingly enough, the two authors of this book are current or retired IBM executives with university connections, and they may be representing the high tech sector of business in particular.

What Business Wants From Higher Education summarizes and promotes one form of higher educational planning that began in the 1990s. It argues that the business community perceives various deficiencies in higher education (apparently in spite of higher education's many accomplishments, not the least of which is providing California and American business with a highly educated work force, as even the authors acknowledge [Oblinger & Verville, 1998, p. 23]). Due to these perceived deficiencies, business desires a series of changes, outlined by Oblinger and Verville, and higher education is implored to adopt them.

Indeed, higher education administrators have been hearing—and responding to—this kind of message throughout the 1990s and up to the present. It has been presented repeatedly in conferences ranging from the administrator-dominated American Association for Higher Education (AAHE) (Marchese, 1998) to the American Council on Education (ACE, which sponsored the book) to conferences organized by university and college chancellors and presidents. The authors of this chapter have also attended some of these conferences. An analysis of the Oblinger and Verville (1998) book will reveal that, for reasons ranging from agreement with to resignation about or accommodation to the messages, administrators have adopted many of the principles this book recommends to reshape higher education, thus increasing the likelihood of conflict between the administration and faculty or students.

This chapter will

1. Show a parallel between the strategic plans previously described as increasing differences between faculty and administration (Wood, 1999) and the strategic plans Oblinger and Verville (1998) assert that business wants higher education to adopt (it should be noted that the earlier discussion was written *before* the authors read Oblinger & Verville, making the two discussions of strategic plans independent of each other).

2. Show that several of the strategic plans are aimed at weakening the faculty, and especially faculty-shared governance of the university, which makes outside business influence considerably easier.

3. Raise questions about whether all these strategic plans for higher education are actually in the interests of business as a whole, even seen from a business perspective.

4. Indicate that the technology sector of business may have unduly influenced university and college administrators to adopt strategic plans that emphasize the widespread use of—and therefore widespread purchase of—expensive technology in higher education, an institution that has a near-captive audience of 14.3 million students.

In the university community, *younger* students alone are said to have a "projected $100 billion-plus annual buying power" (Oblinger & Verville, 1998, pp. 31, 33, 135). This means students have enormous disposable income to purchase millions of computers and related software. In addition, higher education institutions themselves would purchase a parallel number of computers. Given the quick obsolescence of computers, the purchasing of new hardware and software will continue indefinitely, with millions of dollars to be made by technology corporations, also indefinitely.

In the earlier monograph on "the two cultures of faculty and administration," 10 differences between faculty and administration were discussed, again with the list compiled before reading *What Business Wants From Higher Education* (Oblinger & Verville, 1998). That list was derived from synthesizing a decade's worth of faculty and student experience regarding the changes being demanded by university and college administrations, changes often reported throughout the 1990s in *The Chronicle of Higher Education* (e.g., Selingo, 1999), as well as in major media, for example, the *Los Angeles Times* (e.g., Smollar, 1992), *Newsweek* (e.g., Kantrowitz & King, 1992), and the *New York Times* (e.g., Bernstein, 1992) (Wood, 1998a, 1998c; Wood & Valenzuela, 1996, 1997).

RECOMMENDATIONS FROM
What Business Wants From Higher Education

What does the Oblinger and Verville (1998) book recommend to higher education for enhancing its utility to business? A set of recommendations from the book is closely parallel to the issues previously discussed. We will present those recommendations here and take a critical look at them. As will be seen, these are not incidental recommendations. The official journal of the American Association of University Professors (AAUP), *Academe,* has recently taken a serious look at this type of recommendation along the lines to be discussed below (Bellah, 1999; Ohmann, 2000; Schrecker, 2000; see also Zappia, 1999), as have other recent books and articles with titles such as "The Kept University" (Press & Washburn, 2000) and *The Edge of a New Dark Age: The Corporate Takeover of Higher Research and Education* (McMurty, 2000).

Among its many recommendations, *What Business Wants From Higher Education* argues the following:

1. *Higher education should now be learning centered,* with instructional technology (IT) fundamentally changing the traditional "campus-centric model" of higher education to a model of higher education gleaned "from any location," using advanced computer technology, the Web and Internet, and CD-ROMs, all of which are parts of new "electronic communities" (Oblinger & Verville, 1998, p. 95). This kind of recommendation surely furthers the financial interests of companies such as IBM by enabling them to sell millions of dollars of technology to thousands of colleges and universities, serving millions of students (pp. 31, 33, 152). But it is unclear why the rest of the business community would embrace this strategy. Business needs educated workers, not necessarily a workforce trained through distance learning. This is especially the case because distance learning has

many drawbacks, such as lack of direct interaction with professors and other students. Such interaction is a positive attribute of traditional in-class higher education (Farber, 1998). Indeed, David Noble (1998a, 1998b) has shown that universities heavily relying on distance learning will become "digital diploma mills," producing degrees that invite suspicion. Accrediting such degrees has become a national issue, with the AAUP's Committee D (on accreditation) raising serious questions about the quality of distance education programs and the value of degrees awarded. On hearing about the presumed need to widely adopt technology to teach college and university students through distance learning at the aforementioned administration-sponsored conference in 1990, it appeared to many in the audience that technology companies had strong profit motives in gaining support for this kind of higher educational strategic plan, regardless of the quality of education provided (Wood, 1997).

As noted above, the most recent rationale for the widespread use of distance learning in California is to accommodate the increased number of students expected in Tidal Wave II, students who are expected to reflect the increasing diversity of California's population (an analysis of recent California Postsecondary Education Commission [CPEC] data indicates that no one ethnic group in the year 2000, including the white ethnic group, is a numerical majority in California public higher education). However, there has been some backing away from the position of widespread use of distance learning by the CSU Chancellor's Office, which is now also focusing on other ways to accommodate students, such as year-round operation of CSU universities; development of branch campuses such as the National City branch of SDSU; expansion of the workday and workweek to offer Saturday classes and classes beginning at 5 a.m., a program that begins at some Fresno community colleges in Fall 2001; and conversion of other institutions into universities, such as CSU Monterey Bay, which was initially a military base. The creation of new campuses, of course, is yet another way to accommodate the projected students. Because these approaches are derived from standard methods of quality education, using them is more likely to provide continuing quality higher education in California than primary reliance on distance learning as a way to accommodate the students of Tidal Wave II.

2. Accountability measures of student learning are required to assess the benefits of higher education, the book argues (Oblinger & Verville, 1998, p. 108). Why do Oblinger and Verville feel this measurement is necessary? Because, they argue, business needs employees with "the knowledge, skill, and personal attitudes that will enable them to work in the present as well as the future organization" (p. 73). Furthermore, they assert,

> Many executives complain that too many top-level graduates of business schools are good at analyzing textbook cases of business problems but are un-

able to come up with innovative ideas for new business products or services, or ways to stay competitive with similar industries in other countries... there are large gaps between the kind of performance needed for success in a business setting and the kind required for success in schools. (p. 72)

The book assumes that accountability measures of student learning will ensure that these business qualities are learned in college and later retained by the students for use in business. The authors present little actual evidence for this key assumption (see also Shipps, 2000). Especially problematic is the undocumented ability of these measures to indicate a capacity to generate the creative ideas desired by business.

California and American colleges and universities have long since developed extremely rigorous systems of accountability in the tenure review process and in academic reviews of departments. Enormous amounts of evidence about a faculty member's teaching, professional growth and research, and service to the university and larger community are needed before a candidate for tenure actually receives this major academic honor. The Faculty Coalition for Public Higher Education (2000) has detailed the many levels of professional evaluation a professor must experience before getting tenure—and this is years after the same professor had to compete with typically hundreds of other candidates just to get the job in the first place. For departments, accountability is routinely found in detailed academic program reviews, conducted largely by faculty outside the university, in conjunction with the university administration. These reviews can, and often do, have important consequences for changing and improving the departments. These are just two of the rigorous measures of accountability by which faculty and departments are evaluated.

Colleges, universities, and academic professions have many additional mechanisms of accountability in evaluations for research grants and international fellowships, acceptance of books and articles for publication, post-tenure reviews, for merit pay, and promotion to the rank of professor. Each of these is an extremely demanding evaluation and constitutes solid, objective, and well-considered measures of accountability. With recent attention being paid to the extremely important role that teacher quality plays in student learning—teacher quality even being described as the most important factor in student learning—these long-standing measures of faculty quality should continue to be the central benchmarks of faculty, departmental, and college or university accountability.

Finally, it should be noted that the recent focus on accountability measurements is part of an education reform agenda that demands widespread educational improvements without providing necessary additional funding to accomplish this goal. The use of measurements can, therefore, appear to be educational reform without paying the price. However, the actual occurrence of such widespread educational improvements is open to question. There have been reports of cheating and misreporting on high-stakes tests, and a story on the front page of the *Los Angeles Times* ("Crusader Argues," 2000) even argues that this kind of measurement can impede learning. Among other reasons, it forces faculty to "teach to the [objective] test," not develop the creative thinking in students that business is purported to want. Indeed, a proposed widespread use of high stakes testing in the University of Texas system has produced immediate criticism ("Faculty Outcry," 2000).

3. *Universities and colleges should form more strategic alliances with business,* the book asserts. The book states,

> Higher education should establish strategic alliances with potentially important future suppliers—hardware and software vendors, telecommunications companies, publishers, and others. . . . Strategic alliances among higher education institutions will become increasingly common in the future. (Oblinger & Verville, 1998, p. 146)

It might be noted that one colossal failure of such a scheme was CETI, the proposed $300 million deal over several years between the CSU system and four corporations: Microsoft, GTE, Hughes, and Fujitsu. Viewing CETI as a takeover of CSU by business, the faculty, students, and staff of CSU rallied—with the support of corporations not to be included in the deal and the State Legislature, which held a dramatic, standing-room-only hearing in Sacramento—and the project was ended (Wood, 1998b).

A further examination of the kinds of businesses with whom universities and colleges are urged to seek strategic alliances can be instructive. "Hardware and software vendors, telecommunications companies, publishers"—and the vaguely stated "others" (Oblinger & Verville, 1998, p. 146). The three types of companies specifically listed here would all benefit from universities purchasing millions of dollars of computer hardware and software, just the kind of financial arrangement of great interest to technology-oriented companies.

4. *Commodification of higher education is recommended* in several ways by Oblinger and Verville (1998). First, if not foremost, students are now referred to as customers: in the index, under the term *Customers*, it says "*See* College students" (p. 178). Likewise, the entirety of Chapter 3 is devoted to "Profiling the Customer." Indeed, Oblinger and Verville make the same connection found in the earlier monograph between the conception of students as customers and the shift toward "learning-centered" institutions. They say, "Why are colleges and universities changing from an institution-centered to a learning-centered environment? Because . . . it is consumer-oriented" (p. 96).

As those in business know, customers need to purchase products, which in this instance means a college or university education. This also means "learning will occur whenever students can connect to the World Wide Web," making "campus-centered" education obsolete (Oblinger & Verville, 1998, p. 99). These higher educational products come at a cost, which is tuition and related educational expenses, such as outlays for computer hardware and software. Whereas the university receives the tuition, financial outlays for computer hardware and software go to technology companies such as IBM and Microsoft. One of several problems with the CETI arrangement noted above was that universities, students, faculty, and even alumni were to purchase and use only certain computer hardware and software, raising the issue of monopoly, which has followed one of the CETI partners, Microsoft, and its founder, Bill Gates. The end result here is that millions of students, by this model, would be receiving what David Noble (1998a, 1998b) has called digital diploma mill degrees.

5. The importance of intellectual property copyrights over courses remaining with the faculty was cited by Noble (1998b) as a means to slow down the above tendencies toward commodification and commercialization of higher education. Oblinger and Verville (1998) have somewhat different ideas. Whereas they are less dogmatic on ownership of intellectual property than on several other principles, they nonetheless casually refer to teaching as a "work made for hire" (p. 151). As such, the university or college would automatically own the professor's course. However, Oblinger and Verville's main concern is professors working with universities other than their own. The underlying assumption that "deregulation" of higher education is needed (p. 152)—connected to a "free trade" philosophy—may make the authors less sure of who should own professors' courses, because the professors, after all, created the courses, even if they were assisted by universities. Oblinger and Verville conclude, "Most of us are still unsure how to handle quality control and intellectual property management in a medium where one can be a producer as well as a consumer" (p. 52).

6. The large increase in part-time faculty appointments is acknowledged by Oblinger and Verville (1998, pp. 40-41)—without concern about quality of education. They argue that there is no need for all students to have full-time, tenured, or tenure-track professors teaching them (pp. 41, 148-149). Indeed, Oblinger and Verville refer to the "decline of full-time, tenured and tenure-track professors" and to the "increase in part-time faculty," stating that "approximately 35 percent of all faculty are part timers, and more than one-third of the full-time faculty hold term appointments" (p. 40). They indicate that "The traditional faculty ranks may constitute half (or even less) of the profession" (p. 40). Furthermore, they seem to approve of "the new and growing middle category of full-time but non-tenure-track faculty" (p. 40). This is an interesting choice because it addresses the criticism of unattached part-time faculty but still supports significant weakening of faculty-shared governance of the university. In this regard, business should ask higher education leaders why they are supporting policies such as attacks on tenure and reliance on low-paid, insecure part-time faculty, which make a career in higher education unappealing to a younger generation of scholars who would otherwise contribute to the next series of intellectual and practical accomplishments. Again, borrowing the phrase from the widely distributed critique of American high schools in the 1980s, *A Nation at Risk* (1983), it is as though California and American higher education are currently being organized by a foreign enemy.

7. *The abolition of tenure for faculty,* linked to the increasing proportion of part-time faculty, is another policy Oblinger and Verville (1998) support. They state,

> Probably the most visible, incongruous college and university policy to the outside world is tenure. It is under fire across the country. Many view tenure as the single biggest impediment to change in higher education. The tenure system is regularly blamed for many of the perceived ills of higher education. (p. 151)

Oblinger and Verville even argue that tenure is "contrary to the notion of serving customers" (p. 151), going back to their market model of higher education. However, the many benefits of tenure to higher education, students, and the general public have been repeatedly shown (Finkin, 1996; Hohm & Wood, 1998, p. 4). Oblinger and Verville's calls for a "Culture of Change," "Need for Leaders," and "Transformations" (pp. 159-164) all point to attacks on tenure to weaken the faculty role in governing the university. Faculty and students must become aware of this position and organize to meet the challenge (Kuchta, 2000).

Organizations such as the American Association of University Professors (AAUP; 2000), the California Faculty Association (CFA; 2000), and the Faculty Coalition for Public Higher Education (1999) are addressing this national issue, and it has been the focus of recent academic meetings. For example, it was discussed at several recent sessions of the Pacific Sociological Association, organized by Georgie Ann Weatherby. Academic freedom and tenure were the focus of the American Sociological Association's 2000 millennial celebration meeting. Matthew Finkin (1996) published discussions in *The Case for Tenure,* and other views were included in articles for the Special Issue on "The Academy Under Siege" of *Sociological Perspectives* (Hohm & Wood, 1998, p. 4). Again, the business community, along with students and the general public, should insist on increased quality of higher education by supporting an increase in full-time, tenured, and tenure-track faculty instead of lower paid, part-time, temporary faculty. Most successful business leaders in California and America were themselves educated by the full-time, tenured faculty, often at prestigious universities, and thus they should recognize the need to continue this kind of quality higher education for the next generation of students.

8. Increased administrative control, which the book supports, is the goal of attacks on tenure, support for non-tenure-track hiring of faculty, and acceptance of an increasing part-time faculty. In their conclusion to the book, Oblinger and Verville (1998) cite former CSU Chancellor Barry Munitz as arguing that higher education needs more administrators who understand the business model for higher education and are able to enact the business-oriented practices discussed here. He asserts, "The only way we will have leaders who are imaginative, courageous, and professionally trained will be to depart from higher education's traditional view of presidential responsibility as 'the last bastion of amateur management' " (p. 164). The group responsible for bringing about all of these changes, from Oblinger and Verville's standpoint, is these identical "courageous" administrators to whom the book is principally addressed. The authors will likely be comforted by the rise of Ed.D. degrees (Doctor of Education) in higher education administration, which constitutes a significant break with the academically oriented Ph.D. degrees (Doctor of Philosophy) held by the most famous university administrators, such as Clark Kerr, formulator of California's extremely influential Master Plan for Higher Education.

9. The department has been the main organizational focus for academic hiring, retention, and planning in most colleges and universities. Instead, Oblinger and Verville (1998) would prefer academic decision making regarding hiring and related issues to be made by administrators who take business interests directly into account when making such academic decisions (pp. 65, 151-152, 163-164). They indicate, "Reengineering is the order of the day" (p. 5). They even strongly imply that business should help hire faculty (pp. 94, 148-149). In discussing such business needs as training college graduates who can fit into the emerging global economy, Oblinger and Verville assert, "American faculty are . . . unenthusiastic about internationalizing the curriculum . . . it may be difficult for globalization to penetrate higher education" (p. 66). Why is this so? Because "teaching, learning, scholarship, and service [are] highly dependent on the experience base and attitudes of the professoriate" (p. 67). Thus, the administration needs to overrule these faculty tendencies—real or imagined—and hire faculty to implement the policies of business advocated in this book. Interestingly enough, business leaders do not want all of higher education to be turned into business colleges—they even value the breadth of experience characteristic of college graduates (p. 25)—but implementing recommendations like the ones above will likely lead to such a result, creating an ultimate commercial university.

10. Departments that are unproductive for business should be eliminated or significantly diminished, the book implies (Oblinger & Verville, 1998, pp. 62-66, 145, 151-152, 161). The authors question what a bachelor's degree "certifies" these days

(p. 109) and instead call for "accountability . . . to measure [students'] performance" (p. 153). It is instructive to note what is *not* called for: faculty scholarship and development is de-emphasized in importance, even though that is one of the most important focuses for strengthening academic programs (p. 161). Instead, Oblinger and Verville argue, "Demands for increased productivity arise from the common perception that faculty do not teach enough, students do not learn the right things, and administrators are reactive 'firefighters' instead of effective managers" (p. 153). They feel administrators, drawing on an unregulated environment, should move to "add value"—yet another business term—to the curriculum by instituting the skills business needs (Chapter 2).

Oblinger and Verville (1998) finally conclude this line of reasoning with a threat: They argue that to receive necessary funding, higher education needs business support, which will only be forthcoming when higher education makes the drastic changes business considers appropriate (p. 142).

CALIFORNIA'S PROBLEM IN GLOBAL CONTEXT

Several of the strategic plans being applied to California and American higher education have also been appearing in England, Europe, and Third World countries. The threat to tenure in England is substantial, with policies emanating from the government of former Conservative Prime Minister Margaret Thatcher and culminating in the Education Reform Act of 1988, essentially ending the institution of tenure (Court, 1998). One predictable result of the virtual ending of tenure in England has been the replacement of full-time faculty with large numbers of low-paid and insecure part-time instructors, a related strategic plan just discussed (Court, 1998). As previously noted, Great Britain's Open University has also been pioneering yet another strategic plan, the large-scale use of distance learning. Similarly, there have been serious budget problems for higher education there, with faculty implored to become more practical and bring in outside money—for example, by working on projects for business—to fund university activities. Indeed, a faculty member at the University of London has commented that trends such as these, if continued, will refocus British higher education on producing widgets instead of educated people.

At Canada's York University, an elaborate plan was constructed to institute distance learning with little or no faculty input (Noble, 1998a). The faculty quickly perceived this move as an attempt by the administration to drastically change working conditions on the campus without consultation of faculty or students.

Similar attacks on faculty and students at other universities have become sufficiently serious that university disturbances have lately occurred in various countries. York University was the site of the most celebrated disturbance, which entailed a highly publicized strike and ended with the faculty obtaining a more favorable contract than that in the initial plan, with considerable faculty control over the installation of any distance learning schemes (Noble, 1998a, 1998b). Australia and Mexico have had similar university disturbances over the degrading of higher education, as have such countries as South Africa, Fiji, Senegal, France, Guatemala, and Yugoslavia.[3] Indeed, it appears that the strategic plans of one country are routinely studied by other countries for possible utilization. Thus, the adoption of these plans is already international in scope and will likely increase in pace, with the faculty and students of affected countries needing to monitor, and where possible influence, these developments.

ALTERNATIVE EXPLANATIONS FOR THE PROBLEM

The main alternative explanation for the introduction of these strategic plans has been briefly mentioned above: They have been randomly introduced to solve specific problems, such as student access to higher education, technological currency, and budget shortfalls. Taken separately, it is possible to make an argument for the strategic plans as solutions to each of these and related problems. For example, it is possible to argue that periodic budget problems have forced universities to hire more part-time faculty than usual. However, once budget problems have subsided, the practice of hiring part-time faculty often continues, prompting part-time faculty to organize nationally against the continuation of this plan.

Similarly, distance learning was once thought to be an inexpensive way to educate large numbers of people. Yet, even when the large costs of such programs become apparent, the programs have continued. Some administrators have even suggested using money from regular university budgets to support the still-untested distance learning approach.

The main difficulty with the alternative explanation that these strategic plans are randomly introduced is that all of them appear in Oblinger and Verville's (1998) book. Their book is a highly conscious attempt to restructure higher education with little or no input from the main participants in the institution: faculty and students. Oblinger and Verville are drawing on, summarizing, and advocating many discussions of these plans that occurred throughout the 1990s, with roots of these ideas going back to at least the 1980s. Oblinger and Verville organize the strategic plans together in a highly interrelated fashion that is aimed at promoting the implementation of the plans in California and the United States, along with other receptive countries. There is, therefore, nothing random about the introduction of these plans in higher education. The strategic plans, especially in combination, are aimed at greatly increasing business influence over higher education at the state, national, and international levels.

FUTURE PROSPECTS: WHAT TO DO?

The current, unrelenting attacks on California and American higher education—including the imposition of distance learning schemes—will significantly weaken this pivotal institution if they are not prevented by those in power. The solution to all this? Now is the time for the public to insist on restoring California and American higher education by reversing these negative trends of the 1990s and returning American colleges and universities to their previous position of strength and dedicated public service. The public should insist that elected officials pass laws requiring a large majority of in-class instruction instead of a majority of (or all) distance learning classes for college and university degrees; laws helping protect the intellectual property rights of faculty over their courses, such as Assembly member Gloria Romero's Assembly Bill 1773; laws requiring the hiring of substantial numbers of full-time faculty instead of part-time instructors, such as Assembly Bill 1725, which still needs to be properly funded; laws strengthening tenure, which is at the base of the academic freedom required to produce great innovations in science, economic production, and technology itself—the conceptual breakthroughs for which America is famous; laws mandating minimum levels of national and state budgets dedicated to public higher education; and laws that prevent corporate takeovers and privatization of, or undue influence on, public higher education. These and similar laws would go a long way toward supporting higher education against the current onslaught which, if left unchecked, will damage it beyond recognition.

SOLUTIONS AND SUGGESTIONS: STRATEGIES FOR DISTRESSED CAMPUSES

Especially in economic recessions, universities and colleges are subject to direct attacks such as attempts to eliminate academic departments and tenure (American Association of University Professors, 1993). At least 10 strategies discussed at academic conferences can be and have been pursued, particularly by troubled campuses, to extricate themselves from these difficulties. The strategies are:

1. To form faculty alliances with students and staff

2. To form faculty budget committees to evaluate independently any alleged financial crisis on campus (physics professors have been extremely astute as budget analysts on several campuses and should be consulted)

3. To contact outside organizations such as the American Sociological Association and the AAUP for immediate assistance

4. To set up a legal defense fund

5. To seek assistance from sympathetic politicians, including politicians who can point to their support for higher education as a positive accomplishment in re-election campaigns

6. To get faculty and student advocates elected to university senates and seated on important committees such as those concerning budgets and resources

7. To involve the media of television, radio, newspapers, and magazines

8. To set up computer networks to gain assistance from other colleges, universities, and noneducational institutions

9. To give careful attention to tenure, faculty governance, salary, technology transfer, and distance learning issues, along with the issue of gaining access to university budget information, in all collective bargaining and faculty employment contracts

10. To maintain regular contact with legislatures and politicians at the national, state, and local levels, and to make significant increases in political action contributions from university organizations, conveying this financial necessity to the members

NEWER STRATEGIES

At least two political responses have been considered for the previously mentioned structural legislative problem that state budgets are steadily reducing the proportionate share for higher education. Higher education could be included as a mandated part of state and national budgets, as one of us has previously argued (Wood, 1994a, 1994b; see also "ACA 25," 1994). Alternatively, state constitutions could be changed to reduce significantly the mandated proportions of the budgets, as has been considered by the California legislature and elsewhere (California Business-Higher Education Forum, 1994, p. 10). The current use of a third method to deal with lower budgets for higher education—the 4% compacts between the governor and higher education leaders—will probably not survive the next serious recession. It may be necessary to select one of the other two approaches to deal with the problem before higher education is "de-funded" at the next recession. We had the dubious distinction of witnessing an all-out attempt to institute this de-funding procedure as one of the more insidious aspects of the tenure and departmental elimination crisis at SDSU in 1992, when nine academic departments were threatened with elimination, six others were to be drastically reduced, and 146 tenured and tenure-track faculty were threatened with layoffs (AAUP, 1993).

The conservative mood of the country in the mid-1990s, especially after the national Republican sweep of Congress in 1994, boded ill for higher education. Until the 1990s, higher education usually had bipartisan support, with Republicans often seeing education as supportive of business productivity and Democrats often seeing it as an avenue of economic mobility for all Americans. This changed considerably by the mid-1990s, with Democrats often favoring and Republicans often opposing financial aid to higher education (Cordes & Zook, 1995; "Federal Student Loans," 1995; Healy, 1995; Jaschik, 1995; Zook, 1995). If this trend is to be avoided in the future, the American academy may have to help more than usual in electing politicians who will actively support higher education, including during times of recession (Ladd & Lipset, 1975). Faculty may even need to run for election on higher education platforms ("It May Be Remembered," 1995), following the successful strategy of the teachers' unions in Grades K through 12 (California Teachers Association, 1995a, 1995b). This would involve large numbers of professors, students, staff, administrators, chancellors, and boards of trustees campaigning throughout the year for candidates, participating in voter registration drives, paying increased organizational dues for work by political action committees, creating and enlarging the staffs and functions of these committees, writing specific legislation for higher education, and energetically lobbying for legislation favorable to higher education. Indeed, Professor Gloria Romero of CSU, Los Angeles, recently won election to California's Assembly and soon wrote, and got passed, legislation helping protect professors' intellectual property rights, the aforementioned Assembly Bill 1773.

For the long-term, AAUP is planning to publicize the academic profession more positively through activities such as sponsoring a National Public Radio program, writing newspaper columns—and we would add making television appearances and holding Internet discussions—concerned with academic freedom, tenure, and university contributions to the larger society (Cage, 1995). The academic profession during World War II and the Cold War was essential to national security. Some feel this is no longer the case, with fascism and communism no longer directly confronting the United States. This attitude, of course, overlooks many current political realities and borders on a new isolationism. It similarly overlooks the role of universities in providing the essential skills needed to operate and compete in an increasingly technological world (Kerr, 1991, 1993, 1994a, 1994b). Yet, the lessening of immediate warlike threats to the United States in the mid-1990s produced a short-sighted view of the necessity to fund higher education properly.

It appeared that many state and national legislators had become complacent about the maintenance of a highly trained workforce. A publicized talk by a state governor in the mid-1990s indicated that higher education was not a high priority for him or many of his colleagues in other states. The assumption was that the United States will always have a highly trained workforce no matter how much funding is cut from education. This assumption is highly inaccurate, as indicated by the fact that in California alone, 206,000 college students dropped out during the budget-slashing, tuition-raising period of the early 1990s (CSU Chancellor's Office, 1995; Schmidt & Ristine, 1995). These 206,000 students will not receive the sophisticated training that society needs and that is provided by higher education; they will suffer financial, occupational, and personal losses, as well (Schmidt & Ristine, 1995). Undoubtedly, some of them will even join the institution that public higher education has been forced to compete with for funds since the 1990s: the prison system (Rosen, 1995). The states similarly lose tax revenues by cutting higher education because of the greater taxable earnings of college graduates (West, 1995, pp. 136-138). One study showed that in California, "the state would receive $1.12 in taxes for every $1 in funding added to the CSU" (Ristine, 1993). Thus, for every $1 billion in the CSU budget, California gains $1.12 million in taxes. Another way of seeing this is to note that when the Master Plan for Higher Education in California was being developed, economists indicated that more than 40% of the state's economic growth was due to higher education (Kerr, 1991, p. 24). Indeed, several state legislatures began to see the damage to universities in the 1990s and moved to restore lost funding (Lively, 1995). The 4% compacts were an extension of this logic, and we certainly encourage all such positive college and university funding activities by the governor and legislative bodies.

Further activity and influence at the national and state political levels are becoming essential for higher education. The AAUP, the National Science Foundation, the National Institute of Mental Health, and professional associations such as the American Sociological Association and regional AAUPs must intensify lobbying efforts at their respective statehouses, in Congress, and at the White House. The Clinton administration was much more mindful of the need for a strong higher educational system than Republican administrations in the previous decade. Yet, Congress has been much less mindful. This means our tasks are that much more difficult, but the end results have been worth the effort. Since the late 1990s, higher education has been elevated to the top of many political agendas, with both Democratic and Republican presidential nominees in 2000 providing plans to finance higher education more effectively. Democratic candidate Al Gore, for example, proposed a $10,000 college-expense tax credit for middle-income families to support their participation in higher education.

We have been impressed with the lobbying efforts of teachers' associations for Grades K through 12. University associations should study their strategies and do more to emulate their successes, especially by increasing spending on political elections, increasing the number of legislative bills carried for higher education, and responding quickly to any new attacks on higher education. Students formed a national lobbying organization that was a significant influence for legislation concerning student aid ("Federal Student Loans," 1995). They also wrote a Student Declaration of Educational Rights, which was presented to the California legislature by Susan McEntire (1995). The students also formed a very effective E-mail network to focus on the many national problems of higher education, with E-mail addresses such as: nagps@netcom.com or rosati@gusun.georgetown.edu. The professional associations should do likewise if they do not currently have widespread E-mail linkups with interested educational parties who can communicate with legislators. In recent years, several professional groups, such as AAUP and the California Faculty Association, have done just that.

Students are often seen as a kind of silent majority to politicians and other decision makers. Yet, students can take matters into their own hands and bring a more positive conclusion to serious higher educational problems. A vital solution to many of the problems discussed here is having students register to vote and, most important, actually vote, either by absentee ballot or at polling places. Political involvement does not stop there; it needs to continue by holding elected officials accountable for their positions on higher education issues. Setting up meetings with local elected officials or their educational aides makes a difference. Such meetings are a crucial step in beginning relationships with elected officials and their staff.

Members of the SDSU Associated Students External Affairs Board, for example, met with their local elected representatives to begin a business relationship that would address the student loan issue as well as other pertinent educational problems. Inviting local elected officials to the campus is also a good idea. Similarly, it will increase student political participation, as students will already be at the university.

Registering students to vote prior to a general election serves as a stepping stone for increased student involvement in their educational system. At SDSU, the Young Democrats, the California Faculty Association, and the Associated Students were able to register more than 4,000 students in a combined effort in the 1994 elections. The commitment of registering students did not stop with activity on the campus alone. It continued with an absentee ballot drive, which was a success because it was a convenient way for students to vote, and it also facilitated their actual voting. Absentee balloting should be an increasing focus of future electoral campaigns because students are often registered in locales outside the campus area. They often need to re-register if they are to vote near the campus.

One important educational result of these efforts in 1994 was the election from the San Diego area of three of the most higher education-minded Assembly members in California: Dede Alpert, Denise Moreno Ducheny, and Susan Davis, two of whom were in very close races, their margin of victory influenced by the newly registered students.

Lobbying efforts do not end with visits to local representatives. They continue with making trips to the Capitol, as was done by members of the California State Student Association, an organization that now represents about 350,000 students in the CSU system. In 1992, busloads of students from every higher educational system in California went to the Capitol to lobby their respective representatives and make their voices heard. Also, a student vigil at SDSU lasted 6 months and helped elect many officials in 1992 by registering about 8,000 students statewide (Wood, 1998c). Such activities were essential in getting SDSU and the rest of California higher education through the budget-cutting crisis of 1992-1993. Without these efforts by highly dedicated students, several departments of SDSU could have been dismantled or drastically reduced.

Lobbying efforts are not the only ways students can address problems they face. Participation and involvement in university and college committees is absolutely essential. Committee participation also means activity on other decision-making bodies, such as the board of regents or trustees. Board members' policy begins and changes in meetings held by committee. It is here where vocal student representation is necessary.

The media are also useful means to disseminate information to the local community, which may be detached from the university community. Continuous letter writing to newspaper editors, and participation in talk shows and debates, and other events of interest to the media, should be organized and publicized to highlight student issues and concerns about obtaining an affordable, accessible, quality higher education.

SUMMARY AND CONCLUSION

In conclusion, California's Master Plan for Higher Education faces serious threats, making higher education a significant social problem. Many of these threats are summarized in Oblinger and Verville's (1998) *What Business Wants From Higher Education*. Yet, the book goes beyond summarizing to promote the many trends shown in a monograph that focused on the increasing differences between faculty and administration. Aspects of the business model for higher education—increasing business control over the university, especially by the high technology sector—are central to the Oblinger and Verville book, which argues for the need to institute these divisive tendencies in higher education to support business. More generally, the collapse of the Soviet Union has ushered in a perceived triumph of capitalism, which has decided to exert its influences everywhere, including the university. The university has often pointed to the problematical effects of an unbridled economy on other groups and institutions, such as ethnic minorities (Blauner,

1972; Cox, 1948), the poor (Piven & Cloward, 1971), and the federal government (Domhoff, 1998). Now, the academy must deal with similar effects on the faculty, students, and staff in higher education, while drawing on earlier warnings (Bowles & Gintis, 1976; Ovitz, 1993).

Although Oblinger and Verville (1998) is dedicated to gaining business control over higher education, many of the talents the authors say business wants from college graduates are already obtained in a liberal education: critical and analytical thinking; problem-solving abilities; written and oral communication; interpersonal skills, including working with those of diverse backgrounds; knowledge of computers and technology; ability to analyze data; leadership; and creativity. Given the importance of a liberal education (Bellah, 2000), even when it is not directly connected to immediate business interests, the academy now finds itself in the ironic position of having to "save business from itself" by resisting these business demands on col- leges and universities (Wood, 1998a). The future of the Master Plan, as well as California and American higher education in general, depends on the outcome of this struggle.

NOTES

1. G. Diehr discussed these issues in E-mails widely distributed between June 5 and October 23, 2000, with titles such as "Creative Accounting" and "The Expenditure Gap."
2. See National Commission on Excellence in Education (1983). See also Lund and Wild (1993), which recorded the views and experiences of the education and business leaders who were part of the 10-year effort to improve American high school education.
3. Events in the countries mentioned here were described in *The Chronicle of Higher Education* during the years 1998 through 2001.

REFERENCES

ACA 25 might rescue sinking higher education [Editorial]. (1994, March 17). *The Daily Aztec*.
American Association of University Professors (AAUP). (1993, March/April). San Diego State University: An administration's response to fiscal stress. *Academe*, pp. 94-118.
American Association of University Professors (AAUP). (2000). Assessment, accountability, accreditation [Special issue]. *Academe, 86*(1).
Baker, W. (1995, April 7). *Higher education in California: Prospects for the future*. Paper presented to the 66th annual meeting of the Pacific Sociological Association, San Francisco.
Bellah, R. N. (1999). Freedom, coercion, authority. *Academe, 85*, 16-21.
Bellah, R. N. (2000). The true scholar. *Academe, 86*, 18-23.

Bernstein, R. (1992, June 14). The Yale Schmidt leaves behind. *The New York Times Magazine,* pp. 33, 46, 48, 58, 64.

Blauner, R. (1972). *Racial oppression in America.* New York: Harper & Row.

Bowles, S., & Gintis, H. (1976). *Schooling in capitalist America.* New York: Basic Books.

Cage, M. C. (1995, June 23). Association seeks way to improve the image of the professoriate. *The Chronicle of Higher Education,* p. A16.

California audit faults 6 community colleges for violating law on faculty pay. (2000, October 16). *The Chronicle of Higher Education* [On-line].

California Business-Higher Education Forum. (1994). *California fiscal reform: A plan for action, recommendations and summary.* Oakland, CA: Author.

California Faculty Association (CFA). (2000). What is the future of the CSU? [Special issue]. *California Faculty, 4*(1).

California Postsecondary Education Commission. (1999). *California higher education faculty: Tenure-track vs. non tenure-track.* Sacramento: Author.

California Postsecondary Education Commission. (2000). *Providing for progress.* Sacramento: Author.

California Teachers Association. (1995a, June/July). CTA initiative fund established. *CTA Action,* pp. 3-4.

California Teachers Association. (1995b, January/February). Status of CTA-sponsored legislation. *CTA Action,* p. 5.

Cordes, C., & Zook, J. (1995, May 19). Budget ax is overhead: Congress weighs big cuts in funds for education and science programs. *The Chronicle of Higher Education,* p. A31.

Court, S. (1998). Academic tenure and employment in the UK. *Sociological Perspectives, 41,* 767-774.

Cox, O. C. (1948). *Caste, class, and race.* Garden City, NY: Doubleday.

Crusader argues school reforms hinder learning. (2000, February 22). *Los Angeles Times,* pp. 1, 15.

CSU Chancellor's Office. (1995). *Governor's budget documents.* Long Beach, CA: Author.

CSU Chancellor's Office. (2000). *Student needs and priorities survey.* Long Beach, CA: Author.

DeGroot, G. J. (Ed.). (1998). *Student protest: The sixties and after.* London and New York: Addison Wesley Longman.

Domhoff, G. W. (1998). *Who rules America?* (3rd ed.). Mountain View, CA: Mayfield.

Douglass, J. A. (2000). *The California idea and American higher education: 1850 to the 1960 master plan.* Stanford, CA: Stanford University Press.

Faculty Coalition for Public Higher Education. (1999, April 21). Symposium report: Faculty question corporate model for public higher education. *Newsletter, 3*(4).

Faculty Coalition for Public Higher Education. (2000). What does it take to become a professor and to earn tenure? In *Accountability and tenure: A brief* (Section III, pp. 5-7). San Diego, CA: Author.

Faculty outcry greets proposal for competency tests at U. of Texas. (2000, October 6). *The Chronicle of Higher Education,* pp. 35, 38.

Farber, J. (1998). The third circle: On education and distance learning. *Sociological Perspectives, 41*(4), 797-814.

Federal student loans face $10.4 billion cut under compromise. (1995, July 7). *The Chronicle of Higher Education,* p. A21.

Finkin, M. W. (Ed.). (1996). *The case for tenure.* Ithaca, NY: ILR Press of Cornell University Press.

Governor increases Cal Grant aid. (2000, October 19). *The Daily Aztec,* pp. 1, 3.

Greider, W. (1981, December). The education of David Stockman. *The Atlantic Monthly,* pp. 27-54.

Healy, P. (1995, May 19). House panel votes to phase out aid for arts and humanities endowments. *The Chronicle of Higher Education,* p. A36.

Hohm, C. F., & Wood, J. L. (Eds.). (1998). The academy under siege [Special issue]. *Sociological Perspectives, 41*(4).

Ingram, R. T. (1999, May 14). Faculty angst and the search for a common enemy. *The Chronicle of Higher Education,* p. B-10.

It may be remembered as "the great meeting of '95"! (1995, July/August). *ASA Footnotes,* pp. 1, 3, 8.

Jaschik, S (1995, May 5). GOP presidential candidates bring out long knives when higher education programs are on the table. *The Chronicle of Higher Education,* pp. A28- A29.

Kantrowitz, B., & King, P. (1992, September 28). Failing economics: California built a great higher-ed system. Now it's being dismantled. by Barbara *Newsweek,* pp. 32-33.

Kerr, C. (1964). *The uses of the university.* Cambridge, MA: Harvard University Press.

Kerr, C. (1991). *The great transformation in higher education: 1960-1980.* New York: State University of New York Press.

Kerr, C. (1993). *Preserving the master plan: What is to be done in a new epoch of more limited growth of resources?* Testimony before the California Higher Education Summit.

Kerr, C. (with M. L. Gade & M. Kawaoka). (1994a). *Higher education cannot escape history: Issues for the twenty-first century.* New York: State University of New York Press.

Kerr, C. (with M. L. Gade & M. Kawaoka). (1994b). *Troubled times for American higher education: The 1990s and beyond.* New York: State University of New York Press.

Kuchta, D. (2000). *Reviving the master plan: California's higher education challenges* (Occasional Monograph Series, Number 2). San Diego, CA: Faculty Coalition for Public Higher Education.

Ladd, E. C., Jr., & Lipset, S. M. (1975). *The divided academy: Professors and politics.* New York: McGraw-Hill.

Lively, K. (1995, June 23). Legislators in several states rescue colleges from budget cuts. *The Chronicle of Higher Education,* p. A24.

Lo, C.Y.H. (1990). *Small property versus big government: Social origins of the property tax revolt.* Berkeley: University of California Press.

Lund, L., & Wild, C. (1993). *Ten years after* A Nation At Risk (Report No. 1041). New York: Conference Board.

Marchese, T. (1998, May). Not-so-distant competitors: How new providers are remaking the postsecondary market. *Bulletin of the American Association of Higher Education.*

McEntire, S. (1995, May 4). *Student declaration of educational rights.* Presented to the California State Senate Teleconference on Higher Education. California State Senate Television Program, Sacramento, CA.

McMurty, J. (2000). *The edge of a new dark age: The corporate takeover of higher research and education.* Toronto: Comer.

National Commission on Excellence in Education. (1983). *A nation at risk.* Washington, DC: Author.

National Education Association. (2000, October). They're talking on campus... *NEA Higher Education Advocate, 18*(1), 1.

Noble, D. F. (1998a, February). Digital diploma mills: The automation of higher education. *Monthly Review, 49*(9), 38-52.

Noble, D. F. (1998b). Digital diploma mills, Part II: The coming battle over online instruction. *Sociological Perspectives, 41*(4), 815-825.

Oblinger, D. G., & Verville, A.-L. (1998). *What business wants from higher education.* Phoenix, AZ: Oryx Press.

Ohmann, R. (2000, January-February). Historical reflections on accountability. *Academe,* pp. 24-29.

Ovitz, R. (1993). Adversarial research for resisting the entrepreneurialization of the university. *California Sociologist, 16*(1-2), 133-172.

Piven, F. F., & Cloward, R. A. (1971). *Regulating the poor.* New York: Pantheon.

Press, E., & Washburn, J. (2000, March). The kept university. *Atlantic Monthly,* pp. 39-54.

Region update [Column]. (2000, October 12). *San Diego Union-Tribune,* p. 3.

Ristine, J. (1993, February 27). CSU grads benefit state economy, report claims. *San Diego Union–Tribune.*

Ristine, J. (1995, February 2). SDSU's president to quite in '96: Was asked to step down earlier than he planned. *San Diego Union-Tribune,* pp. A-1, A-11.

Rosen, R. (1995, June 11). In the '90s prisons come before schools. *Los Angeles Times.*

San Diego State University. (1997). *Impacts: San Diego State University's role in the growth of the region.* San Diego, CA: SDSU President's Office.

Schmidt, S., & Ristine, J. (1995, April 10). An education in debt: Escalating college fees push loan burdens even higher. *San Diego Union-Tribune,* pp. A-1, A-19.

Schrecker, E. (2000, January-February). How are we doing? [From the editor section]. *Academe,* p. 3.

Selingo, J. (1999, June 11). New chancellor shakes up Cal State with ambitious agenda and blunt style. *The Chronicle of Higher Education,* p. A-32.

Sherman, H. J., & Wood, J. L. (1989). *Sociology: Traditional and radical perspectives* (2nd ed.). New York: Harper & Row.

Shipps, D. (2000, September 3). The business model won't fix schools. *Los Angeles Times,* pp. M2, M6.

Smollar, D. (1992, August 13). "Sociology 7" symbolizes feeling of betrayal at SDSU. *Los Angeles Times,* pp. A1, A22.

Snow, C. P. (1959). *The two cultures.* Cambridge, UK: Cambridge University Press.

UCSD building on brainpower. (2000, November 18). *San Diego Union-Tribune,* pp. B1-B2.

West, E. G. (1995). The economics of higher education. In J. W. Sommer (Ed.), *The academy in crisis: The political economy of higher education* (pp. 135-169). New Brunswick, NJ: Transaction.

Wood, J. L. (1994a, February 4). ACA 25 is the best budget choice. *The Daily Aztec.*

Wood, J. L. (1994b, February 2). Guaranteeing money for higher education. *The Chronicle of Higher Education,* p. B5.

Wood, J. L. (1997). Distance learning—Pitfalls and potential. *California Faculty, 1*(2), 14.

Wood, J. L. (1998a). The academy under siege: An outline of problems and strategies. *Sociological Perspectives, 41*(4), 833-847.

Wood, J. L. (1998b, February 20). In California, a dangerous deal with technology companies. *The Chronicle of Higher Education,* p. B6.

Wood, J. L. (1998c). With a little help from our friends: Student activism and the crisis at San Diego State University. In G. J. DeGroot (Ed.), *Student protest: The sixties and after* (pp. 264-279). London and New York: Addison Wesley Longman.

Wood, J. L. (1999). *C. P. Snow revisited: The two cultures of faculty and administration* (Occasional Monograph Series, Number 1, November). San Diego, CA: Faculty Coalition for Public Higher Education.

Wood, J. L., & Valenzuela, L. T. (1996). The crisis of American higher education. *Thought & Action: The NEA Higher Education Journal, 12*(2), 59-71.

Wood, J. L., & Valenzuela, L. T. (1997). The crisis in higher education. In C. F. Hohm (Ed.), *California's social problems* (pp. 81-98). New York: Longman.

Zappia, C. A. (1999). Academic professionalism and the business model in education: Reflections of a community college historian. *The History Teacher, 33*(1), 55-66.

Zook, J. (1995, May 26). The battle over subsidies: Federal payment on student-loan interest in jeopardy as Congress sets budget. *The Chronicle of Higher Education*, p. A23.

Part IV

CHALLENGES TO GROWTH AND THE ENVIRONMENT

The final section of this text deals with the challenges facing Californians with regard to continued growth and the effects of development on the environment, issues regarding nativistic responses to immigration, and the effects of and responses to natural disasters.

Chapter 10, "Population Growth and Environmental Degradation," by James A. Glynn and Charles F. Hohm, commences with a brief history of the Golden State. The authors take us through four waves of immigration and growth, beginning in the mid-16th century when Juan Rodriguez sailed into San Diego Bay and the state was populated by about 300,000 native people. The first wave was marked by a slow but steady period of colonization, mostly by Spain. The second wave began with the discovery of gold in 1848, the same year that California was ceded by Mexico to the United States, ending a short war between the two countries.

In 1848, only about 15,000 people of European descent lived in California, and the number of Native Americans had been reduced to about 100,000. But, as gold prospectors flocked to the state, the European population soared to 93,000 by 1850. Over the next several decades, the "American" population grew at a very high rate, whereas the native population dwindled. The two big industries of mining and railroad building encouraged thousands of Chinese to seek their fortunes in the new land, as well.

The "pull factors" of new technology spurred the next two periods of growth. The third wave coincided with the introduction of the motion picture industry in Southern California in the early part of the 20th century. After a brief respite from immigration during the Great Depression, California rebounded because of the expansion of the aerospace industry and the development of the transistor, a device that made electronic products more dependable. By 1963, California had become the most populous state in the union, surpassing New York.

The authors point out that many of today's problems in California are due to the distribution of the state's population and the effect that development has had on the environment. The ubiquity of the automobile helped to create two great megalopolises: one around San Francisco Bay, the other stretching outward from the Los Angeles basin. They believe that such urban sprawl may actually lead to a sense of anti-community, a condition that may threaten life rather than enhance it.

Photochemical smog, largely produced by vehicles with internal combustion engines, causes damage to crops and animal life. California's great Central Valley also must deal with the problem of airborne particulate matter. Although large cities and other parts of the valley get an F for ozone concentrations, a great deal of other pollution is produced in the Mojave Desert. Along with the various problems with air quality, California has experienced constant "water wars," with some private companies now attempting to store fresh water in aquifers beneath the Earth's surface for future sale. Water marketing by certain cities and water exchange endanger the state's biodiversity and cropland. The use of pesticides results in bioamplification, increasing the potency of the poisons as they travel up the food chain.

Comparing California's population to the rest of the United States, the authors emphasize the tremendous diversity of the people of California, from Laotian Hmong to Basque shepherds to professionally trained Koreans and East Indians. The vast array of people, speaking more than 100 different languages, has had a negative impact on bilingual education in the state. And this problem is exacerbated by California's continued growth by natural increase, legal immigration, and illegal immigration.

Looking at the state from a global perspective, the authors show that California has one of the largest economies in the world, and the people enjoy a high per capita income. Yet, there is much poverty as manufacturing jobs are lost to Mexican maquiladoras and other offshore factories. At the same time, there is a shortage of highly skilled workers, particularly in the high-tech corridor of Silicon Valley. But the North American Free Trade Agreement (NAFTA) could have a positive effect on the state if Mexico's economy continues to improve, providing Mexican nationals with more discretionary income for imported purchases.

Many non-Hispanic Californians fear *reconquista,* the conversion of Southern California into a Latino subcontinent because of the phenomenal growth of the Hispanic population. Others are troubled by the transformation from secondary industry (manufacturing) to tertiary industry (services). Environmentalists worry about deforestation and other environmentally damaging practices. They are joined by conflict theorists, who charge that business rapes the land for profit.

Looking to the future, the authors see the continuing decline of the white, non-Hispanic population and corresponding increases among Hispanic and Asian peoples. Meanwhile, the state as a whole will grow to more than 40 million people by

2015 and perhaps even beyond 50 million as early as 2020. However, the state will not grow uniformly, and several figures and tables are presented to show the projections. The authors draw four conclusions: (a) The white population will decline as a percentage of the state's total population; (b) the black population will decline, but only slightly; (c) the Asian population will increase significantly; and (4) the Hispanic population will increase most dramatically.

A four-fold solution is suggested for "smart growth," as California continues to burgeon. First, a state land use plan must be devised to prevent further urban sprawl and to protect cropland. This plan must involve a method of equalizing the cost of housing, or California will develop into a two-tiered society, with rich and poor increasingly separated. Second, a state migration plan is needed to redistribute population. Regional think tanks, such as Carol Whiteside's Great Valley Center, could examine the issues and propose solutions. Third, regional planning can prevent the merger of the two megalopolises and put an end to the creation of bedroom communities, where people can live inexpensively but tie up freeways coming and going during commuting hours. Fourth, a state employment plan must be tied to greater funding for higher education so that California can produce the well-trained experts that it will require as it moves into a future of innovation and continuing change.

Chapter 11, "Propositions 187 and 227: A Nativist Response to Mexicans," by Adalberto Aguirre, Jr., focuses on how an immigrant minority has been relegated to a peripheral position in the culture of California. Specifically, the state's dominant group has used the initiative process to pass two pieces of legislation that have had a negative impact on immigrants and Americans of Mexican origin.

Proposition 187, passed in 1994, bars illegal immigrants from benefitting from public education, nonemergency health services, and cash assistance from state agencies. Furthermore, it requires service providers to report suspected aliens to a state agency, and it makes it a felony to use false documents to try to substantiate citizenship. Proposition 227, passed in 1998, requires all school students to be taught in English only.

Aguirre points out that the United States has had a history of nativistic policies, which discriminated against German immigrants in the 19th century, Japanese in the early 20th century, and Chinese until 1943, when China and the United States were allied during World War II. During the 1940s, Mexican-origin youths became the victims of the Zoot Suit riots in Southern California, as the nativistic movement became a mechanism of social control used against California's nonwhite population.

Mexican Americans, in particular, have become socially distant from the general population because nativistic sentiment has emphasized their foreignness. These attitudes were bolstered by tightened controls over the southern border and the repatriation policies during the Great Depression. In turn, certain assumptions have been fostered: (a) Mexican immigrants come to California for a "free ride," (b) immigrants take jobs away from the native population, and (c) Mexican immigrants are not interested in becoming permanent citizens.

Attitudes like these have increased as California's foreign-born population increased from one sixth in 1980 to one fourth in 1990. California has more foreign-born residents than any other state. In addition, the state draws more than its equal share of refugees and those seeking asylum from various countries. But, in recent years, Asia and Mexico have been the primary countries of origin for immigrants arriving in the Golden State, where jobs in manufacturing and low-paid services are strong pull factors. Consequently, higher wages are paid to native-born workers than to those who immigrate.

Because employers associate immigrants' unfamiliarity with English as a sign that they are not dedicated to obtaining U.S. citizenship, many immigrants are relegated to the section of the labor market that has limited opportunities for upward social mobility. Consequently, high percentages of Mexican immigrants are enrolled in English as a Second Language (ESL) classes at all age levels. Fear that English-speaking people of Mexican origin might replace English-speaking natives, Aguirre speculates, fueled the drive behind Proposition 227, which stigmatized immigrant children in educational facilities. He also feels that Proposition 209 (California Civil Rights Initiative) was based on the belief that white workers were losing their jobs to minorities, women, and immigrants.

The author says that the existence of bilingual education in public schools reinforced the idea that schools were immigrant campgrounds and that eventually bilingual speakers would have an advantage over monolingual job applicants. So Propositions 187 and 227 were endorsed by California voters to control the entry and participation of immigrants.

Aguirre's conclusion is that initiatives like Proposition 187 and 227 place all people of Mexican origin at risk in California by transforming state agencies into tools of a police state. It is ironic, he believes, that people of Mexican origin are becoming more separated from society at the same time that California is becoming more racially and ethnically diverse.

The concluding chapter, Chapter 12, "Disasters," by Harvey E. Rich and Loretta Winters, deals with a topic the authors claim is generally neglected in social problems textbooks. Many factors associated with natural disasters, they contend, are really controlled by the products of social consensus: building codes, funding for disaster relief, and so on. They point out that neither of the presidential candidates in 2000 proposed any programs for disaster preparation or mitigation.

The authors say that, in this chapter, the term *disaster* will apply only to sudden events of a catastrophic nature that affect a major portion of a human community or society and overwhelm the capacity of the community to quickly absorb and recover from the loss. Consequently, occurrences such as droughts or famines will not be considered. Moreover, no distinction is made between natural and technological causes of disasters because (a) there has been a shift from sacred to secular reasoning, (b) the newer approaches focus on social consequences, (c) responses to disaster are similar regardless of cause, and (d) it is increasingly difficult to distinguish between the two.

Rich and Winters say that there is now a general agreement that a disaster is not only a physical happening; it is a social event. Earthquakes, fires, and similar events have social consequences based on pre-, trans-, and post-impact activities. Although such disasters cannot always be prevented, their impacts can be vastly reduced depending on how people and agencies react to them.

The history of natural disasters is lacking in documentation because it was the middle of the 20th century before the federal government began to sponsor research into the topic. Although support declined in the early 1950s, the Disaster Research Center, now located at the University of Delaware, has conducted hundreds of field studies over the past couple of decades. Currently, many other organizations are studying the social effects of disasters.

During the past decade, California has suffered major disasters, from major earthquakes to urban riots. Brush fires, especially in the southern part of the state, have been especially devastating. The Golden State also suffers from air pollution, problems with water delivery, and the consequences of toxic waste sites. During torrential rains, mud slides also pose major problems. According to the authors, these tragedies will likely increase during the 21st century as population increases and the population becomes more densely concentrated.

Globally, many of California's problems are shared with countries that have similar social structures and population densities, such as France or Japan. Any of these countries can experience problems with power grids, transportation, and social disruption due to loss of expected conveniences. Less economically developed countries are more likely to have greater mortality connected to events such as floods and typhoons because of weaker building codes and the lack of adequate support systems.

In looking at explanations for reactions to disasters, the authors point out that consensus theorists note a general lack of conflict during a disaster. In fact, research shows that the behavior of people caught up in these circumstances is often goal directed and very positive for four reasons: (a) the disaster is external to the community, (b) immediate attention is required, (c) the problem is usually unambiguous, and (d) suffering is usually attributed to chance, not social status; thus, a leveling of social distinctions occurs.

The power-conflict paradigm suggests that less conflict may be present during a disaster because of the intervention of government agencies. Also, conflict is not absent, although it may occur less frequently. Such analysis also suggests that the poor and minorities are more often the victims of disaster than the rich and those in the dominant group because advantaged groups have more control over events in their lives and environments. There may also be a kind of institutional bias; for example, during Hurricane Hugo, European Americans received more tangible information and emotional support than did African Americans. Finally, there are cultural differences: Some people assume that disasters can be overcome; others believe that the suffering must simply be endured.

Symbolic interactionists assume that events are disasters only if they are defined as such, and disasters do not occur in a social vacuum. Consequently, they focus on risk perception, communication channels, and interactions among helping agencies. Their conclusions suggest that maintaining good communication with relief and prevention agencies is essential in averting the worst results of disasters.

Rich and Winters say that an integrated, multiorganizational orientation toward disaster management is needed, and a key component of this strategy involves communication to the public about acceptable personal behavior and expectable aid from relief agencies. Four factors must be considered: (a) hearing about the emergency, (b) understanding the warning, (c) perceiving the reality of the situation, and (d) seeing oneself as the target of the communication. Responses to such communication rely on (a) environmental cues, (b) social setting factors, (c) receiver social ties, (d) sociodemographic characteristics, (e) psychological traits of the receiver, and (f) selective perception, which may distort people's understanding of reality.

The public is extremely influenced by messages from the public media about impending disasters. Television provides both visual images and dramatic accounts of what is happening. Problems that can be caused by such coverage include (a) sound bites that neglect the subtleties of risk analysis and (b) lack of control over what is aired. Because television is basically an entertainment medium, it tends to concentrate on panic rather than on goal-directed activity. In fact, during disasters, most people do not abandon their occupational roles after assuring themselves that their families are safe.

During disasters, four types of organizations may become involved in rescue and relief attempts. Type 1 organizations are those established associations that have rigidly defined roles and training, such as police and fire departments. Type 2 organizations are expanding organizations characterized by a large group of volunteers who follow the lead of a small group of permanent members. Type 3 organizations are not organized but attempt to help in any way possible; for example, a restaurant that gives away free meals. Type 4 organizations consist of groups that emerge out of friendship or a sense of neighborhood.

Despite the existence of such organizations, there is no such thing as a risk-free society. Consequently, the authors propose (a) earlier and more consistent planning, (b) high priority for mitigation as well as preparedness, (c) better integration of disaster planning in community development, (d) emphasis on the social rather than technical aspects of disaster, and (e) consideration of the special needs of the poor and ethnic minorities. Observing these proposals will help to convince the public and elected officials that disasters are the product of humans grappling with events that are capable of social solutions.

CHAPTER 10

POPULATION GROWTH AND ENVIRONMENTAL DEGRADATION

James A. Glynn
Charles F. Hohm

More than 100 years ago, the British scholar Lord James Bryce (1838-1922) commented that California was "in many respects the most striking state in the whole Union, and [it] has more than any other the character of a great country, capable of standing alone in the world" (Turner & Vieg, 1964, p. 14). Bryce, known as the "unofficial interpreter of the United States to Great Britain," was impressed by the great variety of social and economic possibilities that he observed in the state: minerals, fertile soil, giant forests, and an ambitious citizenry—all of the elements necessary, at least in the late 19th century, for an independent nation-state to achieve greatness. As we shall see in this chapter, these factors led not only to rapid population growth but also to the degradation of the state's natural resources.

HISTORY OF POPULATION GROWTH AND ENVIRONMENTAL DEGRADATION

California has experienced distinct waves of **immigration** and growth that have made the state the largest in the nation. For the purposes of a short chapter like this,

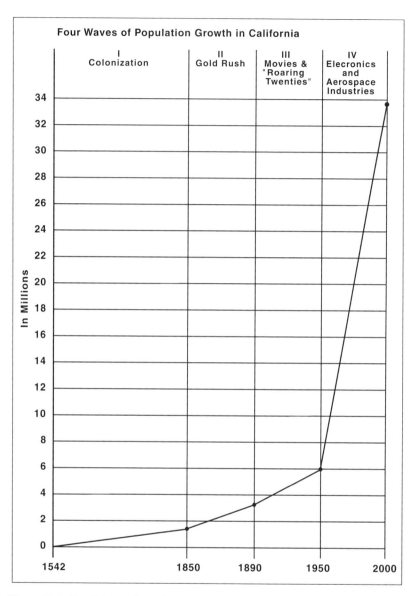

Figure 10.1. Four Waves of Population Growth in California

we will limit our discussion to four historic periods: (a) 16th century to mid-19th century, (b) mid-19th century to early 20th century, (c) early 20th century, and (d) post–World War II (see Figure 10.1).

The first wave of immigration started when Juan Rodriguez, a Portuguese explorer, "discovered" San Diego Bay in 1542. However, it was more than 100 years before a significant number of immigrants followed him. In 1769, the king of Spain ordered settlements to be established in San Diego and Monterey, pronounced Gaspar de Portola governor of the new territory, and granted Father Junipero Serra a license to establish a series of missions. By 1834, 21 missions stretched across California from San Diego to San Francisco, each one day's journey from the other. **Presidios,** or military outposts, were established to provide protection for the little **pueblos,** or towns, that grew up to accommodate the small but growing population (Turner & Vieg, 1964, pp. 2-4).

Gradually, other outsiders established a foothold in the geographic region that was previously occupied by a small number of Native Americans. It has been estimated that 300,000 Native Americans lived in California before the Spanish arrived (Holland, 2000, p. 124). The Hupa inhabited the north; the Maidu, the central section; and the Yuma, the south. Many other tribes occupied specific regions: for example, Pomo in the Sonoma-Mendocino area and Yokuts in the southern part of the Central Valley.

At one point, Sir Francis Drake claimed most of California for England (1579), calling it New Albion. After Mexico won its independence from Spain (1821), it regarded California as part of Mexico. Russians established a trading post at Fort Ross (1812) on a major waterway that is now known as the Russian River. Still, by 1848, there were fewer than 15,000 residents in California (Turner & Vieg, 1964, p. 5), not counting the Native Americans, whose numbers had been reduced to about 100,000 (Holland, 2000, p. 124). That same year was notable for two events that would bring radical changes in a very short time: Gold was discovered at Sutter's Mill, near Coloma, and Mexico ceded the land called California to the United States as part of the **Treaty of Guadalupe Hidalgo,** which officially ended the Mexican-American War. This was the end of the first period of the state's settlement.

The second wave of change was due mainly to internal **migration;** that is, people from other states moving to the region. In 1850, California became a state, and its population expanded to 93,000 non-Native American people, a six-fold increase in just 2 years. Over the next decade, another 287,000 people arrived, a 245% increase, pushing the total population to 380,000 for the 1860 census. Meanwhile, during the 10 years of American rule, 70,000 Native Americans disappeared due to war and European diseases for which the indigenous population had no immunity, leaving a scant 30,000, about 10% of the total who were here before the coming of the Europeans (Holland, 2000, p. 124).

In the 1860s, many Chinese nationals came to California as **sojourners,** single men who intended to make their fortune and then return to China. These men worked in two major industries: mining and railroad building. By the end of the decade, California's Big Four railroad owners—Charles Crocker, Mark Hopkins, Collis P. Huntington, and Leland Stanford, names quite prominent in modern California—had created a rail system that linked the West to the East Coast. During the next several decades, decennial growth ranged between 22% and 54%, so that by the arrival of the 20th century, the state's population stood at nearly 1.5 million (Figure 10.2).

The next two waves were spurred by a number of **pull factors,** conditions that provide a favorable atmosphere for immigration. Certainly, one such factor was the introduction of new technology. In 1907, California's first commercial motion picture, *The Count of Monte Cristo,* was produced in Los Angeles, initiating the third wave. Soon Los Angeles, and in particular the section known as Hollywood, became the movie mecca of the nation. In addition, actors, producers, technicians, and "wannabes" were attracted to this new industry. For millions of Americans, California represented hope as well as glitz. Between 1900 and 1910, the state grew by 60%; during the Roaring Twenties, it grew by an even greater 66% (as shown in Figure 10.2).

The fourth wave followed World War II, succeeded by a significant decline in growth during the **Great Depression** of the 1930s. Part of the reason for the resumed growth of the 1940s (53%) and the 1950s (49%) was the postwar prosperity, the growth of the defense and aerospace industries, and the introduction of more new technology, for example, the invention of the transistor. Before the invention of the transistor, the flow of electricity through radios, televisions, computers, and other electronic devices was controlled by vacuum tubes. These tubes were slow, they generated a great deal of heat, and they were cumbersome: bulky and unwieldy.

In 1947, a team that included William Shockley of Stanford University developed the first transistor, for which the scientists (Shockley, John Bardeen, and Walter H. Brittain) eventually won the Nobel Prize (Burstein & Martin, 1989). More and more applications were found for this new device, which not only replaced vacuum tubes but also increased the speed and reliability of electronic products and led to the miniaturization of circuitry. This new technology attracted new immigrants to the area between San Francisco and San Jose, later known as **Silicon Valley,** after the silicon base of a new generation of *computer chips,* the term used to describe fingernail-size components that now can accommodate millions of transistors.

The new electronics industry, along with a tremendous growth in such aerospace firms as Lockheed, Martin-Marietta, and the Jet Propulsion Laboratory, brought so many people to the state that, by 1963, California surpassed New York as the most populous state in the nation. Achieving this milestone caused some of the state's leaders to reconsider California's history of growth. In 1971, Speaker of the California Assembly Bob Moretti said, "We can no longer accept the proposition that all growth is good." He was concerned that continued growth would put greater stress on the state's ecological system. "What we need to seek," he continued, "is a balance between our resources of air, water, and open space and our population growth" (Bouvier, 1991, p. 3). These concerns will be the topic of our next section.

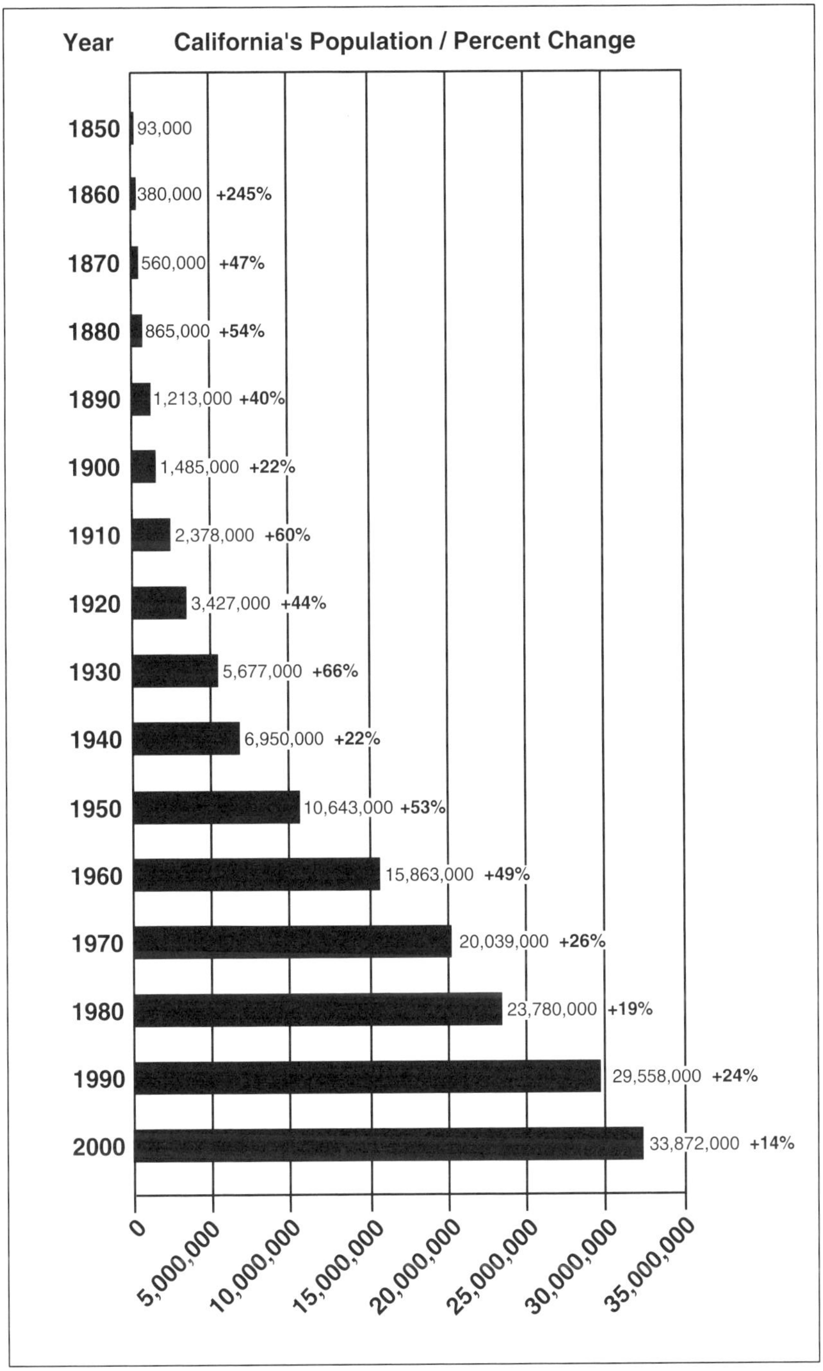

Population Growth and Environmental Degradation ■ 277

Year	California's Population	Percent Change
1850	93,000	
1860	380,000	+245%
1870	560,000	+47%
1880	865,000	+54%
1890	1,213,000	+40%
1900	1,485,000	+22%
1910	2,378,000	+60%
1920	3,427,000	+44%
1930	5,677,000	+66%
1940	6,950,000	+22%
1950	10,643,000	+53%
1960	15,863,000	+49%
1970	20,039,000	+26%
1980	23,780,000	+19%
1990	29,558,000	+24%
2000	33,872,000	+14%

Figure 10.2. California's Population Growth, 1850-2000
SOURCE: U.S. Bureau of the Census.
NOTE: Numbers are rounded to the nearest thousand, percentages to the nearest whole number.

THE PROBLEM IN CALIFORNIA TODAY

Today, California's problem with population seems to be twofold: the distribution of the population and the effect of growth on the environment. At first glance, it would seem that there are perhaps three distinct geographic areas of very dense population: San Francisco, Los Angeles, and San Diego. However, the dramatic effect of the automobile on California's development has caused the problem of **population density** to extend far beyond the political boundaries of these cities. The phenomenon that we have witnessed in the Golden State is the formation of at least two megalopolises. The term **megalopolis** was introduced in 1960 by Jean Gottman to describe the nearly continuous stretch of cities on the nation's northeast seaboard from the southern border of Vermont to central Virginia and from the Atlantic Ocean to the foothills of the Appalachian Mountains (Gottmann, 1961, p. 3).

In recent years, the Office of Management and Budget (part of the Executive Branch of the federal government) has designated such areas **Consolidated Metropolitan Statistical Areas (CMSAs)**, which are huge metropolitan areas (cities and their suburbs, as illustrated by **Metropolitan Statistics Areas [MSAs]**, below) that have merged together (*Intercom*, 1983, p. 3). However, when an MSA is contained within a larger CMSA, it is known as a **Primary Metropolitan Statistical Area (PMSA)**.

Figure 10.3 shows the CMSA that centers on the San Francisco Bay Area in the north and the CMSA that includes Los Angeles but extends into Riverside and San Bernardino Counties in the south. However, it is significant to notice that several MSAs border these areas of heavy population concentration. Only Camp Pendleton provides a buffer between the San Diego MSA and the Los Angeles megalopolis (Glynn, Hohm, & Stewart, 1996, p. 205). Likewise, only San Luis Obispo County (SLO, on the map in Figure 10.3) prevents the **urban sprawl** of the south from merging with that of the north. As this process has proceeded, huge tracts of citrus groves, a wide variety of orchards, and thousands of areas of prime agricultural land have been bulldozed and paved over to accommodate housing for the growing population. As this phenomenon continues to occur, some experts believe that the conditions create a kind of **anti-community**, contributing not to the enhancement of life, but, perhaps, to its demise (Stewart & Glynn, 1985, p. 291).

As mentioned, reliance on the automobile has had a profound effect on the pattern of growth in California and, not incidentally, on some of its worst environmental problems. For example, California leads the world in per person automobile ownership, with estimates ranging as high as 25 million cars registered (Fay, 1991, p. 393; see also Bonfante, 1991, p. 44). The millions of miles covered in daily trips by cars, gas-guzzling sport utility vehicles (SUVs), and light trucks lead to **photochemical smog**, which endangers all life in the state.

Population Growth and Environmental Degradation ■ 279

Figure 10.3. Consolidated Metropolitan Statistical Areas, Primary Metropolitan Statistical Areas, and Metropolitan Statistical Areas in California

NOTE: San Luis Obispo County is the only nonmetropolitan coastal region south of Santa Rosa.

Andrew Goudie (1994) states that photochemical smog "occurs particularly where there is large-scale combustion of petroleum products, as in car-dominated cities like Los Angeles" (p. 332). But the deterioration in **air quality** is not confined to the Los Angeles area. Tom Knudson (1995, p. 19) points out that the southern basin's air pollution is spilling over to the San Joaquin Valley, California's gigantic Central Valley and its most productive agricultural area, where it causes crop damage. There are serious repercussions from this occurrence because the San Joaquin Valley supplies nearly half of the meat, poultry, citrus fruits, and winter vegetables for the rest of the United States (Evans, 1994, p. 24).

Aside from the importation of pollution from the western coastal areas, California's Central Valley has numerous problems of its own. Along with air polluted by the process of photochemical smog, it must also deal with **airborne particulate matter.** In a national survey of deaths caused by fine particle pollution, it was found that the five areas with the most particle pollution deaths are all in California. In order, they are Visalia-Tulare-Porterville (in the Central Valley), with 123 deaths per 100,000 population; Riverside-San Bernardino Counties, 122; Bakersfield (Central Valley), 115; Fresno (Central Valley), 95; and Stockton (Central Valley), 93. In the 239 American cities that were studied, about 64,000 people died prematurely each year from heart and lung ailments caused by the particulate matter ("Outlook," 1996, p. 15).

More air pollution floats to the east of Los Angeles. Ponderosa pines in the San Bernardino Mountains 60 miles away have been extensively damaged by Los Angeles smog, which may reach concentrations of ozone as high as 70 pphm (parts per hundred million). Air is judged to be clean only if its **ozone concentration** is less than 4 pphm (Goudie, 1994, p. 78). According to the "State of the Air 2000" report, 33 of California's 58 counties received an F for their ozone levels. There were 244 unhealthy ozone days in Kern County, 222 in Fresno County, and 189 in Tulare County (Grossi, 2000, pp. A1, A8). All of these counties are in the state's great Central Valley, and each is a major agricultural area.

Interestingly, some of the state's worst polluters are located neither in the major metropolitan areas nor in the agricultural belt that runs down the center of the state. Road dust, for example, is a major cause of airborne particulate matter, and it is more common in rural, nonagricultural regions. In addition, nine of the top polluters in San Bernardino County are located in the Mojave Desert. Figure 10.4 shows the ranking by the Clean Air Trust and the type of activity in which they are involved.

Air quality, although it is a major concern of Californians, is not the only environmental problem. **Water pollution** is also a vital concern. Lake Tahoe in California and Crater Lake in Oregon have been documented as the clearest lakes in the world. According to Tom Stiensta (1995), in the 1960s, both lakes "had about 100 feet of deep blue clarity" (p. A3). Sadly, that has been changing for Lake Tahoe. Boating and other uses by people, logging of forests, and by-products of automotive

Polluter	Activity
Fort Irwin	U.S. Army Training Center
Union Oil Molycorp Inc.	Makes metal ore
Mitsubishi Cement, Lucerne Valley	Makes cement
Riverside Cement Co., Oro Grande	Makes cement
Southern Calif. Gas Co.	Pumps natural gas
North American Chemical	Makes industrial inorganic chemicals
PG&E	Pumps natural gas
US Borax	Makes potash, soda, and borate minerals
S.W. Portland Cement	Makes cement

Figure 10.4. Top Polluters of the Mojave Desert
SOURCE: "7 of 10 Top Polluters," 1997.

combustion have resulted in ever increasing levels of algae in the water. In addition, **acid rain,** another environmental problem largely attributed to automobile exhaust, has deposited high levels of nitrogen in the lake. Phosphorus enters the water as runoff from livestock grazing, golf courses, and fertilizers. When the two elements (nitrogen and phosphorus) mix, algae are produced. Consequently, in Lake Tahoe, algae are growing at the rate of 5% to 7% a year, resulting in decreased clarity (Stiensta, 1995, p. A5).

Water availability is another major factor. Water is essential for manufacturing, residential development, and farming. Its availability has been a center of controversy from the state's earliest days. Both Los Angeles and San Francisco have been criticized for "stealing" water from neighboring territories. San Francisco meets its need for potable water by damming the Tuolomne River and running huge pipelines to the Hetch Hetchy Canyon in Yosemite National Park. Los Angeles drained Lake Owens, turning the Owens Valley into a dust bowl (Knudson, 1995). In the late 1990s, efforts were made to "reclaim" Lake Owens, but the project has been slow to start and is not expected to produce significant results for many years.

Today, there are networks of pipes and pumps throughout the state, diverting water to the Central Valley to be used for agricultural production. Massive channels and aqueducts transport water from the northern part of the state, where population is sparse, to the southern region, where densely packed populations reside. Although northerners often complain about these practices, they are not entirely without blame. Water is a commodity, and it can be sold. Consequently, Yuba City (north of Sacramento) was selling some of its allotment from the New Bullards Bar Reservoir to Southern California in the early 1990s (Linden, 1991). More recently, a privately owned company has been trying to convince the Board of Supervisors of Madera County in the Central Valley to allow it to store more than 400,000 acre-feet of water in the underground **aquifers** beneath Madera Ranch. Farmers, irrigation districts, and other citizens have opposed the idea because they fear that, in time of drought, the water will be sold to the highest bidder (Arcamonte, 2000, pp. A1-A2; Davis, 2000, pp. A1, A16).

Environmentalists are critical of such **water marketing** for two reasons: It can upset an **ecosystem,** and it can deplete the state's **biodiversity.** One way that the shifting of water can affect an ecosystem has to do with **water exchange.** For example, draining the Sacramento River delta siphons water from the San Francisco Bay. The mixed fresh water and salt water is then virtually useless. This, in turn, can have a negative effect on biodiversity, the maintenance of various species within an environment. For example, without a source of fresh water, the agricultural base of the Central Valley would perish. The importance of that possible outcome can only be understood in terms of the effect this would have on the entire nation. The 21 counties that make up the Central Valley produce 250 different crops on 6.7 million acres of irrigated **cropland** (American Farmland Trust, 1995, p. 1).

In addition to endangered cropland, California also has 283 endangered, rare, or threatened species. Two thirds of these continue to decline, despite various state and federal programs to protect them (Linden, 1991). Often, the desires of land developers, commercial entrepreneurs, or even tourists are satisfied at the expense of birds, fish, or other animals. For example, when tourists were disturbed by gnats at Clear Lake, an insecticide called DDD was applied to the lake's surface. The insecticide killed 99% of the larvae that produced the gnats. However, some survived, passing the DDD on to fish. In turn, diving birds, called grebes, ate the fish. **Bioamplification** occurs when a substance like DDD is passed up the food chain, meaning that the dosage becomes more and more fatal. Eventually, dead grebes were found that had 32,000 times the concentration of the pesticide that was in the original application (Goudie, 1994, p. 99).

Although no one may mourn the death of a single variety of bird, it must be remembered that hundreds, possibly thousands of other incidents could be reported. As marshlands are converted to farmlands or housing tracts, many other varieties of fish, birds, and animals are destroyed. Although the creation of federally built dams supplies more than a third of the surface water consumed by irrigated agriculture, "many of them are implicated in species destruction," according to Sandra Postel (1999, p. 121). For example, the Red Bluff Diversion Dam on the Sacramento River blocks the mating route for salmon (Linden, 1991, p. 86). Large development projects also take a toll. The construction of a new campus for the University of California in Merced may destroy the faerie shrimp (McCarthy, 2000, p. A2). As the "Status and Trends of Our Nation's Biological Resources" report states,

> As water supplies were acquired by large development interests with the political and financial ability to move water to the semi-deserts of Southern Cali-

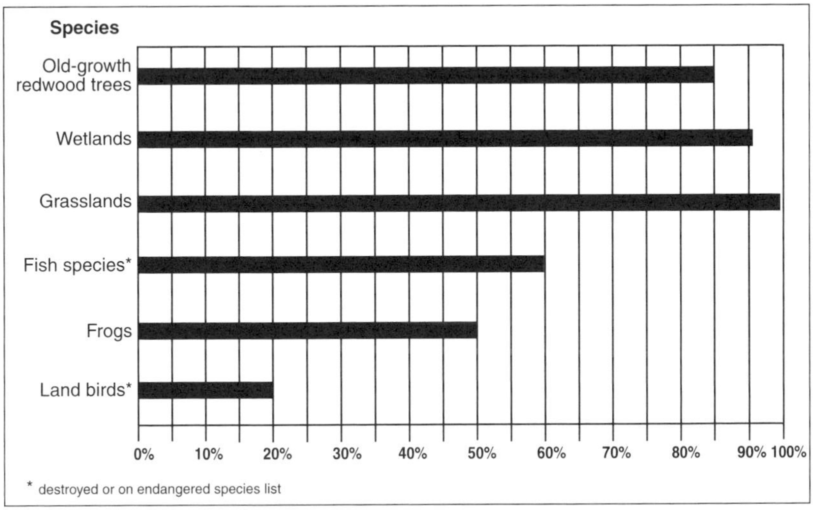

Figure 10.5. Extent of Environmental Destruction in California
SOURCE: Adapted from *Status and Trends of Our Nation's Biological Resources,* cited in Doyle, 1999.

Species (approximate percentages destroyed or on endangered species list):
- Old-growth redwood trees: ~85%
- Wetlands: ~90%
- Grasslands: ~99%
- Fish species*: ~60%
- Frogs: ~50%
- Land birds*: ~20%

* destroyed or on endangered species list

fornia, the growth of cities and agriculture greatly accelerated, resulting in the loss of the incredible richness of the Central Valley. (Doyle, 1999, p. A15)

Even the state's symbol, the grizzly bear, disappeared when the last of the species was shot in 1924 (Knudson, 1995, p. 19). So, California is a state whose symbol, which appears on its flag and many official seals, was a victim of population growth and lack of care for the environment. Figure 10.5 shows the extent of the total devastation.

CALIFORNIA'S POPULATION COMPARED TO THE UNITED STATES

Probably nothing distinguishes California from most of the other states more than its diverse population. Mention has already been made to the Chinese who came to supply labor for the mines and railroads. Japanese came to farm, and they were highly successful until World War II, when residents who were at least one-eighth Japanese were interned in relocation camps. Nearly all of their land was confiscated for nonpayment of taxes or mortgages, or it was lost to unscrupulous merchants, traders, and bankers (McLemore, Romo, & Baker, 2001, p. 189). Sikhs from India arrived in the Sacramento River delta around the beginning of the 20th century. By 1930, there were more Filipinos in Stockton than in any other population center outside the Philippines (Williams, 1993, p. 4). A substantial Basque population is centered in Bakersfield. Basques, who originated in the Pyrenees Mountains between Spain and France, came to tend flocks of sheep that roamed over the landscape of the southern Central Valley. Armenians settled about 100 miles north in Fresno. And Portuguese fishermen found ample supplies of marine life off California's coast.

Hmong people from the mountains of Laos now reside throughout the Central Valley, although they originally settled around Merced. And, in recent years, thousands of them have left the area for factory jobs in the Minneapolis/St. Paul area of Minnesota. Yemeni workers annually leave their farms in the southern portion of the Saudi peninsula in the Middle East to live and labor in migrant camps around Delano. There is a Koreatown in Los Angeles County and a **Little Saigon** (Vietnamese) in Orange County. Professionals from the Middle East and the Asian subcontinent, particularly in the medical field, have filled a great need throughout the Central Valley and in the more rural parts of the state. The same can be said for registered nurses from the Philippines, where the training schools were organized by American doctors.

Schools in Los Angeles and other urban centers must teach children who, collectively, speak literally hundreds of languages, although severe restrictions on **bilingual education** were put into place in the late 1990s by a majority of California voters. But bilingual education originated in California as a recommendation that materialized from a U.S. Supreme Court decision (*Lau v. Nichols*) concerning Chinese school children in San Francisco. That solution then spread to schools with significant populations of Spanish-speaking students. But, with today's matrix of people from all over the world, Californians seem reluctant to use school dollars for multilingual, multiethnic programs.

Between 1990 and 2000, the white population of California declined from 57% to about 47%, whereas its Hispanic population increased from 26% to more than 32% (Free Demographics, 2001). The Kiplinger Editors Group has projected that by 2005, a third of all Californians will be Hispanic (*California Forecasts,* 1996, p. 4). The Asian population of California increased from 9% to 11%, whereas black and Native American populations remained unchanged, at 7% and 1%, respectively (State of California, 2000) (see Figure 10.6). The Kiplinger Editors believe that the Asian population will grow to 16% by 2015 (*The Kiplinger California Newsletter,* 2000a, p. 1)

A little more than half (53%) of the Hispanic growth in California was due to **natural increase** (the difference between births and deaths), but 270,200 Hispanics (47% of the growth) migrated to California during those same years. On the other

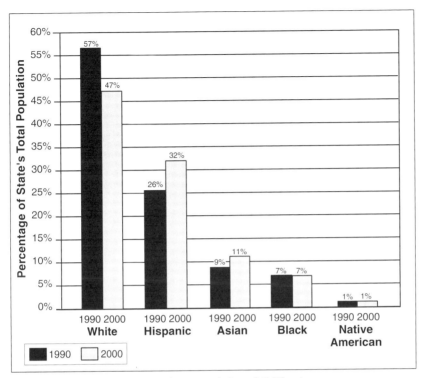

Figure 10.6. Change in Racial/Ethnic Population of California
SOURCE: California State Department of Finance, 2001.

hand, most of the growth of the Asian population (which includes Pacific Islanders) was the result of immigration (62%). Interestingly, during the same time period, California experienced a **net migration** loss that averaged 38,100 white people per year (State of California, 2000, p. 3).

California also absorbs an inordinate number of **illegal immigrants,** also known as undocumented workers. In 1995, the state's Department of Finance estimated that 125,000 people enter California illegally each year (McNeill, 1995). Such estimates are not based on hard data and must be viewed with a degree of suspicion. It is not very likely that any agency knows with any amount of certainty just how many unregistered aliens reside in California at any point in time. However, as shown in Table 10.1, the Urban Institute believes that more than 1.5 million illegal residents were in California in 1997 (Miller, 1997, p. 52).

These demographic dynamics create a California mosaic that is unmatched, not just in the United States, but probably anywhere in the world. One sociological consequence of an extremely diverse population is that it is often difficult for the public to reach a consensus. A variety of cultures means a variety of values. Only agreement

TABLE 10.1 Illegals and Their Children (Estimated Undocumented Immigrant and Student Populations, for States With Largest Numbers of Undocumented Immigrants, 1994)

	Total Undocumented Population	Undocumented as Percentage of Total Population	Undocumented Children Enrolled in Public K to 12	Undocumented as Percentage of Total K to 12 Enrollment
California	1,566,000	5.0	307,000	5.8
New York	490,000	2.7	88,000	3.2
Texas	389,000	2.1	93,900	2.6
Florida	356,000	2.6	97,200	4.8
Illinois	188,000	1.6	16,300	1.4
New Jersey	125,000	1.6	16,300	1.4
Arizona	62,000	1.5	15,000	2.2
Total for seven states	3,176,000	3.0	641,000	

SOURCE: The Urban Institute, Washington, DC. Cited in Miller, 1997.
NOTE: An estimated 5% of California residents and 6% of its public school students are illegal residents of the United States.

among California's tens of millions of people can effect a change in the population's impact on the environment.

CALIFORNIA'S PROBLEM IN GLOBAL CONTEXT

Fulfilling the prophecy of Lord Bryce, if California were an independent country today, it would have the fifth-largest economy in the world, surpassed only by the United States, Japan, Germany, and Great Britain. If the state's growth continues, it could soon surpass Great Britain (Liedtke, 2001). By the turn of the century, the state's gross product was expected to exceed $1.2 trillion (*California Forecasts*, 1996, p. 5). One of the most recent boosts to the economy has come from the establishment of automobile design centers in the Golden State. Daimler-Chrysler designed the PT Cruiser at a studio in Carlsbad; the new VW Beetle originated in Simi Valley; Ford Thunderbird's new look came from a shop in Irvine; and General Motors is looking to California to provide the concept for its Chevrolet Avalanche (*The Kiplinger California Newsletter*, 2000d, p. 2).

At the beginning of the 1990s, California enjoyed the world's fourth-highest **per capita income,** behind Switzerland, Finland, and Sweden (Fay, 1991). The editors of *The Kiplinger California Newsletter* estimate that personal income will grow by 25% to 30% by 2005 (*California Forecasts,* 1996, p. 6). If this projection is correct, California will maintain its position and possibly improve somewhat, despite the fact that there will probably be no significant net increase in manufacturing jobs.

Although some high-tech industries will need more workers, these factory assembly positions will be offset by textile industries moving offshore (meaning that the work is done in other countries), the continuing growth of **maquiladoras** (U.S.-owned factories located south of the U.S.-Mexico border), and heavy reliance on imported goods from countries where labor is relatively cheap (Glynn et al., 1996, pp. 239-241). Despite big corporations' desire to find low-wage manufacturing sites, California still has 4 of the 12 fastest-growing computer chip assembly companies: HCC of Rosemead, and Stanford Microdevices, Silicon Storage Technology, and Marvell Semiconductor, all located in Sunnyvale (*The Kiplinger California Newsletter,* 2000a, p. 2). In addition, the state will need to increase its present workforce in the field of semiconductors by 23%; telephones, by 12%; radio-TV equipment, by 10%; and printed circuit boards, by 9% (*The Kiplinger California Newsletter,* 2000c, p. 2).

High-tech corporations like these increasingly look to the world outside the United States to provide the engineers, technicians, and computer specialists that we lack in this country. In 2000, Congress doubled the number of **H-1B visas,** the work permits that are issued to an educated elite in specified fields. Although the number will rise from 115,000 to 200,000 a year, the fee will probably double, too, from $500 to $1,000 (*The Kiplinger California Newsletter,* 2000c, p. 2). But, desperate high-tech employers in Silicon Valley consider this a small price to pay to fill vacant positions.

Although the Golden State seems to need to import professional workers, it also exports billions of dollars worth of agricultural products. About two thirds of these farm products are shipped to countries that make up the **Pacific Rim** (*The Kiplinger California Newsletter,* 1999). Japan's recently improved economy has spurred recovery in other Asian nations. Figure 10.7 shows that, whereas U.S. farm exports dropped a bit from 1997 levels, they steadily grew again, so that the 2000 shipments reached former levels again. The **North American Free Trade Agreement (NAFTA)** will undoubtedly also have an effect on California's position in the global economy. If the Mexican economy improves, this could mean big increases in U.S. exports south of the border in such areas as electronics, software, machinery, financial services, and transportation equipment (*California Forecasts,* 1996, p. 8). Imports of labor-intensive articles such as clothing will likely increase as well. Whatever the outcome, there is little doubt that California is now part of a global system that includes demographics, the economy, and other superstructures.

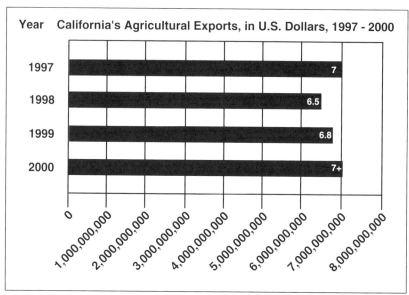

Figure 10.7. California Agricultural Exports, in Dollars, 1997 to 2000
SOURCE: Adapted from *The Kiplinger California Newsletter*, Vol. 35, No. 15, August 4, 1999.

ALTERNATE EXPLANATIONS FOR THE PROBLEM

There are numerous theories to explain the nature of California's development over the years. In this section, we'll examine only four: a **theory of ethnic antagonism**, a form of **sociocultural evolution theory**, conflict theory, and environmental racism.

ETHNIC ANTAGONISM

Today, many Californians fear the possibility of **reconquista**, the reconquest of the Southern California area by Hispanics and its conversion to a Latino subcontinent (Meyer, 1992, p. 32). Such histrionics have historical precedent in the Golden State. For example, Chinese accounted for only 9% of the state's population when California pressured Congress to pass the **Chinese Exclusion Act of 1882.** However, they made up more than 25% of the labor force (Parillo, 1994, p. 274). Because the Chinese, mostly male sojourners, worked for substandard wages, California experienced a **split labor market.**

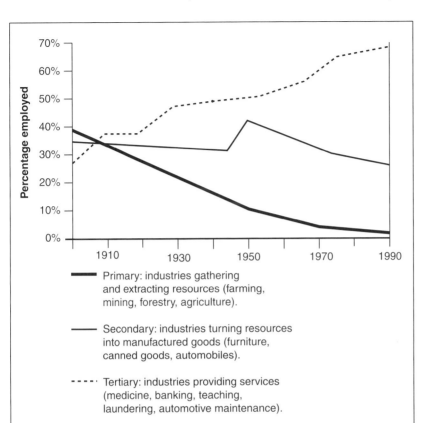

Figure 10.8. Economic Industries in the United States
SOURCE: Adapted from Glynn, Hohm, and Stewart (1996), p. 324.

According to Edna Bonacich (1972), a split labor market exists when two ethnically identifiable groups perform services for different remuneration. When this situation occurs, the higher paid group generally discriminates against the lower paid one, often leading to **exclusion** of the lower paid workers from the market. This theory explains the attitude toward both the Chinese in the late 19th century and illegal immigrants (most of whom are Hispanic) in the late 20th century.

SOCIOCULTURAL EVOLUTION

A form of sociocultural evolution theory holds that societies tend to transform from less complex to more complex forms. This can be applied to the type of economic industry that is predominant. Figure 10.8 shows that **primary economic industry** employed the greatest percentage of people in the nation a hundred years ago. **Secondary economic industry** peaked in the middle of the 20th century. **Tertiary economic industry** has steadily increased.

In the early days of statehood, California needed many people to work in the primary sector of the economy: miners, farmers, foresters, fish catchers. Those industries still exist, but a much smaller percentage of the population is needed to produce greater yields. Some mining is now done by exploding sections of mountains and then sorting the debris on conveyor belts. Agribusiness is buying up family farms. **Deforestation** is now accomplished by a controversial method called clear cutting. And fishing now involves both the use of huge drag nets and **aquiculture** (where the fish are commercially bred).

Although secondary industry has been declining nationwide since the 1950s, it grew slightly until the 1990s in California, mainly because of the expansion of the electronics and aerospace industries. However, the closure of military bases, the shutting down of automobile assembly factories, and the use of maquiladoras in Mexico has negatively affected secondary industry in California.

Yet, the loss of such jobs has been more than compensated for by the increased employment of people in the service sector. According to *California Forecasts* (1996), more than half of future growth in employment will be in personal services, business services, consulting and legal services, entertainment, and health care services. Another 20% will be in wholesale and retail sales. Similar gains will be made in finance, real estate, transportation, and public utilities. But, the report emphasizes, there will be no big gains in manufacturing jobs (p. 6).

CONFLICT THEORY

Conflict theorists would probably be joined by environmentalists in charging that business has raped the land for profit. The exploding of whole mountains and the clear-cutting of entire forests would not sit well with either group. Each would also point out that the manufacture of computer chips dumps millions of gallons of pollutants into the waters and onto the land every year. A relatively small percentage of the population benefits economically from these ecologically damaging activities.

As Dan Walters (1999) points out, as California enters into its third economic incarnation (tertiary industry), it tends to be a two-tiered society: one well educated and affluent, the other poorly educated and poverty stricken. For example, whereas the urbanized areas (San Francisco, Los Angeles, and San Diego) have experienced increases in employment, double-digit unemployment is still common in the Central Valley, an area that still relies on primary-industry workers (Torres & Torres, 1999).

A question of continuing concern to state leaders is how can the nation's largest state control the growth of its various economic enterprises when it seems certain that its population will grow substantially well into the 21st century.

ENVIRONMENTAL RACISM

In the preceding sections of this chapter, population growth was seen as a major culprit explaining environmental degradation. However, environmental degradation cannot be laid simply at the door of population growth. Environmental sociologists, many of whom take a Marxist view, see environmental problems stemming from unbridled capitalism (see, e.g., Goldblatt, 1996; O'Connor, 1984, 1987; Redclift & Benton, 1994). These scholars assert that the manner in which societies organize their economic activities is profoundly related to how much environmental degradation exists and what segments of the population are most likely to be adversely affected. For example, in California, much of the farming that is done in the Central Valley is conducted by huge agricultural conglomerates that use vast amounts of water, pesticides, and herbicides. These agricultural practices result in soil salinization and the runoff of pesticides and herbicides into streams and rivers as well as underground water supplies. Environmental racism exists when certain ethnic or racial groups are disproportionately affected in a negative way by the above. For example, most of the farm workers who work in the Central Valley of California are poor Mexican migrants. There is evidence that these people are exposed to pesticides and herbicides at such a high rate that their health is at risk. Studies have shown cancer clusters in many farm worker communities. So, although environmental pollution and degradation can negatively affect all Californians, minorities are much more likely to be affected.

FUTURE PROSPECTS

California has always had a love-hate relationship with population growth. In 1994, 60% of the state's voters, most of whom were white, cast ballots in favor of **Proposition 187,** which made illegal immigrants ineligible for public educational services, nonemergency health care, and other publicly financed social services. Undoubtedly, some of these voters feared the possibility of reconquista. It is probably also true that many voters saw the proposition as a way of limiting the future growth of the state, especially the growth of the Hispanic population. Figure 10.9 shows that nearly one third of the state's population today is Hispanic, and that percentage will continue to grow into the foreseeable future. Around 2040, roughly half of all people in the Golden State will be of Hispanic heritage.

Meanwhile, the white population will decline from its current 49.8% to less than one third over the next 40 years. During the same time, the Asian population will slowly advance to about 12%, and blacks will decrease slightly to about 6% (Fay, 1995). All the while, the state itself will continue to grow. The *California Almanac*

292 ■ CHALLENGES TO GROWTH AND THE ENVIRONMENT

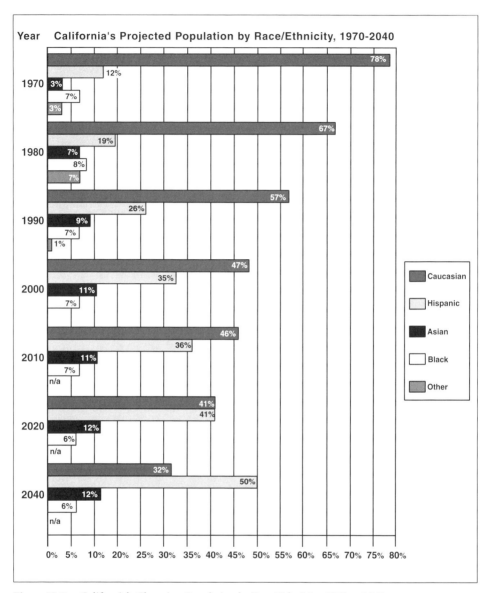

Figure 10.9. California's Changing Population by Race/Ethnicity, 1970 to 2040
SOURCE: Adapted from Fay, 1995.
NOTE: Data is projected for 2010-2040.

(Fay, 1995) estimates that the population will reach 38,385,460 in the year 2005, the editors of *The Kiplinger California Newsletter* (2000e, p. 4) believe that population will be 40.8 million by 2010, and the Population Reference Bureau (1999) believes that the Golden State will have slightly more than 41 million by 2015. Leon Bouvier, now retired but formerly vice president of the Population Reference Bureau, esti-

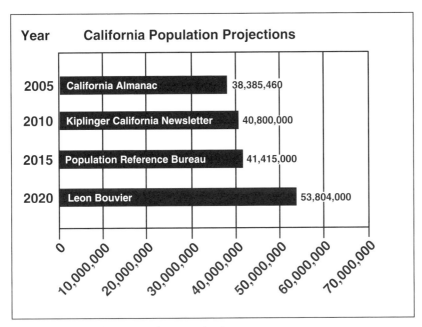

Figure 10.10. California Population Projections

mates that the state could reach an astounding 53,804,000 population as early as 2020 (Bouvier, 1991, p. 27). (The various estimates are summarized and compared in Figure 10.10.)

If any of the projections in Figure 10.10 are correct, California would be a postindustrial state within one of the most economically developed countries of the world that is growing like a less developed country. The accuracy of the projections is not as important as the fact that all of the sources consulted predict that California will continue to grow at a rapid pace. However, the state will not grow uniformly (see Figure 10.11).

The areas that will add the most people will be in the southern part of the state, the southern part of the Central Valley, and the eastern sector of the San Francisco Bay Area. Moderate growth will be observed in the Central Coast region, the northern part of the Central Valley, the San Francisco area, and the areas south of San Francisco and north of Marin County. The fewest people will be added to the northern counties and those along the east-central border. However, we should not confuse the addition of numbers of people with percentage change. Figures 10.12 and 10.13 demonstrate the difference. Figure 10.12 shows that some of California's largest cities have added significant numbers of people, but Figure 10.13 shows that the greatest percentage gain has been among some of the state's smallest cities.

Notice, also, that two of California's largest cities, Los Angeles and Long Beach, actually lost population in the mid-1990s (Figure 10.14). This trend is consistent

294 ■ CHALLENGES TO GROWTH AND THE ENVIRONMENT

Figure 10.11. Population Growth Trends in California Counties, 1995 to 2005
SOURCE: Adapted from *The Kiplinger California Newsletter*, 1995.

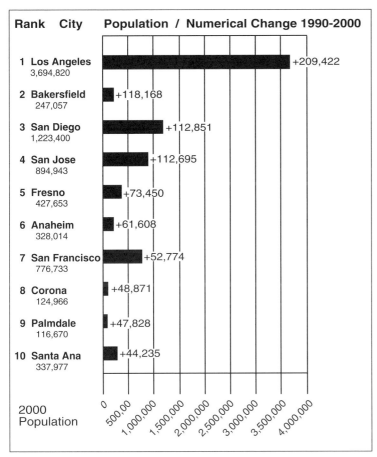

Figure 10.12. 10 Fastest-Growing California Cities Based on Numerical Change
SOURCE: California Department of Finance, 2001.

with other data from around the United States and elsewhere in the world. For example, a great number of highly industrialized countries in Europe will actually grow smaller in population during the next decade, including Bulgaria, Latvia, Ukraine, Russia, Belarus, Hungary, Estonia, Romania, Germany, Italy, Austria, and Slovakia (Kent, 1999).

Probably the best conclusions that can be drawn from these data are (a) the white population of the state will decline as a percentage of California's total population, eventually becoming a numeric minority; (b) the Hispanic population will increase quite dramatically, eventually becoming a numeric majority; (c) the Asian population will increase significantly, perhaps constituting about one eighth of the total; and (d) the black population will decline, but only slightly. In addition, California's largest cities will not increase very much, if at all. But our smaller cities will burgeon, perhaps creating more urban sprawl if they are located near already large metropol-

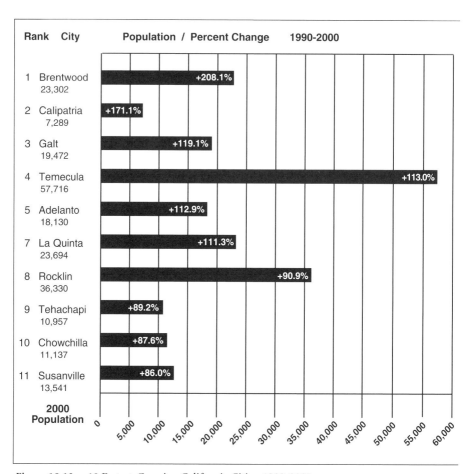

Figure 10.13. 10 Fastest-Growing California Cities, 1990-2000
SOURCE: California State Department of Finance, 2001.

itan areas. Facing projections such as these, the citizens of the Golden State need to come up with very creative solutions.

SOLUTIONS TO THE PROBLEMS

Because problems of environmental degradation are intricately linked to the rapid growth of California's population, it is impossible to consider the solutions of one issue without the other. Growth is a certainty, but what Californians need to consider is **smart growth.** The state can continue to expand its population and protect its environment if people and their leaders act intelligently. We suggest four action plans: (a) state land use, (b) migration, (c) **regional planning,** and (d) employment.

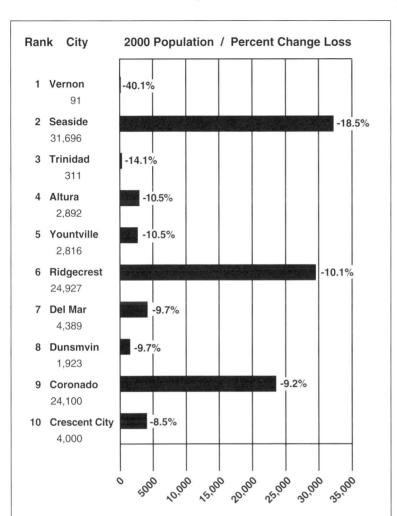

Figure 10.14. Cities With the Largest Percentile Loss in Population in California, 1990-2000.
SOURCE: California State Department of Finance, 2001.

STATE LAND USE PLAN

One of the most significant issues facing the state is the possibility that urban sprawl will encroach on prime agricultural land. Over the next 40 years, the population in areas that hold some of California's most productive croplands will grow at frightening levels. In the great Central Valley, for example, Kern, Madera, Merced, Stanislaus, and Tulare Counties will grow by more than 300%. All of the other counties that make up that agricultural sector of the state will grow between 200% and 300% (American Farmland Trust, 1995). Part of the reason for this population expansion is tremendous differentials in the cost of housing. In large urban areas, the cost of an average house has increased to the point where only the richest residents can consider the possibility of purchase. For example, in Palo Alto (located in the electronic mecca called Silicon Valley), a small, one-bedroom house was listed in 2000 for $495,000. However, prospective buyers eventually bid the price up to $750,000. The median price of a detached home was a whopping $847,000 (McAllister, 2000). Even that pales by contrast to Hillsborough, where the median-priced house costs $1,782,500; to Los Altos Hills, $2,286,000; and to Atherton, $3,175,000 (*The Kiplinger California Newsletter*, 2000b, p. 2). By contrast, the median price of a three-bedroom, two-bath home in Modesto (in California's Central Valley) was $125,000 (Doo, 1999).

One possible solution would be to rely on the State Legislature to enact some kind of regulation to equalize the cost of housing throughout the state. This, of course, causes major philosophical problems in a country and a state that put a high value on free enterprise and a market-controlled economy. The problem is complicated by the existence of a **two-tiered system of stratification** in the state, where the poor and the rich are separated by greater margins as the years go by.

Of course, the availability or lack of water will play a major role in any land use policy. Various **water wars** throughout the state have pitted farmers against real estate developers, rural areas against urban centers, and environmentalists against most pro-growth movements. But, it seems that the population/environment problem will not be solved unless various elements within the state can devise some sort of smart growth plan.

STATE MIGRATION PLAN

In the early 1990s, California experienced a net loss in migration, meaning that more people moved out of the state than moved into it. But according to the Center for the Continuing Study of the California Economy (cited in Dortch, 1995), that trend is reversing itself because of a strengthening economy. The center projects that net migration will be positive until 2005, and possibly until 2010, at perhaps 172,000 per year (Dortch, 1995, p. 4). However, there may be a countervailing trend.

California's elderly population, described as those who are 65 years of age or older, is growing rapidly. In 1993, this group accounted for 3.3 million of the state's population, but by 2020, that number will double to 6.6 million ("Elderly Residents," 1996). As more people move onto fixed incomes, they may seek a place of residence where taxes are lower, the price of housing is more moderate, and the overall cost of living is more affordable, in states like Louisiana or Mississippi. For example, between 1990 and 1998, California lost 1,624,284 people, but Florida gained 853,933; both Georgia and North Carolina each had a net migration in excess of 400,000 (Westphal, 1999). This phenomenon is fueled by the fact that retirees can sell their expensive houses in California's urban areas, take their equity, and live comfortably in less expensive parts of California or in other less expensive states.

Although some may choose to leave the state, others may simply move to more affordable, and less populated, areas within the state. For example, only 36.2% of people with a median income can afford to buy a house in San Diego, 27.5%, in San Jose; and only 11.9%, in San Francisco. However, 63.5% of median income earners can afford a house in Modesto; 63.7%, in Riverside-San Bernardino; and 73.3%, in Bakersfield (Nax, 2000). In fact, between 1990 and 1998, almost 24,000 people moved from Alameda County to less expensive cities in San Joaquin and Stanislaus Counties. Another 24,000 left Santa Clara County (the heart of Silicon Valley) to live in San Joaquin, Stanislaus, and Merced Counties (Sbranti, 1999, p. A1). Interestingly, this influx sent thousands of residents of those counties further inland and into the mountainous areas of Mariposa, Tuolumne, Calaveras, and Amador Counties (Sbranti, 1999, p. A20).

All of this migration, however, seems to be occurring without any planning. Historically, city councils and county boards of supervisors have resisted any attempt by state agencies to dictate policy. But, in recent years, there have been attempts to develop regional planning. One example is **Great Valley Center,** a think-tank type of organization presided over by Carol Whiteside, former mayor of Modesto. She sees the possibility of building consensus among the various counties that constitute the Central Valley (Glynn, 2000). Similar plans could be undertaken for the west San Francisco Bay Area, the Inland Empire (including Riverside and San Bernardino), and other areas that share commonalities.

REGIONAL PLANNING

Regional planning involves decision-making processes that are shared among city and county administrations. Working together, governments can overcome the burden that urban sprawl has placed on the people of the state and on California's environment. The building pattern that has dominated the California landscape for decades is the erection of single-family homes in the suburbs adjacent to cities and, increasingly, the building of commercial strip malls to accommodate the residents of the single-family homes. This has resulted in severe traffic problems, which are escalating faster than the growth of the population. Affordable housing in the far-flung suburbs has drawn families from the central cities and the original suburbs. However, most of the employment is in the cities, and many Californians spend several hours each day commuting in stop-and-go traffic.

Examples may be found all around the state's urban centers. Many families have moved from San Diego County to buy more affordable homes in Temecula, which is located in the southern part of neighboring Riverside County. Most of the folks who moved to Temecula still work in San Diego, and, as a result, Interstate 15 (which connects the two counties) has turned into a virtual parking lot during peak hours.

The same phenomenon can be found in the San Francisco Bay Area, where many families commute several hours a day between their jobs in Silicon Valley and the reasonably priced homes that they have purchased in Modesto, in the Central Valley. Still others are commuting between San Jose (which is expensive, but which supplies many jobs) and Hollister (which is less costly).

Smart growth calls for the preservation of open space in the hinterland and the placement of affordable housing and commercial buildings in a high-density, mixed format in downtown areas of cities. Such a configuration, with housing and commercial buildings next to one another, puts more people close to their jobs, alleviating the need for so many people to use the freeways to commute to and from work. Also, high density urban housing would encourage the growth of mass rapid transit systems, such as trains, light-rail cars, and buses.

STATE EMPLOYMENT PLAN

Employment is a pull factor for migration, one that encourages people to move into an area. Historically, the Golden State possessed many pull factors, as stated earlier in this chapter. However, during the 1980s, manufacturing jobs declined by 18% (Bonfante, 1991, p. 48). In the 1990s, at least 60,000 jobs were lost because of the closure of military bases (Evans, 1994, p. 22). On the other hand, opportunities in services are increasing dramatically. There is a never-ending need for teachers. Local control of school districts has created a kind of **academic headhunting,** with each district attempting to steal teachers away from other districts (Mehas, Garcia, & Buster, 1999). The only true solution is to train more students and attract them to the teaching profession.

The same can be said for hospitals. By 2010, California will need an additional 43,000 nurses. Several pieces of legislation have been introduced to increase the output of the state's 96 nursing programs. But, a lack of resources is a serious hindrance. In 1997, the California State University system had to deny admission to 44% of qualified candidates for nursing programs, and the same situation applied to at least 5,000 students at the community colleges ("Gov. Davis," 1999).

In addition, in the early 2000s, the movie industry, which spurred California's third wave of growth, may well provide a subsequent pull factor. It is in the midst of tremendous technological change, particularly in digitally produced special effects, which is drawing new people into the field. The Kiplinger Editors say, "Since California is the leader in creative talent as well as technology, it is certain to benefit from this revolution" (*California Forecasts,* 1996, p. 15). This statement seems to fit well with Lord Bryce's comment, made nearly a hundred years ago, "the real question will not be about making more wealth or having more people, but whether the people will then be happier or better than they have been hitherto or are at this moment" (Bouvier, 1991, p. 2). If the people of the state take advantage of these opportunities, the answer will be in the affirmative.

REFERENCES

American Farmland Trust. (1995). *Future urban growth in California's Central Valley: The bottom line for agriculture and taxpayers* [Summary report]. Washington, DC: Author.

Arcamonte, M. S. (2000, August 10). M(adera) I(rrigation) D(istrict) opposes Madera Ranch groundwater bank. *Madera Tribune,* pp. A1, A2.

Bonacich, E. (1972). A theory of ethnic antagonism: The split labor market. *American Sociological Review, 37,* 547-559.

Bonfante, J. (1991, November 18). The endangered dream. *Time,* pp. 42-49.

Bouvier, L. (1991). *Fifty million Californians?* Washington, DC: Federation for American Immigration Reform.

Burstein, J. S., & Martin, E. G. (1989). *Computer information systems.* Chicago: Dryden.

California Department of Finance. (2001). http://www.dof.ca.gov/html/demograp/Table1.xls

California Forecasts. (1996). Washington, DC.: The Kiplinger California Newsletter.

Davis, J. (2000, April 25). Madera farmers oppose water banking. *Fresno Bee,* pp. A1, A16.

Doo, J. (1999, October 7). Modesto's median home now $125,000. *Modesto Bee,* p. C1.

Dortch, S. (1995, November). California's next decade. *American Demographics,* pp. 4-6.

Doyle, M. (1999, September 17). State's natural features rubbed out, study finds. *Fresno Bee,* p. A15.

Dreazen, Y. J. (2000, December 31). Census shows gains in South and West. *The Wall Street Journal,* pp. A2, A18.

Elderly residents in California expected to double by 2020. (1996, May 21). *Bakersfield Californian,* p. A3.

Evans, R. (1994, February). Golden State's fall from grace. *Geographical,* pp. 22-25.

Fay, J. S. (Ed.). (1991). *California almanac* (6th ed.). Santa Barbara, CA: Pacific Data Resources.

Fay, J. S. (Ed.). (1995). *California almanac* (7th ed.). Santa Barbara, CA: Pacific Data Resources.

Free Demographics. (2001). http://www.freedemographics.com/AllocateOnline.d22

Glynn, J. A. (2000, April 3). Are you part of the problem or part of the solution? *Madera Tribune,* p. A4.

Glynn, J. A., Hohm, C. F., & Stewart, E. W. (1996). *Global social problems.* New York: Harper-Collins.

Goldblatt, D. (1996). *Social theory and the environment.* Boulder, Co: Westview.

Gottmann, J. (1961). *Megalopolis.* New York: Twentieth Century Fund.

Goudie, A. (1994). *The human impact on the natural environment.* Cambridge: MIT Press.

Gov. Davis should focus on nurse shortage. (1999, September 20). *Fresno Bee,* p. A11.

Grossi, M. (2000, May 23). Central Valley air graded an "F". *Fresno Bee,* pp. A1, A8.

Holland, G. (2000, June). Two dreams of California. *Harper's Magazine,* pp. 122-124.

Intercom. (1983). Washington, DC. Population Reference Bureau.

Kent, M. M. (1999). Shrinking societies favor procreation. In *Population today* (pp. 4-5). Washington, DC: Population Reference Bureau.

The Kiplinger California Newsletter. 1999. Vol. 35, No. 15, Aug. 4.

The Kiplinger California Newsletter. (2000a). Vol. 36, No. 9, May 3.

The Kiplinger California Newsletter. (2000b). Vol. 36, No. 10, May 17.
The Kiplinger California Newsletter. (2000c). Vol. 36, No. 11, June 7.
The Kiplinger California Newsletter. (2000d). Vol. 36, No. 14, July 19.
The Kiplinger California Newsletter. (2000e). Vol. 36, No. 16, August 16.
Knudson, T. (1995). Is it too late to save California? *California Journal, 26*(4), 18-22.
Liedkte, M. (2001, June 16). California's economy in world's 5th. *Fresno Bee,* p. A21.
Linden, E. (1991, November 18). Gobbling up the land. *Time,* pp. 83, 86-87.
McAllister, S. (2000, February 6). Small houses sell for big prices in popular Palo Alto. *Fresno Bee,* p. F6.
McCarthy, C. (2000, April 29). Faerie shrimp threatened by shrimp. *Fresno Bee,* pp. A1, A12.
McLemore, S. D., Romo, H. D., & Baker, R. G. (2001). *Racial and ethnic relations in America.* Boston: Allyn & Bacon.
McNeill, D. (1995, Winter). California continues slow population growth. *California Demographics,* pp. 1, 8.
Mehas, P. G., Garcia, C., & Buster, W. (1999, December 4). K-12 Master Plan is needed to rescue California schools. *Fresno Bee,* p. B-14.
Meyer, M. (1992, November 9). Los Angeles 2010: A Latino subcontinent. *Newsweek,* pp. 32, 35.
Miller, B. (1997, October). Educating the "other" children. *American Demographics,* pp. 49-54.
Nax, S. (2000, January 9). Affordable homes a matter of view. *Fresno Bee,* p. F1.
O'Connor, J. (1984). *Accumulation crisis.* New York: Basil Blackwell.
O'Connor, J. (1987). *The meaning of crisis.* New York: Basil Blackwell.
Outlook. (1996, May 20). *U.S. News and World Report,* p. 15.
Parillo, V. N. (1994). *Strangers to these shores.* New York: Macmillan.
Population Reference Bureau. (1999). *1999 United States population data sheet.* Washington, DC: Author.
Postel, S. (1999). *Pillar of sand: Can the irrigation miracle last?* New York: Norton.
Redclift, M., & Benton, T. (Eds.). (1994). *Social theory and the global environment.* London: Routledge.
Sbranti, J. N. (1999, January 1). Moving through the '90s. *Modesto Bee,* pp. A1, A20.
7 of 10 top polluters in San Bernardino County. (1997, May 31). *Bakersfield Californian,* p. B4.
State of California. (2000). *Race/ethnic population estimates: Components of change for California counties, April 1990 to July 1997.* Sacramento, CA: Department of Finance.
Stewart, E. W., & Glynn, J. A. (1985). *Introduction to sociology.* New York: McGraw-Hill.
Stiensta, T. (1995, May 7). Algae destroying Lake Tahoe. *Los Angeles Times,* pp. A3, A5.
Torres, A. S., & Torres, R. (1999, November 26). Valley's dual society of haves, have-nots has to be addressed. *Fresno Bee,* p. B6.
Turner, H. A., & Vieg, J. A. (1964). *The government and politics of California.* New York: McGraw-Hill.
Walters, D. (1999, December 27). Two-tier economy becomes a reality across California. *Fresno Bee,* p. A9.
Westphal, D. (1999, September 13). People moving south. *Modesto Bee,* pp. A1, A16.
Williams, P. (1993). Spotlight on . . . San Joaquin County. *California State Census Data Census Newsletter, 2*(3), 3-4.

CHAPTER 11

PROPOSITIONS 187 AND 227
A Nativist Response to Mexicans

Adalberto Aguirre, Jr.

Is it coincidence that as people of Mexican origin increased their representation in California society during the 1990s, they became the targets of nativist attacks?[1] The passage of **Proposition 187** in November 1994 and **Proposition 227** in June 1998 were stages in an escalating attack on the state's Mexican-origin population, especially the state's Mexican immigrant population. The approval of these propositions by a predominantly Anglo electorate was one outcome of a **nativist mind-set** in California's Anglo population.[2] Nativist feelings in the Anglo population were fueled by a sense of urgency that something had to be done to stem the tide of increasing numbers in the state's Mexican-origin population.[3] The approval of Propo- sitions 187 and 227 was, thus, a nativist response to fears in the Anglo population that Mexican immigrants were filled with an insatiable hunger for resources the Anglo population was unwilling to share.

HISTORY OF THE PROBLEM IN CALIFORNIA

The image of an immigrant culture that is used to characterize and romanticize the social fabric of U.S. society is under attack in California. The immigrant culture has

Author's Note: This chapter benefitted from the comments offered by Hisauro Garza, Ruben Martinez, and David Baker. My graduate students have provided assistance with this project: Shoon Lio has offered insight on the chapter's direction, and Gilbert Garcia enabled me to archive research material. My heartfelt thanks to Glenn Tsunokai and Michele Adams for helping this one-finger typist speed up his work. Any shortcomings I claim as my own.

been relegated to a peripheral place in mainstream society and has lost most of its romantic appeal because the new immigrants are non-Europeans from the Third World. Historically, the immigrant culture has served as a bridge for the absorption of European immigrants into U.S. society. For Mexican immigrants, however, the immigrant culture reinforces their differences in mainstream U.S. society. As a result, the immigrant culture is not a bridge for Mexican immigrants in California society. In this context, then, Propositions 187 and 227 are an attack on the immigrant culture because they rob Mexican immigrants of voice and presence in California society.

The purpose of this chapter is to examine the context surrounding Propositions 187 and 227 as a nativist response to the presence of Mexican immigrants in California society. Mexican immigrants have not been the only victims of nativism in California. The history of California society is interspersed with nativist responses to the presence of people of color (Perea, Delgado, Harris, & Wildman, 2000; Takaki, 1993). It will be argued in this chapter that Propositions 187 and 227 are the outcome of a nativist mind-set that portrays Mexican immigrants as threats to the economic and educational infrastructure of California society. Implicit in the argument is the idea that Propositions 187 and 227 are a tool of a political economy in which the political state, or government, promotes the interests of one social group over an entire population; thus, the political state is responsible for maintaining cultural hegemony between political culture and civic society (Lears, 1985). Propositions 187 and 227 are thus a social control mechanism nested in the state's political culture, a mechanism that protects the interests of a dominant group (i.e., Anglo population) from a perceived threat (i.e., Mexican immigrants) to its control over civic society.

THE PROPOSITIONS

Although it is not possible to conduct a detailed analysis of Propositions 187 and 227 in this chapter, it is necessary to provide the reader with a descriptive summary of the propositions. In November 1994, California voters approved Proposition 187, the Save Our State (SOS) initiative, by a vote of 59% to 41%. There are five major sections to the proposition:

1. Illegal aliens are barred from the state's public education system, and education institutions are required to verify the legal status of students and their parents.

2. Providers of publicly paid, nonemergency health services are required to verify the legal status of people seeking their services.

3. People seeking cash assistance and other benefits are required to verify their legal status before receiving services.

4. Service providers are required to report suspected illegal aliens to the State Attorney General and the Immigration and Naturalization Service (INS).

5. It is a state felony to make, distribute, and use false documents that conceal a person's legal status to obtain public benefits or employment (Garcia, 1995; Schuler, 1996).

Proposition 227, the English Language in Public Schools initiative, was approved by California voters in June 1998 by a vote of 61% to 39%. Proposition 227 dismantled bilingual education in California's public schools by requiring all students to be taught academic subjects in English. Proposition 227 mandated that teachers at public schools must teach all subjects to non-English speaking children in sheltered immersion programs. The proposition also made school personnel, especially teachers, liable for attorney's fees and damages if they do not follow an English-only curriculum and if they fail to report suspected immigrant children to the proper school authorities (Gullixson, 1999; Johnson, 1999).

NATIVISM IN CALIFORNIA AND THE UNITED STATES

Nativistic reaction to immigrants is not new in either the United States or California.[4] The Know-Nothing political party of the 1850s used **nativism** successfully in its political platform to win support among voters (Carlson, 1989). The Know-Nothing political party called for restrictive immigration laws and laws that would prevent the foreign born from serving in political office. Irving (1993) has noted that nativist attitudes at the end of the 19th century reflected the fear of an immigrant surge, especially the extension of working-class immigrant women beyond their tenement housing. The German and German American population became targets of nativist feelings during World War I (Perea, 1992). Nativist attacks on Germans included public floggings and lynchings. During the early 19th century, nativist attacks against parochial schools in the United States were rooted in the belief that immigrants were a threat to American culture and the English language (Ross, 1994). The perceived threat of aliens owning land in California resulted in the passage of the Alien Land Law in 1913, prohibiting aliens ineligible for citizenship from owning real property. The Alien Land Law was primarily a "discriminatory law directed against the Japanese" (Ferguson, 1947, pp. 61-62). The Alien Land Law was, then, a nativist response to aliens because it protected the interests of white European immigrant land owners in California (McGovney, 1947).

From the beginning, nativistic fears were a driving force in shaping the character of California society. The cession of California to the United States in 1848 was a catalyst for Anglo immigration into California. As the Anglo population increased its numbers during the 1850s, it petitioned Washington, D.C., for legislation that would bar Indians from owning land, especially mining operations, and from residing near Anglo settlements (Scafidi, 1999). An Anglo-controlled legislature passed laws in the 1880s that banned Chinese immigration, prohibited Chinese laborers from competing with Anglo laborers for work, and prevented Chinese from owning land (McClain, 1984). During the 1850s, Anglos challenged in court the land grants held by Mexican families in California (Cameron, 1998). The creation of a commission by Congress in 1851 to review land grants in California was the first step in displacing Mexican families from their land (Barrera, 1979). Thus, nativism in early California's Anglo population was anti-immigrant and anti-minority.

Mexican immigrants in California have been frequent targets of nativist sentiment. During the Great Depression, for example, Mexican immigrants as well as Mexican Americans became the targets of nativist sentiment, which portrayed them as threats to the state's and country's economy. Mexican-origin people were portrayed as parasites on social welfare programs and as depriving "true" Americans of jobs during the economic depression. As a response to public outcry, thousands of Mexican immigrants and Mexican Americans became targets of a movement fueled by nativist sentiment, and implemented by the INS, which promoted their deportation to Mexico (Hoffman, 1974; McWilliams, 1968). On February 21, 1931, the INS launched a raid on Mexican-origin people in La Placita in downtown Los Angeles, resulting in the deportation of hundreds of families to Mexico (Olivo, 2001). By 1940, more than 1 million Mexican-origin people had been deported from the United States. Most of those deported, about 60%, had been born in the United States (Balderrama & Rodriguez, 1995).

Ironically, with the entry of the United States into World War II and the resulting reduction of American workers in the labor pool, much of the public sentiment that had supported the deportation of Mexican-origin people during the Depression became focused on encouraging Mexican laborers to work in the agricultural fields of the southwestern United States. The creation of the **Bracero Program** by Congress in 1942 opened the door to allow thousands of Mexican laborers to enter the United States as legal seasonal workers in a growing agricultural industrial complex in the southwestern United States (Aguirre & Turner, 2001). Not surprisingly, 12 years later in 1954, the INS launched **Operation Wetback** to apprehend Mexican laborers in the United States who had not entered legally as braceros. Operation Wetback, however, was an attempt by the U.S. government to appease public beliefs that Mexicans were streaming uncontrollably across the border and that braceros were taking jobs away from Americans. By 1959, the INS had deported 3.8 million people to Mexico (Grebler, Moore, & Guzman, 1970). Operation Wetback reinforced the perception held by Mexicans living in the United States that theirs was a marginal and tenuous presence.

In the 1940s, the **Zoot Suit riots** in Los Angeles focused on attacks by U.S. military personnel on Mexican-origin youth who were wearing zoot suits (Mazon, 1984). Although it was not exclusive to Mexican-origin youth, the zoot suit was a distinctive style of dress often associated with deviant subcultures. The zoot suit was popularized in American culture by members of the jitterbug cult. In general, the zoot suit was exaggerated clothing consisting of a long coat, with or without a vented back and wide lapels, and high-rise trousers with wide legs pegged at the ankle. Mexican-origin youth wearing zoot suits were perceived by Anglos as deviants and threats to American patriotism. On June 3, 1943, U.S. military personnel attacked Mexican zoot suiters in an attempt to clean up Los Angeles. The efforts by Anglo authorities, especially the police, to cover up the attacks by U.S. military personnel on zoot-suited Mexican-origin youth were nested in a nativisit perception that Mexican-origin youth were un-American.

The general observation can be made that nativist sentiment in California history has primarily targeted non-white immigrant populations. The social context that fostered nativist sentiment in California's white population is described by Almaguer (1994): "The cultural division of the world into different categories of humanity led white, European Americans in California to arrogantly privilege themselves as superior to non-European people of color" (p. 7). White European immigrants, consequently, built an infrastructure for social institutions to promote and protect their interests. Nativist movements in California arose, then, when non-white immigrants posed a threat to the infrastructure. The nativist movements also became a social control mechanism over non-white immigrants in California.

MEXICAN IMMIGRANTS AND NATIVISM

The general public's perception in California that Mexican immigrants are streaming uncontrollably into the state has resulted in increased nativistic sentiment (Chavira, 1990). Nativistic reaction to Mexican immigrants differs from previous nativist movements in California history because it focuses attention on their "foreignness" in U.S. society (Chang, 1993). Mexican immigrants are foreigners; but their foreignness is perceived as a threat to the social fabric. As such, the element of foreignness reinforces the perception that Mexican immigrants do not belong and increases their social distance from California society.[5] For example, nativist sentiment focused on the foreignness of Mexican immigrants in 1920s Los Angeles is typified in a *Saturday Evening Post* article, which describes

> the endless streets crowded with the shacks of illiterate, diseased, pauperized Mexicans, taking no interest whatever in the community, living constantly on

the ragged edge of starvation, bringing countless numbers of American citizens into the world with the reckless prodigality of rabbits. (quoted in Sanchez, 1993, p. 96)

The perceived foreignness of Mexican immigrants has inflamed nativist sentiment by fostering the belief that all Mexican-origin people are immigrants.[6] Despite the fact that the majority (64%) of all Latinos in the United States are born in the United States, Latinos are generally perceived by the public as immigrants (Enchautegui, 1995). Nativist sentiment is reflected in an increased feeling among Californians that the U.S. border with Mexico needs tighter control and that immigration policies such as repatriation need to be implemented (Jost, 1995). According to a *Los Angeles Times* poll conducted on December 4, 1993, a sizable number of Californians consider illegal immigration from Mexico to be the state's major problem (78%); many think that the National Guard should be used to guard the U.S.-Mexico border (58%). The majority (65%) of Californians also favor an amendment to the U.S. Constitution that would bar citizenship to children born in the United States to illegal immigrant parents. Accordingly, a *Los Angeles Times* poll conducted on August 3, 1996, found that a majority of Californians (57%) believed that greater job creation in Mexico and a tougher border policy would reduce immigration from Mexico.

As nativist responses to Mexican immigrants in California, Propositions 187 and 227 were based on a set of assumptions regarding the presence of Mexican immigrants in California society. The first assumption is that Mexican immigrants come to California for a "free ride," that is, to live at public expense. The second assumption is that Mexican immigrants take jobs away from U.S. workers, especially Anglo workers.[7] The third assumption is that Mexican immigrants are not interested in establishing permanent residency in the United States. In general, these three assumptions are contradicted by research showing that Mexican immigrants in California perform jobs regarded as undesirable by U.S. workers, do not depend on welfare services for a livelihood, and intend to remain permanently in the United States (Aguirre, 1993; McCarthy & Valdez, 1986; Portes & Rumbaut, 1990; Ronfeldt & Ortiz de Oppermann, 1990). These three faulty assumptions are instrumental in supporting a mind-set that is nativistic in its character and, as a result, hostile toward Mexican immigrants.

CALIFORNIA'S IMMIGRANT POPULATION

According to Parker (1994) and Muller (1993), California is a magnet for immigrants. In 1990, one in four California residents was foreign born, compared to one in six in 1980 (Vernez, 1994). The General Accounting Office (1993) estimates that more than half of the legal and undocumented immigrants in the United States settle in California. In 1997, there were 25.8 million foreign-born people in the United States, 9.7% of the U.S. population (Schmidley & Alvarado, 1998). Almost one fourth of the foreign-born population in 1997 resided in California. About 27% of the foreign-born were born in Mexico. Interestingly, the number of legal immigrants settling in California declined between 1992 and 1998, from 336,663 to 170,126 (Rolph, 1992; U.S. Department of Justice, 1999). Almost 40% of the legal immigrants who settled in California were located in the Los Angeles area and were from Mexico (McCarthy & Vernez, 1998).

California is also the state where the majority of refugees/asylum seekers arriving in the United States settle.[8] Immigrants who arrive in the United States as refugees/asylum seekers are people who fear persecution in their own country. Between 1980 and 1990, the majority of immigrants (70%) who arrived as refugees/asylum seekers in the United States came from Asia, and 52% of them settled in California (Rolph, 1992). One result was that, in California, the Asian population grew at twice the rate of the Mexican population between 1980 and 1990. Between 1982 and 1990, 54% of all immigrants who qualified as applicants under the Immigration Reform and Control Act (IRCA) settled in California. IRCA applicants are those aliens offered legal status under the act passed in 1986; they must have resided continuously and unlawfully in the United States since January 1, 1982 (Baker, 1990). The majority of immigrants (88%) who qualified as IRCA applicants came from Mexico. Regarding their settlement patterns, the majority of IRCA applicants (67%) and refugees/asylum seekers (63%) in California settled in the Los Angeles-Long Beach-Santa Ana-Anaheim metropolitan corridor.

Some general observations about California's immigrant population can be inferred from the preceding statistical portrait. First, Asia and Mexico are the primary countries of origin for immigrants arriving in California. Second, California's immigrant population is a mosaic of sorts, in that it draws from the various pools of immigrants, such as refugees/asylum seekers and IRCA immigrants, arriving in the United States. Third, the immigrant population in California is concentrated in the Los Angeles-Long Beach-Santa Ana-Anaheim metropolitan corridor.

IMMIGRANTS AND WORK

Nativist feeling among Californians peaks regarding the topic of work. Such nativist feeling ranges between the two ends of a spectrum. At one end of the spectrum, immigrants are regarded as contributors to work and wages (Fix & Passel, 1994; Fix & Zimmermann, 1993; Martin, 1993; Simon, 1989). Because immigrants are eager to work, they often accept work at below average wages. Sometimes, these wages are anywhere between $2 and $3 per hour below the minimum wage. As a result, immigrants are instrumental in keeping jobs in California, such as manufacturing furniture and apparel, that would otherwise go elsewhere, probably to an underdeveloped country. At the other end of the spectrum, immigrants are blamed for increasing unemployment. Because the urban areas in which immigrants settle are very often economically depressed, immigrants' entry into these urban areas does not counter the economic depression but rather it enhances it (Briggs, 1984; Enchautegui, 1993). That is, the types of jobs that immigrants are likely to find are not associated with wages that promote upward social mobility. Although the wages earned by immigrants transform them into the working poor, the money is insufficient to alter the social context in which immigrants reside.[9]

In an examination of Mexican immigrants in the California economy, Vernez (1993) found that California's demand for labor, especially in manufacturing and service industries, is a major factor in attracting immigrants from Mexico and Asia. According to Vernez, between 1960 and 1980, Mexican immigrants in California increased their distribution across all occupational sectors of the California economy. However, in 1980, the majority of Mexican immigrant workers (75%) were in manufacturing. In contrast, the representation of U.S.-born workers in agriculture and manufacturing decreased between 1960 and 1980 but increased in service industries during the same period.

Regarding occupational categories, the representation of U.S.-born workers in white-collar occupations increased 12% from 39% to 51% between 1960 and 1980 but remained unchanged (12%) for Mexican immigrant workers (Vernez, 1993). Although the representation of U.S.-born workers in operative/laborer occupations decreased 10% between 1960 and 1980, from 44% to 34%, the representation of Mexican immigrant workers in these occupations increased 8%, from 73% to 81%. Regarding wages, U.S.-born male workers had the largest gain in hourly wages between 1960 and 1980, from $2.77 to $9.45. They were followed, in order, by U.S.-born female workers ($2.50 to $7.48), Mexican-immigrant male workers ($1.97 to $6.75), and Mexican-immigrant female workers ($1.95 to $6.43).

The following observations can be made regarding Mexican immigrant workers in the California economy. First, if industry's need for labor is a pull factor for Mexican immigrants, then the increased representation of Mexican immigrant workers between 1960 and 1980 in manufacturing industries reflects the growth of these industries in the California economy.[10] Second, Mexican immigrant workers are not a threat, vis-à-vis their occupations, to the middle class in California. The majority (51%) of U.S.-born workers in 1980 were found in white-collar occupations, whereas the majority (81%) of Mexican-immigrant workers were in operative/laborer occupations. If anything, Mexican immigrant workers have become segmented as operatives/laborers in California society.

SEGMENTATION OF MEXICAN IMMIGRANT WORKERS

The lack of language skills in English is regarded as a primary reason for the segmentation of Mexican immigrant workers into a secondary labor market (Chiswick, 1990; Hamel, 1993; Hayflich, 1992; Meisenheimer, 1992). Their lack of English language skills is also used to argue that Mexican immigrants are not interested in identifying with U.S. society. The secondary labor market consists of low wage, high risk, and high turnover jobs, which are in most cases unattractive to U.S. workers. As a result, the majority of Mexican immigrants find themselves in a labor market that has limited socioeconomic mobility and that does not require workers to have English language skills (McManus, 1990; Muller & Espenshade, 1985). That is, Mexican immigrant workers can perform their work tasks without the need to speak or write the English language. It is not surprising, then, that the general public believes immigrants are unwilling to acquire language skills in English.

In an attempt to examine the issue of immigrants and their learning English, the author surveyed a population of Mexican immigrants enrolling for the first time in an English as a Second Language (ESL) adult education program in Riverside County. A total of 608 students who listed their country of origin as Mexico were enrolled in 10 ESL classes. Of the 608 students, 330 were males and 278 were females. The descriptive features for the sample of students follow:

1. About 76% had less than 5 years of formal schooling, 17% had completed the equivalent of a U.S. high school degree, 2% had earned a technical/vocational certificate, and 5% had completed college.

2. About 51% were enrolled in beginning ESL, 28% were enrolled in intermediate ESL, and 21% were enrolled in advanced ESL.

3. Students gave the following reasons, in decreasing importance, for enrolling in the ESL class: get a better paying job, get more education, learn English to help children with school, communicate better in U.S. society.

An examination of the association between ESL class placement and students' educational background was found to be significant ($\chi^2 = 64.95, p < .001$); increased educational attainment is associated with higher ESL class placement. Interestingly, the test results also showed that students were distributed across three ESL levels: beginning, intermediate, and advanced. That is, the students demonstrated various levels of English proficiency rather than only one level (e.g., beginning). In particular, those students with a high school degree or more were more likely to have test scores placing them in intermediate and advanced ESL levels.

The association between ESL placement and students' educational attainment suggests that people in this sample of Mexican immigrants may have arrived in California with language skills in English. If this is the case, then one may speculate that these students may regard ESL classes as a means to establish permanency within an occupational world that is rapidly becoming insensitive to non-English speaking workers (see Cacas, 1995; Califa, 1989; Fink, Robinson, & Wyld, 1996; Murphy, Barlow, & Diane, 1993). This scenario takes on greater meaning if one considers that a shrinking job market may force white English-speaking workers to compete with Mexican immigrants for the same jobs, especially in the service and manufacturing industries. In these circumstances, an employer would be constrained in using English-language ability as a criterion for excluding Mexican immigrant workers from the workplace.

It could be asserted that the fear that Mexican immigrants would learn English as a way of displacing Anglo workers fueled Anglo support for Proposition 227. On the one hand, Proposition 227 became a vehicle with which Anglo voters buttressed Proposition 187's mandate to deny public services to immigrants. On the other hand, Proposition 227's ban on bilingual instruction upped the ante on how non-English and limited-English speaking immigrant children would be educated in California's schools. In a sense, by denying immigrant children access to bilingual instruction, the chances are increased that the public school will not be able to offer them the same opportunities available to English-speaking children. As a result, language becomes a discriminating dimension between immigrant and Anglo children in school.[11]

On the surface, Proposition 227 appears to be a reasonable approach for educating immigrant children by teaching them only in English. If instructed only in English, supporters of Proposition 227 argued, immigrant children would improve their educational outcomes in California's public schools (Farquharson, 1999; Zabetakis, 1998). However, Proposition 227 amends the state's education code to deny instructional services (e.g., bilingual instruction) to immigrant children. That is, Proposition 227's objective is to constrain, not enhance, the educational context for immigrant children. How, then, does a restrictive educational context improve educational outcomes for immigrant children? Ironically, Proposition 227 ends up stigmatizing immigrant children in California public education instead of improving their access to the opportunity structure in California society. As Vargas (1999) observes,

> By singling out a particular group and proscribing them from speaking in their chosen language, the measure implies that speaking in any language other than English is inappropriate. Perhaps those who do not speak English are de facto non-citizens; or implicit in their inability to speak English is disloyalty; perhaps they are unintelligent; perhaps they are conspiring against or talking about the Anglo majority. (p. 531)

Thus, an Anglo electorate used Proposition 227 to tell immigrant children, our language (English) is the only one spoken in school.

THE NATIVISTIC CONTEXT

Proposition 187 creates a "state-run system to verify the legal status of all persons seeking public education, health care and other public benefits, and adds public education to the list of services for which unauthorized aliens are ineligible" (Martin, 1995, p. 255). Proposition 227 reinforced Proposition 187 by requiring school personnel to report and identify suspected immigrant children and encouraged outsiders (e.g., parents of school children) to identify school personnel not teaching an English-only curriculum. The promoters of Propositions 187 and 227 convinced voters that tighter control over public service expenditures and instructional services would free the state from its economic troubles and unfavorable educational outcomes. In particular, tighter control over "unauthorized aliens" and their children would guarantee a solution to the state's ills. Both propositions targeted immigrants as a suspect category, thus increasing their identifiability for public officials in California society. Where Proposition 187 made immigrants suspect in the delivery of public services, Proposition 227 stigmatized immigrant children in public education. Both propositions thus aimed to reduce the participation of Mexican immigrants and their children in California's civic culture.

Harold Ezell, former director of the INS's West Coast office, argued that Proposition 187 was necessary in California because "the people are tired of watching their state run wild and become a Third World country" (quoted in Wood, 1994, p. 1). Sharing Ezell's perspective, one of Proposition 187's drafters, Barbara Cox, warned that an immigrant invasion from Mexico would result in a "Mexico-controlled California [that] could vote to establish Spanish as the sole language of California, 10 million more English-speaking Californians could flee, and there would be a statewide vote to leave the Union and annex California to Mexico" (quoted in Rodger & Doyle, 1994, p. A1).

The architect of Proposition 227, Ron Unz, observed that his own grandparents "came to California in the 1920s and 1930s as poor European immigrants ... who came to work and become successful ... not to sit back and be a burden on those already here" (quoted in Baraba, 1997, p. B2). The implicit characterization in Proposition 227 of Mexican immigrants as a burden on those already here appealed to an Anglo electorate that was anti-immigrant and anti-minority (Attinasi, 1998; Escobedo, 1999). In particular, Proposition 227 reinforced the perception that immigrant children were a burden on the state's public schools. Ironically, anti-immigration activists joined the anti-Proposition 227 campaign because the proposition did not go far enough in removing immigrant children from public schools (Puente, 1998).

AN ANTI-MEXICAN MIND-SET

The passage of Proposition 187 is characterized by Martinez (1995) as "aimed at restricting the rights of illegal immigrants," following in the footsteps of "the Operation Wetback of the '50s, . . . the Repatriation of the '30s, . . . the Greaser Laws of the late 19th Century" (p. 2). The context characterized by Martinez for Proposition 187 can be extended to Proposition 227, because both the spirit and intent of the propositions are the same: to control Mexican immigrants so as to protect Anglo interests. What is interesting about Martinez's characterization of Proposition 187 is that it links the proposition with a set of other nativist movements that targeted Mexican immigrants and Mexican Americans. For example, Operation Wetback targeted braceros for deportation at time when U.S. labor organizations saw immigrant labor as a threat to wage levels for U.S. workers and the American public perceived braceros as taking jobs away from Americans (Aguirre & Turner, 2001). Similarly, during the repatriation movement of the 1930s, Mexican immigrants and Mexican Americans were targeted for deportation because public officials saw them as taking jobs away from American workers. The Greaser Laws of the late 19th century were anti-Mexican legislation that sought to limit and control the presence of Mexicans in U.S. society. During the gold mining boom in California, vagrancy laws were passed in mining areas that targeted Mexicans. Because Mexicans were prohibited from owning mines, their presence in a mining town would result in their arrest as vagrants. As a result, Mexicans generally avoided mining towns in California.

Operation Wetback, the repatriation movement of the 1930s, and the Greaser Laws were events nested in nativist sentiments within the white population. The white population regarded Mexican-origin people as a threat to the infrastructure they built, especially in the Southwest. The resultant nativist movements transformed Mexican-origin people into **scapegoats.**[12] Similarly, Propositions 187 and 227 can be regarded as an indicator of the white population's frustration with, and growing intolerance of, Mexican-origin people and Mexican immigrants. Both propositions exhibit an anti-Mexican mind-set rooted in the belief that Mexicans are a threat in California society to the economic and educational resources controlled by the Anglo population.

SOCIAL FORCES

Since the late 1980s, California's economy and system of public education have been undergoing noticeable changes. One of these changes has been the downsizing of the aerospace and related industries. Aerospace companies have reduced the number of California employees, or they have relocated to other states offering lower production costs. A direct result of this change in the aerospace industry has been the loss of white-collar and skilled labor jobs. An analysis of how the job losses have affected the California population shows that the job losses have displaced white workers to a larger degree than other workers (e.g., see California State Legislature, 1985; Vernez, 1993).

As the number of white-collar and skilled labor jobs has been reduced, the number of service and manufacturing jobs has increased. It has been noted in this chapter that California's demand for workers in the service and manufacturing industries serves as a social force in attracting Mexican immigrants. It has also been noted that the strength of this demand has resulted in the segmentation of Mexican immigrant workers in the service and manufacturing industries. Not surprisingly, then, as the number of service and manufacturing jobs has increased, so has the number of Mexican immigrant workers in California.

In addition, the reduction of white-collar and skilled labor jobs in the California economy has enhanced the visibility of job growth in the service and manufacturing industries. As occupants of service and manufacturing jobs, Mexican immigrant workers have become much more visible, that is, *identifiable.* Interestingly, because people displaced from white-collar and skilled labor jobs are not likely to pursue jobs in the service and manufacturing industries, the presence of Mexican immigrant workers becomes more noticeable. As a result, the proportionate representation of Mexican immigrant workers appears to increase as the job sector shrinks. Especially, it would appear as if the presence of Mexican immigrant workers in California is at the expense of white workers. It is not surprising, then, that the California Civil Rights Initiative, Proposition 209, has been called the offspring of Proposition 187 because the former is based on the belief that white workers are losing their jobs to minorities, women, and immigrants (Schrag, 1995).

The entry of Mexican immigrants into California's job market has increased the size of the state's Mexican-origin population, as well as enhancing the profitability of the service and manufacturing industries. As an outcome of the entry of Mexican immigrant workers, for example, into the textile, apparel, and manufacturing industries, "California employers have seen their labor costs decline relative to those of employers elsewhere in the United States and yet have not lost their productivity advantage" (McCarthy & Vernez, 1998, p. 31). In other words, the availability of Mexican immigrant labor has increased the profit-making activities in service and manufacturing industries.

Although the service and manufacturing industries enjoy increased profits made possible by cheap immigrant labor, these industries are usually the first ones to downsize by laying off workers or slowing production during times of economic instability, such as a recession (Miller, 2001). In addition, if the industries encounter higher than expected production costs, they are likely to shield their profitability by downsizing. Soaring energy costs in California are placing many of these industries at risk. In turn, thousands of Mexican immigrant workers will be put at risk. If the industries decide to counter higher energy costs by lowering production levels, then fewer workers will be needed. Because many of these workers are likely to be Mexican immigrants, their displacement by these industries will increase their presence and identifiability in California society. Will the enhanced presence of displaced Mexican immigrant labor result in nativist perceptions that they are a burden on California society?

Because a large number of Mexican immigrants eventually relocated their families from Mexico to California, the 1980s and 1990s witnessed a dramatic increase of Mexican immigrant children enrolling in California's public schools (Macias, 1993). The noticeable presence of immigrant children in the schools caused supporters of Proposition 227 to argue that the schools were becoming immigrant campgrounds. The availability of bilingual instruction in California's public schools reinforced the perception that schools were immigrant campgrounds. The perception was fueled by an increasing fear that the schools would be transformed into non-English speaking environments. As a result, immigrant children were perceived as a threat to the social fabric in California society. Proposition 227 proposed to reduce the threat by requiring that all academic subjects be taught in English. In the end, the fear was not so much that immigrant children would displace white children in the public schools. After all, their immigrant parents were limited participants in the electoral process and were ignored by local school boards and school personnel. Rather, the fear was that bilingual immigrant children would have an unfair advantage over monolingual English-speaking children in a California society that was becoming increasingly culturally and linguistically diverse. In this sense, Proposition 227 promotes the nativist fear that diversity is a threat to California's social fabric.

SOCIAL CONTROL

The perceived threat of Mexican immigrant workers to white job holders and white children in public schools is a social force that shaped a nativist context for Propositions 187 and 227. The noticeable presence of Mexican immigrants in California society has resulted in nativist beliefs that characterize Mexicans as the source of California's economic troubles and as the cause of low test scores in the public schools. By targeting Mexican immigrants, Propositions 187 and 227 seek to control the entry and participation of Mexican immigrants in California society.

On the one hand, Propositions 187 and 227 are symbolic tools used by the Anglo population to express their power over Mexican immigrants. Because Mexican immigrants are limited in their right to vote in U.S. elections, Anglo voters can create social policy to distance themselves from immigrants. In this sense, Propositions 187 and 227 heighten the meaning of *difference,* of foreignness, for Mexican immigrants in California society. On the other hand, Propositions 187 and 227 silence Mexican immigrants and their children in California society by making them scapegoats. Proposition 187 and Proposition 227 create the impression that Mexican immigrants must be controlled to protect California's economic and educational resources. In this sense, Propositions 187 and 227 are a visible expression of the power held by white people to silence the Mexican population in California society.

CONCLUDING REMARKS

Propositions 187 and 227 have created a context in California society that places all Mexican-origin people in California at risk. In the public's mind, all Mexican-origin people are perceived as foreigners and intruders in California society. As a result, the propositions emphasize the tenuous relationship of Mexican-origin people to California society. To avoid criticism, the California legislature has suggested that a card identification system be developed and enforced by the Immigration and Naturalization Service (INS) to facilitate the identification of Mexican immigrants and Mexican Americans. An identity card, however, would polarize people by nativity in the Mexican-origin population in California, and it would likely distance the Mexican-origin population from Central American immigrants. One can only speculate that a growing and polarized Latino population, consisting primarily of Mexican-origin people and Central American immigrants, would be attractive to political interests interested in maintaining a colonized relationship between Latinos and Anglos in California society. The attraction becomes stronger for Anglo society if it can continue to use the political process, such as the passage of initiatives, to express its will over a growing Latino population.

The implementation of Propositions 187 and 227 promotes the establishment of screening processes that are characteristic of a police state. The propositions transform public agencies and educational institutions into police agents by requiring public officials, especially school officials, to report suspected aliens to authorities. Consequently, the personal freedom and liberty of all Mexican-origin people is at risk. The repatriation movement of the 1930s is a good example of how nativist ideology can result in political initiatives that create victims out of innocent bystanders in society. It must not be forgotten that the repatriation movement resulted in millions of Mexican American people being misidentified by the INS and being deported to Mexico. It is not Mexican immigrants who are a threat to California society. Rather, it is the nativist beliefs that promoted Propositions 187 and 227 that are a threat to California society because they do not support the establishment of a multicultural environment. Ironically, the propositions increase the social distance of Mexican-origin people from a California society that is rapidly becoming more linguistically and culturally diverse in its social fabric.

Finally, it is foolish to believe that the passage of Propositions 187 and 227 will remove Mexican immigrants and Mexican Americans from California society. As long as the service and manufacturing industries continue to depend on inexpensive labor, Mexican-origin people will remain active participants in California's economy. However, their participation may be curtailed by the problems faced by the service and manufacturing industries. The power crisis faced by California society, and especially by the service and manufacturing sector, is likely to affect everyone. But will it affect everyone in the same manner? If the service and manufacturing industries decide to deal with escalating energy costs by furloughing or firing Mexican immigrant workers, will the workers be perceived by the public as victims of the legislature's ill-planned strategy to deregulate the state's energy industry? Will their displacement be perceived by the public as an indicator that there are too many Mexican immigrants in the state? Will the public respond to the presence of displaced Mexican immigrant workers in California society by using the electoral process to implement initiatives that promote their removal from California society? If past activity in California society serves as a basis for answering these questions, then one can assume that Mexican immigrants will continue to be portrayed by nativist beliefs as social problems in California society.

NOTES

1. California's Latino population increased steadily over the last two decades. Using a conservative estimate, the Mexican-origin component of the Latino population is about 70% (Aguirre & Turner, 2001). As a result, the bulk of the growth in California's Latino population over the last two decades occurred in the Mexican-origin component, primarily as a result of increased levels of Mexican immigrants. Interestingly, close to 50% of all immigrants in California are from Mexico (McCarthy & Vernez, 1998). Between 1990 and 1999, the U.S. Census Bureau estimates that California's Latino population increased by 32%, from 7.8 million people to 10.5 million (Nelson & O'Reilly, 2000; Verdin, 2000).

2. In the November 1994 election, 81% of the voters were white and 17% were minority (Latino, black, and Asian). The majority (63%) of white voters approved Proposition 187, whereas the majority (61%) of minority voters voted against the proposition. In the June 1998 election, 69% of the voters were white and 10% were minority. The majority (67%) of white voters approved Proposition 227, whereas the majority (53%) of minority voters did not approve the proposition.

3. I suspect that Proposition 209, California's Civil Rights Initiative, was also the outcome of nativist fears in the state's Anglo population. Even though Proposition 209 was a direct attempt, by banning affirmative action, to restrict minority access to social

opportunity, I believe that the nativist fears (e.g., anti-immigrant) in the Anglo population that were the political undercurrent for Proposition 187 shifted over to Proposition 209. The anti-immigrant mind-set that fueled support for Proposition 187 may have been generalized into an anti-minority mind-set among Anglo voters regarding Proposition 209. Summaries and interpretive analyses of Proposition 209 can be found in Doherty (1999), Gotanda et al. (1996), and Volokh (1997).

4. Historically, immigrants have been regarded as a threat to the Anglo-Saxon core in U.S. society (Aguirre & Turner, 2001). In retaining their mother tongue and their cultural orientations, immigrants are perceived as a threat to the Anglo-Saxon core's language (English) and cultural orientations (e.g., individualism, democracy, etc.). For summary reviews of the research literature regarding nativist responses in U.S. society to immigrants, see Cornell and Bratton (1999), Hernandez-Truyol (1996), Wade (1996), and Vargas (1998).

5. The perceived foreignness of Mexican immigrants, in turn, increases their identifiability in California society. In some cases, their level of identifiability has detrimental outcomes. For example, on June 12, 1995, the home of a Mexican family was set on fire in the city of Palmdale, California, by suspected white supremacists, according to police. White power slogans were spray-painted on the home's walls, and the word *Mexico* was painted on a wall with an X through it (Arredondo, 1995).

In another incident, a white female was overheard by bakery workers making comments about her hatred toward Mexicans on September 1, 2000, in Van Nuys, California. According to witnesses, the woman left the bakery and entered her car, which was parked in the bakery's parking lot. At the time, a Mexican male was loading boxes of bagels into his car in the bakery's parking lot. Witnesses stated that the woman drove her car directly into the Mexican male numerous times before dragging him onto the street with her car. The woman then returned to the bakery, bought a bagel, and locked herself in her car (Liu, 2000).

6. A national public opinion poll, Attitudes Towards Immigrants Survey, conducted by Knight Ridder on May 2, 1997, found that one third of the adult respondents viewed Mexican immigrants as unfavorable when compared with immigrants from other countries and as creating more problems in U.S. society than immigrants from other countries.

7. Mexican immigrant labor has changed the character of the U.S. labor force not by displacing Anglo workers but by assuming jobs that have increased the profitability of some occupational sectors, most notably, jobs in the service/manufacturing industries. Mexican immigrant workers have moved into jobs—at fast food restaurants, poultry farms, car washes, supermarket deliverymen, and so on—that are shunned by most Americans (Greenhouse, 2000; Katz, 1996). The low wages paid by employers to immigrant workers in these jobs have allowed employers to increase their profitability (Bustos & Mathis, 2000).

8. The official statistics on refugees/asylum seekers are sometimes confusing. For example, according to the INS (1999), 2,441 refugees/asylum seekers were granted permanent resident status in 1998 and resided in California. That number represents only 4% of the total number of refugees/asylum seekers granted permanent resident status in the United States in 1998. However, the number of refugees/asylum seekers who resided in California in 1998 was 22,610, or 43% of the total number of people requesting asylum in the United States. It may be that the population of refugees/asylum seekers residing in California who are not granted permanent resident status is much larger than the population of refugees/asylum seekers granted permanent resident status.

9. Mexican immigrants are caught in a precarious position. The jobs they assume do not have opportunities for upward mobility. Washing cars or working on a poultry farm offer few, if any, opportunities for occupational advancement. In addition, the wages paid in those jobs are so low that Mexican immigrants are unable to accumulate the capital necessary for starting a business or becoming venture capitalists. In contrast, European immigrants coming to the United States in the early 19th century could use low wage jobs as a way of moving up; that is, jobs then had more opportunity for upward mobility because the occupational sectors were emerging in an expanding social fabric. As a result, although they work long hours and more calendar days than the average American worker, Mexican immigrants are still trying to catch up with the income growth enjoyed by Californians in the 1990s (Dreazen, 2000). In addition, although Mexican-origin households in California, including immigrant households, reduced their debt burden in the 1990s, they also saw their household net worth drop by 24% (Walsh, 2000).

10. McCarthy and Vernez (1998) estimate that the number of immigrants in the service and manufacturing industry quintupled between 1960 and 1990.

11. The precursor to Proposition 227 was Proposition 63. In 1986, a predominantly Anglo electorate approved Proposition 63, which amended the state constitution to declare English the official language of California. Proposition 63 has often been used by opponents of bilingual instruction to argue for the denial of educational services to non-English and limited-English speaking children in California's public schools.

12. Higham (1955) defines nativism as "intense opposition to an internal minority on the ground of its foreign (i.e., "un-American") connections [which] translates . . . into a zeal to destroy the enemies of a distinctively American way of life" (p. 4). To survive as a social force in society, nativism must identify targets (e.g., people or groups) it can blame for problems or misfortunes in society. These people or groups thus become scapegoats. For example, during Operation Wetback, Mexican-origin people were labeled economic burdens in U.S. society, when in reality U.S. society simply wanted to divest itself of Mexican labor that had served its purpose for American businesses and agricultural interests. There are other examples of the use of scapegoats. During the 1980s recession, U.S. auto workers blamed the Japanese and, unfortunately, Asians in general for declining sales of U.S. automobiles. The welfare cheat, instead of organizational problems associated with an ever-expanding bureaucratic structure, is blamed for problems in the welfare system. The homeless are blamed for being alcoholics or drug addicts; no examination is made of how social service agencies fail to meet their needs. Finally, according to Calabresi (1994), scapegoats are important because they make it possible to blame someone for a problem in society rather than addressing the problem itself.

REFERENCES

Aguirre, A., Jr. (1993). Communication media and Mexican social issues: A focus on English-language and U.S.-origin communication media. *International Journal of Comparative Sociology, 34,* 231-243.

Aguirre, A., Jr., & Turner, J. (2001). *American ethnicity: The dynamics and consequences of discrimination* (3rd ed.). New York: McGraw-Hill.

Almaguer, T. (1994). *Racial fault lines: The historical origins of white supremacy in California.* Berkeley: University of California Press.

Arredondo, M. (1995, June 12). Incendio destruye vivienda en Palmdale (Fire destroys Palmdale house). *La Opinion,* p. 3A.

Attinasi, J. (1998). English only for California children and the aftermath of Proposition 227. *Education, 119,* 263-284.

Baker, S. (1990). *The cautious welcome: The legislative programs of the Immigration Reform and Control Act.* Santa Monica, CA: RAND and the Urban Institute.

Balderrama, F., & Rodriguez, R. (1995). *Decade of betrayal: Mexican repatriation in the 1930s.* Albuquerque: University of New Mexico Press.

Baraba, M. (1997, August 31). GOP bid to mend rift with Latinos still strained. *Los Angeles Times,* p. B2.

Barrera, M. (1979). *Race and class in the Southwest: A theory of racial inequality.* Notre Dame, IN: University of Notre Dame Press.

Briggs, V. (1984). *Immigration policy and the American labor force.* Baltimore, MD: Johns Hopkins University Press.

Bustos, S., & Mathis, D. (2000, May 23). Small towns shaped by influx of Hispanics. *USA Today,* p. 10A.

Cacas, S. (1995). The language of hate: Discrimination complaints are growing as the workplace becomes more bilingual. *Human Rights, 22,* 30-33.

Calabresi, G. (1994). Scapegoats. *Quinnipiac Law Review, 14,* 83-89.

Califa, A. (1989). Declaring English the official language: Prejudice spoken here. *Harvard Civil Rights-Civil Liberties Law Review, 24,* 293-348.

California State Legislature. (1985). *The undereducation of minorities and the impact on the California economy.* Sacramento: State of California Joint Publications Office.

Cameron, C. (1998). One hundred fifty years of solitude: Reflections on the end of the history academy's dominance of scholarship on the Treaty of Guadalupe Hidalgo. *Southwestern Journal of Law & Trade in the Americas, 5,* 83-107.

Carlson, A. C. (1989). The rhetoric of the Know-Nothing Party: Nativism as a response to the rhetorical situation. *Southern Communication Journal, 54,* 364-384.

Chang, R. (1993). Toward an Asian American legal scholarship: Critical race theory, poststructuralism, and narrative space. *California Law Review, 81,* 1243-1323.

Chavira, R. (1990, November 19). Hatred, fear, and vigilance: A flood of illegal Mexican immigrants fuels a wave of nativism and a tense face-off on the Tijuana border. *Time,* pp. 12-16.

Chiswick, B. (1990, Spring). *Language in the labor market: The immigrant experience in Canada and the United States.* Paper presented at the Conference on Immigration, Language, and Ethnic Issues: Public Policy in Canada and the United States, Washington, DC.

Cornell, D., & Bratton, W. (1999). Deadweight costs and intrinsic wrongs of nativism: Economics, freedom, and legal suppression of Spanish. *Cornell Law Review, 84,* 595-695.

Doherty, B. (1999). Creative advocacy in defense of affirmative action: A comparative institutional analysis of Proposition 209. *Wisconsin Law Review,* 91-117.

Dreazen, Y. (2000, March 15). Racial wealth gap huge. *The (Riverside, CA) Press Enterprise,* p. A1.

Enchautegui, M. (1993). *The effects of immigration on the wages and employment of black males.* Washington, DC: Urban Institute.

Enchautegui, M. (1995). *Policy implications of Latino poverty.* Washington, DC: Urban Institute.

Escobedo, D. (1999). Propositions 187 and 227: Latino immigrant rights to education. *Human Rights, 26,* 13-15.

Farquharson, M. (1999). Proposition 227: A burning issue for California's bilingual students. *Boston University Public Interest Law Journal, 8,* 333-359.

Ferguson, E. (1947). The California Alien Land Law and the Fourteenth Amendment. *California Law Review, 35,* 61-90.

Fink, R., Robinson, R., & Wyld, D. (1996). English-only work rules: Balancing fair employment considerations in a multicultural and multilingual healthcare workforce. *Hospital & Health Services Administration, 41,* 473-483.

Fix, M., & Passel, J. (1994). *Immigration and immigrants: Setting the record straight.* Washington, DC: Urban Institute.

Fix, M., & Zimmerman, W. (1993). *After arrival: An overview of federal immigration policy in the United States.* Washington, DC: Urban Institute.

Garcia, R. (1995). Critical race theory and Proposition 187: The racial politics of immigration law. *Chicano-Latino Law Review, 17,* 118-154.

General Accounting Office. (1993). *Illegal aliens: Despite data limitations, current methods provide better population estimates.* Washington, DC: Government Printing Office.

Gotanda, N., Bayati, J., Berkman, S., Lanier, C., McMillan-Delaney, H., Tate, S., & Yoshida, J. (1996). Legal implications of Proposition 209—The California civil rights initiative. *Western State University Law Review, 24,* 1-55.

Grebler, L., Moore, J., & Guzman, R. (1970). *The Mexican-American people: The nation's second largest minority.* New York: Free Press.

Greenhouse, S. (2000, September 4). Major change in labor force. *The (Riverside, CA) Press Enterprise,* p. A1.

Gullixson, K. (1999). California Proposition 227: An examination of the legal, educational, and practical issues surrounding the new law. *Law and Inequality, 17,* 505-536.

Hamel, R. (1993). No job. No home. No English. *American Demographics, 15,* 42-43.

Hayflich, P. F. (1992). Why don't they speak English? *Training, 29,* 75-78.

Hernandez-Truyol, B. (1996). Natives, newcomers, and nativism: A human rights model for the twenty-first century. *Fordham Urban Law Journal, 23,* 1075-1136.

Higham, J. (1955). *Strangers in the land: Patterns of American nativism, 1860-1925.* New Brunswick, NJ: Rutgers University Press.

Hoffman, A. (1974). *Unwanted Mexican Americans in the Great Depression: Repatriation pressures, 1929-1939.* Tucson: University of Arizona Press.

Immigration and Naturalization Service. (1999). *1998 statistical yearbook.* Washington, DC: Author.

Irving, K. (1993). Gendered space, racialized space: Nativism, the immigrant women, and Stephen Crane's "Maggie." *College Literature, 20,* 30-44.

Johnson, C. (1999). The California backlash against bilingual education: *Valeria G. v. Wilson* and Proposition 227. *University of San Francisco Law Review, 34,* 169-195.

Jost, K. (1995). Cracking down on immigration. *CQ Researcher, 5,* 97-120.

Katz, J. (1996, November 11). For determined field hands, hope quickly fades. *Los Angeles Times,* p. A1.

Lears, T.J.J. (1985). The concept of cultural hegemony: Problems and possibilities. *American Historical Review, 90,* 567-593.

Liu, C. (2000, September 7). Driver charged with hate crime in death. *Los Angeles Times,* p. B5.

Macias, R. (1993). Language and ethnic classification of language minorities: Chicano and Latino students in the 1990s. *Hispanic Journal of Behavioral Sciences, 15,* 230-257.

Martin, P. (1993). The missing bridge: How immigrant networks keep Americans out of dirty jobs. *Population and Environment, 14,* 539-566.

Martin, P. (1995). Proposition 187 in California. *International Migration Review, 29,* 255-263.

Martinez, R. (1995, September 3). Bringing back an L.A. hero: Ruben Salazar gave a passionate voice to city's Chicanos. *Los Angeles Times Book Review,* pp. 2, 11.

Mazon, M. (1984). *The Zoot Suit riots: The psychology of symbolic annihilation.* Austin: University of Texas Press.

McCarthy, K., & Valdez, R. B. (1986). *Current and future effects of Mexican immigration in California.* Santa Monica, CA: RAND.

McCarthy, K., & Vernez, G. (1998). *Immigration in a changing economy: California's experience—questions and answers.* Santa Monica, CA: RAND.

McClain, C. (1984). The Chinese struggle for civil rights in nineteenth century America: The first phase, 1850-1870. *California Law Review, 72,* 529-568.

McGovney, D. (1947). The anti-Japanese land laws of California and ten other states. *California Law Review, 35,* 7-60.

McManus, W. (1990). Labor market effects of language enclaves: Hispanic men in the United States. *Journal of Human Resources, 25,* 228-252.

McWilliams, C. (1968). *North from Mexico: The Spanish-speaking people of the United States.* New York: Greenwood.

Meisenheimer, J. (1992). How do immigrants fare in the U.S. labor market? *Monthly Labor Review, 115,* 3-19.

Miller, G. (2001, February 23). Minorities may be first to feel pain if recession hits America. *The (Riverside, CA) Press Enterprise,* p. C1.

Muller, T. (1993). *Immigrants and the American city.* New York: New York University Press.

Muller, T., & Espenshade, T. (1985). *The fourth wave: California's newest immigrants.* Washington, DC: Urban Institute.

Murphy, B., Barlow, W., & Diane, D. (1993). English-only rules might not violate Civil Rights Act. *Personnel Journal, 72,* 24.

Nelson, S., & O'Reilly, R. (2000, August 30). Minorities become majority in state, census officials say. *Los Angeles Times,* p. A1.

Olivo, A. (2001, February 25). Ghosts of a 1931 raid: A random INS roundup set the tone for decades of ethnic tension. *Los Angeles Times,* p. B1.

Parker, T. (1994). The California story: Immigrants come to California as a result of federal—not state—policies. *Public Welfare, 52,* 16-20.

Perea, J. (1992). Demography and distrust: An essay on American languages, cultural pluralism, and official English. *Minnesota Law Review, 77,* 269-373.

Perea, J., Delgado, R., Harris, A., & Wildman, S. (2000). *Race and races: Cases and resources for a diverse America.* St. Paul, MN: West Group.

Portes, A., & Rumbaut, R. (1990). *Immigrant America: A portrait.* Berkeley: University of California Press.

Puente, M. (1998, May 27). Proposition 227 unites some unlikely teammates. *USA Today,* p. 4A.

Rodger, P., & Doyle, M. (1994, January 9). War of words. *Fresno Bee,* p. A1.

Rolph, E. (1992). *Immigration policies: Legacy from the 1980s and issues for the 1990s.* Santa Monica, CA: RAND.

Ronfeldt, D., & Ortiz de Oppermann, M. (1990). *Mexican immigration, U.S. investment, and U.S.-Mexico relations.* Santa Monica, CA: RAND and Urban Institute.

Ross, W. (1994). *Forging new freedoms: Nativism, education, and the Constitution, 1917-1927.* Lincoln: University of Nebraska Press.

Sanchez, G. (1993). *Becoming Mexican American: Ethnicity, culture, and identity in Chicano Los Angeles, 1900-1945.* New York: Oxford University Press.

Scafidi, S. (1999). Native Americans and civic identity in Alta California. *North Dakota Law Review, 75,* 423-448.

Schmidley, D., & Alvarado, H. (1998). *The foreign born population: March 1997* (Update, Current Population Reports P20-507). Washington, DC: U.S. Bureau of the Census.

Schrag, P. (1995). Son of 187: Anti-affirmative action propositions. *New Republic, 212,* 16-19.

Schuler, K. (1996). Equal protection and the undocumented immigrant: California's Proposition 187. *Boston College Third World Law Journal, 16,* 275-312.

Simon, J. (1989). *The economic consequences of immigration.* Cambridge, MA: Basil Blackwell.

Takaki, R. (1993). *A different mirror: A history of multicultural America.* Boston: Little, Brown.

U.S. Department of Justice. (1999). *Legal immigration, fiscal year 1998.* Washington, DC: Immigration and Naturalization Service.

Vargas, S. (1998). Deconstructing homo[geneous] Americanus: The white ethnic immigrant narrative and its exclusionary effect. *Tulane Law Review, 72,* 1493-1596.

Vargas, S. (1999). Judicial review of initiatives and referendums in which majorities vote on minorities' democratic citizenship. *Ohio State Law Journal, 60,* 399-555.

Verdin, T. (2000, August 30). Minorities become majority in state. *The (Riverside, CA) Press Enterprise,* p. A1.

Vernez, G. (1993). *Mexican labor in California's economy: From rapid growth to likely stability.* Santa Monica, CA: RAND.

Vernez, G. (1994). *Undocumented immigration: An irritant or significant problem in U.S.-Mexico relations?* Santa Monica, CA: RAND.

Volokh, E. (1997). The California civil rights initiative: An interpretive guide. *UCLA Law Review, 44,* 1335-1402.

Wade, D. (1996). The conclusion that a sinister conspiracy of foreign origin controls organized crime: The influence of nativism in the Kefauver Committee. *Northern Illinois University Law Review, 16,* 371-409.

Walsh, M. (2000, March 25). Latinos' net worth shrinking despite boom times. *Los Angeles Times,* p. A16.

Wood, D. (1994, June 1). Ballot vote on illegal immigrants set for fall in California. *Christian Science Monitor,* p. 1.

Zabetakis, A. (1998). Proposition 227: Death for bilingual education. *Georgetown Immigration Law Journal, 13,* 105-128.

CHAPTER 12

DISASTERS

Harvey E. Rich
Loretta I. Winters

Social problems affect social structures rather than just involving personal concerns. They affect society rather than selected individuals and require structural and cultural solutions. Only at the point where private issues associated with **disaster** are translated into public concerns will general support for disaster prevention and mitigation measures emerge. This transformation will occur, in part, through the influence of **claims makers,** individuals or groups who attempt to publicly define situations as harmful and to promote action to change them (Stallings, 1995). Scientists are among the major claims makers for natural disasters, yet, as we suggest in the section on the media, they are without real power. As Stallings (1995) indicates,

> Whether the risk [of earthquakes] is primarily a threat to life or potential economic disaster, whether it is widespread or confined to one area, whether it is more important or less important than other public issues, and so forth, depends on what risk promoters are able to accomplish rather than on self-evident "facts" about the earth. (p. 2)

Social problems textbooks have typically neglected the topic of disasters, with the exception of the first edition of Merton and Nisbet's *Contemporary Social Problems* (Fritz, 1961). Although this work was frequently cited in disaster research, the chapter was dropped in later editions of the book due to lack of agreement on whether the subject matter was a social problem, a subset of collective movements, or a branch of social organizations.

We believe there is a high degree of arbitrariness to subject placements. This chapter attempts to show that social activities concerning disasters, such as the design and enforcement of building codes, disaster relief, and funding for earthquake prediction, are all based on societal consensus. Due to highly visible media attention and a growing list of natural and technological disasters, societal awareness and the belief that something can be done has reached a high level in California, indicating that this topic clearly deserves treatment in a social problems context. In the present chapter of the second edition of *California's Social Problems,* we are again following this approach.

Rossi, Wright, and Weber-Burdin (1982) suggest that despite the view by sociological researchers that disasters be considered social and political issues, most people see them as private troubles. Studies indicate that "individuals may not take significant action to prepare themselves for a disaster because they see mitigation and prevention as primarily a governmental responsibility" (Dynes, 1992, p. 72).

Yet, in the 1992 presidential election, candidates, civil rights organizations, consumer groups, and environmentalists lacked earthquake reduction platforms in their programs (Stallings, 1995). In the 2000 election, only one candidate even made environmental issues a priority, and no candidate said anything about disaster preparation or mitigation. Grassroots-level efforts to generate mitigation programs are not common because these efforts are viewed as the responsibility of higher levels of government. But, higher levels of government perceive these programs to be the responsibility of local government. The lack of commitment and urgency on the part of individuals and political actors to place disasters in the realm of general welfare may contribute to little community change until people experience firsthand the effects of a disaster.

DEFINING DISASTER

In this chapter, we will focus on natural and technological disasters. However, before discussing disasters in any detail, we should carefully define our terms. Politicians, the public, and some researchers have attached the term *disaster* to a wide variety of situations that threaten human life or property. We will adopt the definition that *disaster* applies only to sudden events of a catastrophic nature that affect a major portion of a human community or society and overwhelm the capacity of the community to quickly absorb and recover from the loss (Bates & Peacock, 1987). Events that may overwhelm the community or the society but that occur slowly, such as droughts or famines, will remain outside the purview of our narrowed definition. Furthermore, disasters elicit consensus behaviors and are therefore distinguishable from conflict crises such as riots or terrorist acts (Quarantelli, 1999b).

Barton (1969) added a useful perspective to the dialogue by using the concept of **collective stress.** A collective stress situation is one that occurs when many members of a social system fail to receive expected conditions of life from the system, such as safety, protection from attack, food, shelter, and income (Wenger, 1987).

We will exclude from consideration as disasters those events that affect only a few individuals or a neighborhood. These might be referred to as calamities, but they do not occur on a sufficient scale to overwhelm the capacity of the community to deal with them. Furthermore, when communities can simply use their traditional response organizations, we might prefer to call such events *emergencies* (Wenger, 1987). This distinction is significant because the few organizations that plan for disasters fall short in their efforts because they do not acknowledge the differences between emergencies and disasters (Quarantelli, 1994b).

In this chapter, we make no distinction between natural and technological causes of disasters for four reasons:

1. The worldwide shift from religious values toward secularism changed perceptions of the causation of disasters from other-worldly to mechanistic rationale.

2. Analysis of the disaster phenomenon moved from concentration on the physical aspects to socially oriented approaches.

3. The manner in which individuals and society prepare for, respond to, and recover from natural or technological events is very similar.

4. It is increasingly difficulty to distinguish between the two types of disasters, as technological and geographical incidents become more intertwined.

We elaborate on these points in the following paragraphs.

Although the popular conception of natural disasters continues to be that they are "acts of God," over which humans have little control, the spread of secularism has brought a different explanation for natural disasters. Physical forces are no longer viewed as acts of God by scientists, but are viewed as natural events made understandable using the scientific method. That is, scientific epistemologies have gained prominence over nonscientific ones for understanding the forces behind natural events. Earthquakes are now considered a result of the movement of tectonic plates beneath the earth's surface, which might in the future be predictable. Scientific understanding conveys a perception of control, hence an active stance rather than a passive one toward some aspects of the environment leading to disaster. This orientation creates a new arena of activity for scientists, preparedness in addition to response. In fact, the United Nations proclaimed the 1990s to be the International Decade for Natural Disaster Reduction.

Many physical scientists and some social scientists, in particular geographers, have made the distinction between disasters caused by natural agents and technological ones created by people. They have primarily defined such events in physical terms (Quarantelli, 1994a). However, since the earliest research on disasters in the 1920s, sociological researchers have generally conceptualized these occurrences in social terms, ignoring the physical/social difference. Agreement now exists among social researchers that there is no difference between a natural and a technological disaster because all disasters are the product of human actions. A "disaster is not a physical happening, it is a social event" (Quarantelli, 1993b, p. 24).

Although the origins of the two types might have different categories of agents, the trans-disaster period and the post-disaster recovery of both types of events have a great number of similarities and can be described primarily by their social origins, processes, and results (Quarantelli, 1999b). In general, disaster prevention, preparedness, and recovery processes are fairly generic. The reactions of political entities and social structures, as well as the behaviors of individuals, tend to be quite similar whether one is dealing with massive dislocation from a hurricane, a 100-year flood, collapsing structures in a 6.8 magnitude earthquake, or a toxic chemical release (Quarantelli, 1999b).

Floods, fires, earthquakes, and other disasters have social consequences only due to pre-, trans-, and post-impact activities of individuals and communities. Thus, large brush fires, such as those that occurred in the Los Angeles area in 1993, are not inherently disasters; they are nature's way of revitalizing the local ecological community. They only become defined as dreadful because humans persist in building large numbers of highly flammable structures in the midst of dense brush habitats. In general, damage is inversely related to precautions taken by builders, use of materials to construct fire-resistant structures, and generous ground clearances. Even so, it might be argued that buildings should not be placed in high danger areas. Similarly, by preventing building in known flood plains, flood damage could be minimized and, by uniformly enforcing and requiring updating of building codes, earthquake damage could be substantially reduced. Buildings might shake, but if nothing were broken, an earthquake would not be a disaster. The point is not that such natural events can be prevented but that their impact would be much reduced if community perceptions and norms were different. Thus, all disasters, natural or otherwise, are social events with social precursors, social activities, and social consequences. Quarantelli (1999c) affirms this perspective, stating, "it is a misnomer to talk about *natural disasters* as if they could exist outside of the actions and decisions of human beings and their societies" (p. 11).

In the past, disasters were primarily caused by natural agents, but due to the enormous complexities of contemporary technology, the integration of modern technological systems, and the growth and concentration of people in large metropolitan areas, the incidence of technological disasters is increasing (Quarantelli,

1999c) and promises to provide society with some of the most disruptive events of the future (Akimoto, 1987). Furthermore, as mentioned above, it will become increasingly difficult to distinguish between natural and technological disasters because natural disaster agents now have the potential to generate or magnify concurrent technological disasters. This is expected to occur at a greater rate because of the proliferation of hazardous substances and the increasing vulnerabilities of social structures (Quarantelli, 1993a, 1996, 1999c).

HISTORY OF DISASTER RESEARCH

Although the first study of disasters occurred in 1920,[1] writings in disaster research were few until the middle of the 20th century (Darley & Gilbert, 1985; Quarantelli, 1992, 1994b). This is when social researchers began to view natural disasters as a product of physical agents and the social setting. Control over environmental impact is an important aspect of this approach because people are considered active participants in the prevention, mitigation, preparedness, response, and recovery from a disaster. The federal government initiated the first natural disaster research endeavors during the 1950s; sociologists were the primary group of investigators involved at this time. The U.S. military was interested in understanding peacetime disasters to apply results to conditions of warfare. Numerous studies were conducted at universities; however, the National Opinion Research Center (NORC) at the University of Chicago undertook the most important research between 1950 and 1954.

At the end of the 1950s, government support declined and was replaced by support from civilian agencies. In 1952, the National Academy of Sciences' National Research Council appointed a Committee on Disaster Studies, which focused on the social psychology of mass emergencies. The Disaster Research Group formed in 1957 to continue disaster research but dissolved in 1961, with the Disaster Research Center (DRC) at Ohio State University taking over its work in 1963. In Ohio, an individualistic orientation to disaster research yielded to a purely sociological approach. In 1985, the DRC moved to the University of Delaware. By the start of the 21st century, the DRC had conducted several hundred field studies on preparedness, response, and recovery from natural and technological disasters as well as civil disturbances (Webb, 1999).

The 1970s and 1980s marked dramatic increases in disaster research. In the mid-1970s, women began to enter this area in greater numbers (Tierney, 1998). During this period, a variety of efforts to codify the research were attempted (Quarantelli, 1994b). These include Mileti, Drabek, and Haas's (1975) summarization of research findings of 191 published studies on emergencies in *Human Sys-*

tems in Extreme Environments. Quarantelli and Dynes of the DRC school produced another major review of the literature in 1975 in the *Annual Review of Sociology.* These discussions clearly revealed the variety of myths about disasters pervasive in the popular culture, which will be discussed later in the chapter.

The Research Committee on Disasters within the International Sociological Association was established in 1982 and includes a substantial number of constituents outside the field of sociology. Members come from more than 30 countries. The association now publishes a journal, the *International Journal of Mass Emergencies and Disasters.*

By the 1990s, disaster research had spread beyond the DRC to a variety of American universities. At present, nonacademic institutions such as the Oak Ridge National Laboratory and the Battelle Human Research Institute also engage in social science research on disasters.

The latest research includes work by geographers, political scientists, psychologists, anthropologists, medical researchers, and public administrators. The primary focus had been on natural disasters; however in the late 1970s and in the 1980s, inquiry broadened to include investigations into technological disasters, with non-American research an increasingly important part of the field.

THE CALIFORNIA EXPERIENCE

During the past decade or so, California has seen such a variety of natural disasters and human-created devastations that many would agree the glow has left the Golden State. In 1987, a 6.1 magnitude earthquake struck Whittier; in 1989, the 6.7 magnitude Loma Prieta quake sent a portion of the upper level of an Oakland freeway crashing down onto the lower level, and the Marina district in San Francisco was destroyed; in 1992, riots/uprisings in the Los Angeles inner-city areas by African Americans and Latinos claimed 54 lives (Lozano, 1994) and left $1 billion in property damage (Lee, 1995).

In 1992, a 7.5 magnitude earthquake was centered in the Big Bear region, far enough away from population centers to do minimal damage but of such magnitude that it reminded everyone of their vulnerability to the forces of nature. The winter of 1993 brought massive brush fires to Southern California, destroying 220,000 acres of brush and forest (Hal, Hurst, & Zamichow, 1993) and leaving more than $1 billion in property damage (Platte, 1994). The devastation denuded hillsides, resulting in flooding during the winter and spring and the dislocation of thousands of individuals. In January 1994, a 6.8 magnitude quake centered in Northridge became the largest natural disaster to strike the United States in the 20th century, causing an estimated $15 billion to $20 billion in property damage, from which California has yet to recover. To these events should be added the daily bombardment of chemicals from air pollution; toxic waste sites, such as Casmalia and Stringfellow, which have not been cleaned up; and a myriad of other chemical explosions, water purification failures, brush fires, floods, and the like.

Are these events occurring more frequently, and if so, why? Are we perhaps more susceptible to disasters in the second half of the 20th century, or are we just more sensitized to these events because of the ubiquity and speed of the modern media? We intend to address these questions later in the chapter.

CALIFORNIA DISASTERS COMPARED TO THE UNITED STATES

The type of technological disaster will vary depending on the industrial production facilities that happen to be located in a particular community or state. Most states have highly diversified economies, and therefore, a variety of potential **catastrophes** can occur. Examples would include chemical plant fires, refinery explosions, nuclear or gas-fired power plant accidents, accidents involving highly explosive or nuclear materials during transport, and fire in a high-rise structure. Numerous other possibilities exist in our highly complex industrial infrastructure. Recall that, in this chapter, we are specifically excluding terrorist acts or other acts of violence that may result in outcomes lacking consensus.

However, the geographic distribution of natural disasters varies considerably by locale. Hurricanes are common along the East Coast and in the South, tornadoes in the Midwest, fires and earthquakes in the Far West. Although the processes in the pre-, trans-, and post-disaster periods are similar for all disasters, California has to deal primarily with three specific types of disasters: earthquakes, brush fires, and rain- and mud-induced erosion and flooding.

California consistently has had to deal with destruction wrought by earthquakes and fires. Major earthquake fault lines run throughout the state, exposing large metropolitan areas as well as rural communities to potentially devastating destruction. As a result, the state has developed a statewide emergency-planning system. The Office of Emergency Services was used extensively during the Southern California fires of 1993 and during the Northridge earthquake of 1994, the most costly catastrophe the United States has yet encountered. The 1994 earthquake put this system to the test; the state agencies responsible for implementation worked fairly effectively with community and private organizations during the recovery phase (Nigg, 1995).

Although the frequency of disaster occurrence is going to increase everywhere in the 21st century and the extent of damage will be greater than ever, this trend is likely to be more pronounced in California than in most states due to its large and growing population, with highly dense concentrations along the coast from San Francisco to San Diego. Contributing to the potential for increasing rates and extent of disaster damage is cutting edge technology, California's trademark, which is so easily damaged, and the enormous diversity of its people.

Highly complex, integrated industrial infrastructures concentrated in major metropolitan communities, typical throughout California, are more susceptible to damage from major events. The variety of ethnic groups in the state, with more than 100 spoken languages, makes it likely that communication devoted to preparedness and recovery will be impeded when a disaster strikes. Furthermore, the disparity in income between the richest and poorest sectors in California portends uneven allocation and use of resources in dealing with disasters. The poorest segments of society are more likely to show the greatest impact and experience the slowest recovery from a disaster.

CALIFORNIA, THE DEVELOPED WORLD, AND THE DEVELOPING WORLD

California, as the most populated state in the United States, shares many features of other states and developed countries. It has a highly urbanized population, most citizens residing in three major metropolitan areas. In Southern California, these areas stretch for hundreds of miles. Complex, highly integrated technological systems are susceptible to major problems. Electric power generation and distribution illustrate that vulnerability. Electricity in the West is distributed over one large power grid, which can easily be interrupted as we saw in 1984, when nearly 2 million people in 10 Western states, including California, were blacked out from a power disruption in Washington, before California could isolate itself from the grid (Lerten, 1984). A similar massive power shutdown occurred in 1994, when outages struck California and seven other Western states (Hecht, 1994). Transportation systems disintegrate when major arteries are shut down, as happened with freeway overpass collapses during both the 1989 Loma Prieta and the 1994 Northridge earthquakes. Fires started by people and through natural causes quickly spread from brush area to brush area and have the capability of disabling major portions of a large metropolitan community, as occurred in Southern California in 1993.

Add to this list the destruction that might take place if a large toxic chemical compound were to become air- or water-bound, if destructive biological agents were released into a city water system, or if the unimaginable catastrophe of a partial or complete nuclear plant meltdown were to occur. Even our information systems have become so centralized and cybernated that a computer virus might cause communication gridlock or perhaps the loss of hundreds of millions of dollars in financial transactions. These potentialities for disaster are shared by other advanced industrialized countries, especially those possessing densely populated centers with highly integrated technological systems, such as those in Europe or Japan.

Less developed countries that are industrializing at a rapid rate, such as South Korea, Brazil, Mexico, Singapore, and Taiwan, share many vulnerabilities of modern industrialized states. However, large sectors of their populations live in rural areas, less exposed to modern system breakdowns.

Deaths from natural disasters occur on an enormous scale far more frequently in less developed countries, despite the lack of integration of the various sectors of their technological and social systems. This illustrates how modern technology has decreased the likelihood that citizens will die from a massive catastrophic event. The United States has had only three natural disasters in its history that have killed more than 1,000 people. That number pales into insignificance when compared with such figures as the 300,000 and 139,000, respectively, who died in typhoons in Bangladesh in 1970 and 1991 (The Disaster Center, 2000). In Tangshan, China, in 1976, an earthquake killed more than 300,000 individuals, and a flood in Bangladesh in 1987 left 25 million homeless. An earthquake in 1989 in Armenia killed 30,000 people.

In recent years, we can see this pattern enduring as death on a massive scale occurs regularly, with small- and large-scale devastations continuing to strike frequently. In 1997, a single landslide during Hurricane Pauline killed more than 375 people in Mexico, when a river of mud buried homes on the outskirts of Acapulco (Ferriss, 2001; Riley, 1999). In 1998, as many as 10,000 people died in Latin America, and the economies of Honduras and Nicaragua were decimated, as a result of Hurricane Mitch (Riley, 1999). A 7.4 magnitude earthquake in Turkey in 1999 is estimated to have killed as many as 40,000 people (*Los Angeles Times*, 1999). In the first 2 months of 2001 alone, three major earthquakes took an enormous toll in lives. In January, a 7.9 temblor, which devastated the state of Gujarat in India, may have left half a million people homeless and more than 50,000 dead ("Earthquake in India, 2001; Li & Dahlburg, 2001). Two massive earthquakes, with magnitudes of 6.6 and 7.6, struck El Salvador in January and February leaving over 1.3 million people homeless out of a population of 6 million, and at least 1,100 dead (Nordwall, 2001; "Supporting El Salvador," 2001).

The differences in mortality between California and countries of the developing world can be attributed entirely to stronger building codes, housing regulations, and land use regulations in California (Quarantelli, 1994a), as well as deforestation, soil erosion, and poverty in the developing nations (Ferriss, 2001). The negative consequences for California and other highly developed areas are primarily massive economic losses and social disruptions. The Loma Prieta earthquake killed fewer than 90 individuals, yet the direct economic losses are estimated at near $7 billion and the indirect losses at more than twice that amount (Mulligan & Kristoff, 1994).

SOCIOLOGICAL PERSPECTIVES FOR EXAMINING DISASTER

CONSENSUS THEORY

Because theoretical perspective has a large impact on the questions asked and the evidence selected for analysis, it is important to examine the various extant disaster perspectives. One view, consensus theory, draws on the structural functional paradigm. It dominated sociological research following World War II. Because of a large increase in disaster research conducted by sociologists in the following decade, much research since then has followed this paradigm.

Stallings (1988) points out that, from a consensus perspective, "disasters are treated as independent variables that provoke both planned and emergent social responses" (p. 579). This approach views responses to disasters as restoring a society's equilibrium. The restoration is manifest in the cooperation that rests on agreed-on norms that shape behavior directed at recovery. Recovery ultimately works toward maintaining the status quo. Thus, high levels of agreement on social goals and the means for achieving them, which arise during the emergency phase of disasters, emphasize within the community the existence of "strong pressures to restore rather than change previous social structures and processes" (Stallings, 1988, p. 573). This perspective is supported by evidence that conflict is virtually nonexistent during the emergency period of natural disasters (Quarantelli & Dynes, 1976). In fact, much of early disaster research found that rational, goal-directed, rather than chaotic, behavior predominated during a disaster crisis (Fritz, 1961). Furthermore, emergent groups or social structures tend to bring out the best, rather than the worst in people (Stallings, 1988).

Four assumptions operate under the consensus model of disasters (Dynes & Quarantelli, 1971; Quarantelli & Dynes, 1976; Wenger, 1987). First, the disaster agent is external to the community and confronted by all of its members. Second, immediate attention to the problem is warranted. Third, the event presents the community with clear and nonambiguous problems. Finally, the suffering that arises is perceived as due to chance and not social status.

Normative consensus arises in response to these assumptions. Operating norms stress the responsibility of individuals to community over self (Quarantelli & Dynes, 1976). Furthermore, goal-directed behavior values activity benefiting the community as a whole rather than a small number of individuals. Thus, a leveling of social distinctions occurs. Setting aside individualism for communalism during the emergency period is the backbone of the consensus approach. Stallings (1988) elaborates, "There is consensus that no actor—individual, family, business, political party, or government—should be 'better off' as a result of the disaster. Only the community as a whole should emerge a 'winner' " (p. 573). **Emergent norms** of assistance to those in need of help further discourage partisan political conflict and inhibit various forms of economic hostility. They also reduce the probability of conflict over private property by discouraging looting and clarifying the distinction between public property and private property.

CONFLICT THEORY

Another view of disasters is derived from the power-conflict paradigm, which emphasizes power differentials and antagonisms in society. Stallings (1988) examined the research conducted by both consensus and conflict proponents and provided a detailed analysis of the literature. He reached a number of conclusions. The first is that disaster conflict theory is not as developed as consensus theory, and therefore, little research in this area flows from this perspective. Actually, much analysis carried out using a conflict orientation relies on studies conducted by researchers using the consensus perspective. However, those researchers provide alternative interpretations for the absence of social hostilities during the emergency phase.

Conflict theorists attribute the lack of disagreement both to the intervention of the state in the disaster process and to the recognition by participants of common interests in restoring prior structural arrangements. They view disaster as "both an independent variable responsible for triggering various types of citizen and government responses and as a dependent variable directly influenced by both preinterruption of social patterns and by what government does or does not do before impact" (Stallings, 1988, p. 580).

Stallings (1988) enumerates various criticisms of the consensus approach that conflict theory addresses. First, he points out that social conflict is not actually absent during the emergency phase of disasters but occurs relatively less frequently. He also notes that the normative consensus found in the emergency phase appears to apply in the United States but not necessarily in other societies because of the greater political consensus that exists in this country. Furthermore, this generalization characterizes the emergency phase but not necessarily other stages (i.e., planning, restoration, and recovery). Actually, most research on conflict during disasters focuses on the later post-impact phases (restoration and recovery). High levels of social disagreement tend to mark these stages, as commonly held consensus evaporates when early tensions reassert themselves over time. Greater agreement about the immediate causes and solutions for natural rather than technological disasters exists due to the fact that blame and scapegoating are common during and after technological mishaps.

The idea that disasters only exist when there are victims is important to a power-conflict analysis. It follows that such events will have the greatest effect on the poor and politically disenfranchised. Susman, O'Keefe, and Wisner (1983) argue that a marginalization process places the poor in a more vulnerable position than the more affluent when calamity strikes. They state, "The disasters resulting from such a process are not 'Acts of God.' Quite the reverse; The poor, instead of inheriting the Earth, are being eaten up by it."

Advantaged groups have more control in the event of a disaster because they have the resources to mediate its effects. They avoid residency in the most vulnerable areas and can afford housing that is designed to withstand the devastating effects of disasters. Although California is vulnerable to earthquakes, the affluent are better able to afford newer homes, earthquake proofing, and earthquake insurance. They are also more able to recover from economic setbacks of repair and relocation. Homes built in the 5 years prior to the Northridge earthquake were less vulnerable to quake damage than older homes, as the dramatic collapse and flattening of the Northridge Meadows apartments clearly demonstrated.

Tierney (1999), in discussing the link between imposed risk and social inequality, points to prodevelopment interest organizations, landlords' associations, and the real estate lobby as groups that tend to oppose earthquake hazard mitigation for economic reasons. These groups attempt to defeat proposed measures and weaken adopted measures. Members participate in these activities while enjoying a buffer to the harmful effects of disasters suffered by the poor but largely avoided by the affluent. Furthermore, Quarantelli (1999a) reveals that the higher people's social class the more likely they are to recover to the pre-impact level after a disaster.

Minority groups are also in a vulnerable position in the United States and are more prone to suffer from a variety of harmful environmental and technological mishaps. Bullard (1993) describes the environmental plight of people of color.

> Communities are not all created equal. In the United States, for example, some communities are routinely poisoned while the government looks the other way. Environmental regulations have not uniformly benefited all segments of society. People of color (African Americans, Latinos, Asians, Pacific Islanders, and Native Americans) are disproportionately harmed by industrial toxins on their jobs and in their neighborhoods. These groups must contend with dirty air and drinking water—the by-products of municipal landfills, incinerators, polluting industries, and hazardous waste treatment, storage, and disposal facilities. (p. 15)

Kaniasty and Norris (1995) found that European American victims of Hurricane Hugo received more tangible, informational, and emotional support than African American. Furthermore, they found that the discrepancy between the amount of social support received by African Americans and European Americans widened as the economic loss increased.

Conflict theorists argue that disaster relief maintains the current economic arrangements of society. Emergent norms act to restore previous social structures and processes (Stallings, 1988). After the emergency phase of a disaster, there is little tolerance for changing existing social arrangements. As an example, local officials from a small southwestern city withdrew their request for aid following a flood because a federal official proposed building new facilities for low-income Hispanic flood victims outside the flood plain, where Causcasians already primarily resided.

The literature on ethnic minorities and disasters, although sparse, does indicate a difference in the way various groups view, plan, and respond to natural disasters. For example, ethnic minorities tend to show the smallest reaction to warnings. This has the potential to be a meaningful issue in Southern California with its large and growing proportion of minorities. An illustration of the consequences of this factor is that, because evacuation response time is important in reaching safety from flash floods, minority group members will suffer greater adverse consequences as a result of flash flooding.

Personal response to disasters is also influenced by group membership characteristics. As Quarantelli (1996) argues, some ethnic and minority groups assume hazards can be overcome, whereas others assume people have to accept and adjust to the threats of disasters. This shows that differing beliefs about disasters affect mitigation and prevention efforts as well as emergency preparedness. Some minority group members place an emphasis on extended family that provides extensive support during a disaster crisis. When family resources are low, lack of trust placed in outsiders could prevent members of such groups from seeking outside help.

NATURAL DISASTERS IN CALIFORNIA TODAY

Much of the current sociological writing about disasters takes a social constructionist view, one that draws on the symbolic interactionist paradigm. The underlying assumptions of this approach are, first, events are disasters only if defined as such and, second, disasters do not occur in a social vacuum. It is not the event that is important but understanding how it is perceived and acted on and how it affects consequences that are significant. Thus, we need to examine the major social actors, what these individuals or groups have to say, and reactions to statements made by them.

Risk perception depends on the images created by **risk promoters** via communication channels such as television, newspapers, movies, and government hearings. An oil refinery in flames next to private houses is a very powerful image of consequences of inadequate zoning laws, whereas flames engulfing homes in the middle of high brush may illustrate lack of enforcement of housing placement regulations.

Risk promoters have varying concerns. The following quotes from a brochure prepared by the Southern California Earthquake Preparedness Project are illustrative:

> Earth scientists unanimously agree on the inevitability of major earthquakes in California. The gradual movement of the Pacific Plate relative to the North American Plate leads to the inexorable concentration of strain along the San Andreas and related fault systems. . . . (Stallings, 1995, p. 24)

and

> Strong ground shaking, which is the primary cause of damage during earthquakes, often extends over vast areas . . . Casualties and property damage occur principally because of the failure of manmade structures. (Stallings, 1995, p. 24)

Emphasis here is clearly placed on the physical setting and natural causes of destruction, with a nod toward engineering solutions to structural failure.

In California, as elsewhere in the United States, the principal natural hazards claims makers are physical scientists and engineers. Representatives from these professions create the majority of input to governmental agencies and commissions. Perception of the dangers created by natural hazards centers on information given by these groups to the public through the media. The preference of the media and policy groups is for statements emphasizing the power of nature and of engineering ways to lower the danger to individuals once an event takes place, rather than underlying social and political processes that contribute to disaster impacts (Stallings, 1995).

Earth scientists have a stake in defining the temblor danger as human inability to predict earthquake occurrence. In large measure, this is because federal grant-making agencies are predisposed to confer grants for the study of physical phenomena such as the frequency and impacts of ground movement and earthquake prediction capability. Interestingly, local politicians and businesses occasionally find this approach to be a threat in its own right, citing the fact that community revenue losses from such predictions would be greater than assistance generated by federal intervention after a quake. This provides an example of counterclaims makers, who disagree with the usefulness of the quake prediction approach taken by the claims-making scientific establishment (Stallings, 1995).

Nonetheless, the aftermath of the 1971 Sylmar earthquake illustrates the dominance of the political process by scientific and technocratic claims makers. Three of the half dozen or so hearings on the subject involved individuals associated with one of these groups. In one hearing, earth scientists and engineers were in the spotlight for much of the time and made several points, among them (a) that more money was needed for basic research to understand the dynamics of earthquakes, (b) that future earthquakes in urban areas would produce even worse damage, and

(c) that failure to use existing knowledge in the placement of structures away from active fault lines was a cause of unacceptably high risk (Stallings, 1995). We will discuss the role of the media in granting claims makers legitimacy later in the chapter.

PLANNING

The normal or predisaster period defines community readiness before an event occurs. Preparedness measures are undertaken prior to a disaster to mitigate the negative consequences of the occurrence of the event (Perry, 1987).

Most communities have organizational structures already in place, which are set up to deal with the advent of crises, such as the police, fire department, an emergency management agency, the Red Cross, and the national guard. These organizations define their institutional missions as routinely responding to emergency tasks. Most take the disaster management view that disasters are just big emergencies. They do not regard a disaster as qualitatively different from ongoing routine emergencies. That is, during a disaster, traditional organizational boundaries need to be temporarily altered to permit a more coherent and integrated effort by a disaster response network. California, during the past decade, has been almost alone in setting up a structure, the Office of Emergency Services, to deal with the overarching need to work with a multitude of organizations during major disasters. This statewide organization, with a 1995-1996 budget of only $15 million, is charged with the immediate response and coordination of the state's efforts to deal with significant emergencies, which include earthquakes, floods, fires, riots, and major toxic spills (Reich, 1995). However, an integrated multiorganizational orientation toward disaster management is not typical of counterdisaster approaches, particularly at the community level (Britton, 1987).

Planning and preparedness require a variety of activities, which encompass (a) hazard vulnerability surveys, (b) assessment of potential disaster threats, and (c) the design of preparedness programs. These programs include warning systems, evacuation plans, citizen sheltering plans, and hazard awareness activities. A key component of such plans involves communication and public education from authorities about behavior and expectations when the time comes (Perry, 1987).

COMMUNICATION

Efforts to communicate information to the public fall into two broad categories. The first, hazards communication, is concerned with the nature of the risk associated with activities that have the potential to become critical emergencies. Examples are living near a nuclear power plant, near a large chemical complex, in a flood plain, or in a risk-prone brush area. The second, warnings communication, addresses risk when an emergency/disaster is imminent. Research on hazard awareness communication has been contradictory, with some studies suggesting that education can enhance risk perception whereas other studies show no or limited results. The following are four factors that affect the process of receiving risk warnings:

1. *Hearing about the emergency*: Typically, this will be through public warning systems or the media.

2. *Understanding the warning*: Meanings vary among people, and what is perceived may not agree with the message intended by the sender. For example, some people may leave their houses when hearing a brush fire warning, whereas others may believe that the wind is blowing away from their location and stay home watching events on television.

3. *Reality perception*: People will be encouraged to act on the warning if they believe that it is real and that the contents of the message are accurate.

4. *Self as target*: Individuals must see themselves as the intended targets of the information before they will act on it.

Thus, people must first hear the warning, understand what is stated in the message, believe that the warning is real, and view themselves as a target of the message before they will act as emergency organizations hope. Confirmation is an important part of the process of believing the validity of the message and personalizing it, which is why telephone lines become saturated so rapidly at the time of crisis. People place calls to relatives and friends to confirm the information and to view their role in it (Fitzpatrick & Mileti, 1994).

It has long been established in communication research that sender characteristics affect the communication process. Briefly, the following characteristics of the sender affect how the person receiving the message will react: the source of the information; message consistency, accuracy, and clarity; level of certainty about events and what to do; level of detail; frequency of repeating the message; specification of location; and number of distribution channels for the message.

As might be expected, responses to potential or real disasters also vary with the situational and personal characteristics of the receiver of hazard warnings. These situational/personal characteristics fall into six general categories:

1. *Environmental cues* or physical characteristics of an emergency setting.

2. *Social setting* factors: for example, family togetherness at the time of the event.

3. *Receiver social ties*: Greater family cohesion influences the decision to respond.

4. *Sociodemographic characteristics* of the receiver: Gender, social class, and ethnicity influence hearing, understanding, believing, personalizing, and responding.

5. *Psychological traits* of the receiver: In illustration, people with an internal locus of control, who are self-determined, act as if they have control over their behavior. On the other hand, people with an external locus of control, who are characterized by fatalistic views of the world, act as if their fate is in the hands of others and are less likely to respond to warnings.

6. *Selective perception* plays a role in how individuals process warning information. People tend to filter information to conform with existing views. If they do not believe that floods are particularly dangerous, then they are likely to ignore most warnings although the warnings may be directed at them. (Fitzpatrick & Mileti, 1994)

Similarly, many people living near nuclear power plants do not believe that the probability of problems is very great and ignore communication about evacuation plans, whereas others believe that leaks from such plants cause a variety of maladies and therefore tune into every report of environmental toxicity from radiation, no matter how minor.

The preceding discussion indicates that emergency preparedness authorities must pay attention to the social psychological processes at work in preparing education programs. If the listed characteristics of both senders and receivers are not given careful consideration, the impact of the messages may be lost or, worse, completely misinterpreted.

Dynes (1992) points out that scientists inform the public through the media, and neither the scientists nor the media alone are enough to move the public to acknowledge the importance of mitigation measures. Information received from the media is discussed with family, friends, and neighbors. Discussion, used to clarify and evaluate media reports, contributes to personal earthquake preparation. Palm, Hidgson, Blanchard, and Lyons (1990) found no correlation between perceived and actual physical risk, which might be expected since perceived risk is mediated by discussion of such risk. Furthermore, as Stallings (1995) indicates, a discrepancy exists between the advocacy of scientific experts and safety officials and the public's apparent lack of concern about earthquakes. These findings indicate that communication from the scientific community does not generally appear to be received adequately or that is does not generate behavioral change toward natural disaster mitigation by individuals.

THE MEDIA

Not surprisingly, public perception of disasters is highly influenced by the messages sent through the media, particularly television, where people receive most of their news, especially about breaking stories. Three aspects of television influence the way information is purveyed. The first is the emphasis on visuals, pictures that typify and dramatize an event. The second is the need for breaking news, usually involving dramatic action stories, and the third is the time compression element of reporting. Stallings (1995) indicates that earthquake threat stories usually run for a few days after an earthquake when coverage is winding down and reporters are seeking new angles. Stories about earthquake disasters emphasize themes of human tragedy and heroism. These stories highlight the drama and give emphasis to individual effort to survive or recover from the disaster. People interviewed on camera often are victims, rescue workers, and high-ranking officials. Authoritative interviews are usually with earth scientists and, occasionally, disaster preparedness experts. Stories about the earthquake threat are frequently about prediction.

These aspects of the television medium shape the type of story reported. The most obvious factor is that with many competing events, there is little time for each story. Thus, time rarely exists for elaboration of complex points. Sound bites play best on air, and the subtleties of risk analysis are lost. Second, claims makers have virtually no control over what is aired and what is not. Therefore, even experts feel left out of the communication process because their interviews are just as likely as any others to be cut to a sound bite or excluded from the day's reporting entirely. Because of the compression aspect of television news, interviews with important disaster authorities are likely to run next to interviews with completely uninformed individuals, confusing the public. To the untrained public eye, all interviews take on nearly equal moral authority (Stallings, 1995).

The dominant theme conveyed in the media is that nature has taken its toll or, in the case of a technological event, a controllably engineered part of the system has failed, but survival is up to the heroic individual. Technology and science will prepare us to deal with similar events in the future.

As mentioned earlier, the media perpetuate the dominant view that scientists who provide technical solutions to specific problems are the primary authorities to focus on when dealing with disasters. Thus, in the case of earthquakes, earth scientists are interviewed about quake magnitudes, frequency of occurrence, ground shaking, and area vulnerability. For brush fires, botanists discuss types of chapparal, fire prevention authorities display the latest fire fighting equipment, and structural engineers examine the fire resistant capacities of various types of structures. It is unusual to see a story examining the political infighting regarding location of housing or fire resistant materials, the logic of building codes, the political alliances that allow structures to be placed in their present locations, or the implications of disaster preparedness for minorities.

The media also help to perpetuate various myths about consensus disasters, whether natural or technological. These myths include (a) the occurrence of panic, (b) widespread looting and criminal behavior, (c) the existence of shock and long-term psychological trauma, and (d) considerable helplessness among victims. The preponderance of research has shown that these myths are substantially wrong. Panic flight is minimal, existing only during very specific situations, such as when it is perceived that the danger is immediate danger or that limited escape routes might be closing. Most behavior during and after the emergency phase is guided by substantial rationality and altruism. People are not immobilized. In fact, they are likely to take the initiative in acting to deal with the event. People are more likely to converge on the affected area to do what they think has to be done in the crisis. They do not sit by passively and wait for help from formal relief and emergency organizations. Anti-social and criminal behavior is similarly insignificant during the emergency stage. Although some shock does occur during the disaster, most research indicates the lack of long-term psychological problems resulting from disasters (Dynes, Tierney, & Fritz, 1994; Quarantelli, 1999b; Wenger, 1987).

EMERGENT ROLES AND GROUPS

The concept of role is an important bridge between the group and the individual. Sociologists use roles in a variety of ways; the most common have focused on either predictable behavior patterns or expectations about appropriate behavior. In either case, organizations are nothing more than collections of integrated sets of roles. We can gain substantial insight into disaster behavior by examining the normative pressures individuals feel depending on their group memberships. Thus, many researchers had predicted that occupational roles would be abandoned in large numbers during disasters because of member loyalties and ties toward the family (Killian, 1952). More recent research suggests that role conflict is not a problem during emergencies (Dynes, 1987). In a variety of studies, it appears that virtually no one abandoned his or her work role. Some individuals engaged in search behavior for a family member, usually in conjunction with their occupational role. Those who were at home reported to work shortly afterward, if needed. Several explanations have been advanced for the lack of occupational/family role conflict. The first is that family roles are not rigidly separated from work. Frequently, family members engage in emergency responsibilities as a unit. Second, disaster alters the normal role divisions within a community and creates an emergency consensus, which prioritizes roles. That is, family, community, and employer agree on which roles are

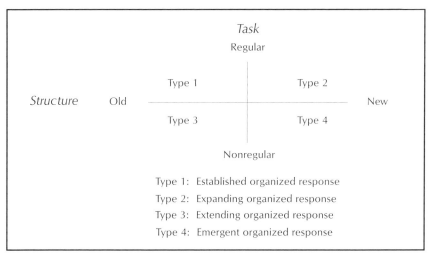

Figure 12.1. Disaster Research Center Typology of Organized Responses to Disaster
SOURCE: Webb, 1999.

important in the short run. One might abandon a family vacation or dinner at home to cope with the situation. The problem now appears to be not how to avoid conflict among multiple memberships but rather how to help organizations incorporate into their structures the large number of people who seek relevant roles during the emergency.

The occurrence of a disaster assaults the community with numerous simultaneous problems, just when the communication and organizational structures that normally deal with varying aspects of social life are in turmoil. Organizations that are set up to deal with emergencies will typically respond, but their members may have role conflicts with other groups to which they belong. Other organizations may not define their goals as emergency management but may, for one reason or another, find themselves working on such tasks. The Disaster Research Center (DRC) has produced a typology of group responses in the aftermath of a disaster. This approach lists four types of groups functioning after the crisis event (Webb, 1999) (see Figure 12.1).

Type 1 consists of established organizations that have traditional roles in a disaster, which provide guidance for members: Police and fire departments provide the clearest examples of this, with their rigidly defined role expectations. These organizations become involved quite early because they are set up to deal with emergencies, can mobilize significant resources quickly, and are able to deal with the situation with less ambiguity than are other organizations. They are usually quite effective in the early stages of the crisis.

Type 2 consists of expanding organizations, those with a small permanent membership that is supplemented by a large group of volunteers. The permanent members may continue to play traditional roles, or they may take on new ones along with the volunteers. The influx of many new people creates flexibility to deal with unusual circumstances, which is frequently lacking among established organizations. Examples of this organizational type are hospitals, shelters for the homeless, and food distribution centers.

Type 3 consists of existing organizations that extend themselves to disaster responsibilities (extending organizations), although they are not prepared for this role. These organizations may enlist in this effort entirely, or they may dedicate substantial portions of organizational resources to the endeavor. Relief efforts of the Boy Scouts, free meals offered by a restaurant, and the activities of church groups are examples of this type of organization. Involvement of these groups is frequently limited, and difficulties develop around interfacing with other more traditional organizations.

Finally, Type 4 groups emerge out of friendship patterns or formal organizations that did not exist prior to the disaster (emergent organizations). The participants generally possess substantial resources, skills, and experience, which they bring to bear on the emergency tasks. These groups are ephemeral, existing only during the crisis, although some organizations may exist in a noncrisis stage. Examples are ad hoc groups formed to force cleanup of a toxic disposal site, to prevent the location of a nuclear reactor near a community, and to distribute meals to the temporarily homeless.

The unique aspect of a disaster is that such groups arise spontaneously. Because their roles are not specified, Type 4 groups tend to conflict with the first two types of groups, which have more clearly defined rules governing what they can and should do. The fencing of the perimeter of California State University, Northridge, in the aftermath of the 1994 Northridge earthquake provides an example of this conflict. Relations were strained between the university administration and campus police on the one hand, and ad hoc groups of faculty and staff who wished to aid in the reconstruction process on the other. Some among the administrative staff viewed faculty as a nuisance and "in the way" of reconstruction, rather than as a positive force for additional relief.

DISASTERS AND THE FUTURE

We in California and the rest of the developed world have had the view for the past 100 years that science and technology have improved and will continue to enhance the quality of our lives. In many ways, this statement has been true. Unfortunately, there are risks associated with a modern industrialized state. Life always has had its share of danger from natural causes; floods demolish homes and crops, earthquakes destroy most built structures, fires wreak havoc on undeveloped and developed land, and hurricanes disrupt entire communities. All these events, of course, result in loss of life and disorder, sometimes on a very large scale. Industrialization has resulted in technological improvements that mitigate the worst aspects of these disasters, and this trend is likely to continue. On the other hand, the forces of modernization also produce vulnerabilities that did not exist previously and create potentialities for new large-scale disasters.

Because of the development of prediction and warning techniques, we are better able to prepare for large-scale natural disasters and there has been a relative decline in deaths. For example, the employment of satellite tracking systems that give days of warning moderates the ravages of large storms and the effects of flooding. Dams can withstand greater flood pressures, and modern structural engineering of buildings prevents large loss of life from most earthquakes and fires. However, a variety of factors are present in the modern industrial world that tend to increase the probability that disasters of the future will be more frequent and their effects more widespread.

The global increase in population density in many areas in the 20th century, coupled with migration to urban areas, has contributed to the creation of highly complex interrelated systems that are easily broken. These systems have enormous potential impact on large numbers of people in a limited locale. The haphazard development of most cities, placing substantial risk factors from technological events in the midst of residential and working populations, has exacerbated the brittleness of urban communities. Supporting systems such as water, electricity, gas, transportation arteries, and information networks are increasingly interrelated. A major catastrophe is likely to affect systems in a chain reaction effect, paralyzing an entire city.

These factors coupled with the escalation in use of complex technologies such as vast chemical complexes, nuclear power plants, and the widespread employment of radioactive materials in health centers throughout the state and country has enhanced the probability that the occurrence of small accidents or natural events will result in major emergencies. Unfortunately, these trends toward complexity and vulnerability of larger populations are likely to increase over time rather than abate (Akimoto, 1987; Quarantelli, 1993a, 1996).

COPING WITH DISASTERS

Despite everything that can be done, it is not realistic to think that a risk-free society is possible. The consequences of social and political changes, outlined previously, coupled with the sheer complexity of most modern technical systems, make it likely that disasters of the future will continue and will be more frequent and potentially worse than what we have seen in the past.

Because disasters are social creations, we believe that society has the means to lessen the dangers and deal more effectively with the effects from severe natural and technological misfortunes. The following are five suggested means for coping with disasters in the future:

1. Earlier and more consistent planning. People must view preparation for disasters primarily as a social response. The technical aspects of disaster preparation are very important, but whether such methods will be used, how they will be employed, for whom they will work, and the degree of preparation are political and social questions. Governments and organizations must focus on social solutions: communication, organization, integration of public and private organizations, citizen involvement, effective involvement of the media, and so on, for disaster preparation.

Planning must involve both mitigation and preparedness features. *Mitigation* involves measures taken prior to a disaster impact, whereas *preparedness* concerns activities taken when the disaster is imminent. Examples of mitigation measures would be uniform enforcement of earthquake building codes, brush clearance laws, and roof flammability standards, along with strict penalties for standards violations for refineries, chemical plants, and nuclear materials factories or power plants. Siting potentially hazardous production facilities in locations and ways that minimize the danger to large populations and enforcing zoning ordinances that keep the dangers to a minimum are further examples.

When a disaster is at hand, elements of effective communication and organizational coordination must be worked out prior to the event and must be practiced over and over so that all personnel in the affected areas will understand what they might encounter. Disaster personnel also need to understand that flexible processes are more important than detailed disaster plans, which often are overlooked and unusable at the crucial moment.

2. Give mitigation as high a priority as preparedness, response, and recovery. Society functions at its highest alert when crisis is near or present. At such times, it is typical to de-emphasize other social roles and to divert resources from other areas to the pending disaster. As the crisis stage passes and community life begins to take on a semblance of normality, there is generally a limited period of intense individual and organizational interest in the specific crisis or in crises generally. Passing and enforcing laws dealing with mitigation, such as retrofitting, flood measures, construction, and building codes, must remain high lists of social priorities to minimize future disasters.

3. Better integration of disaster planning in societal and community development planning. One prime reason mitigation and preparedness are not central to organizations and individuals is that planning, when it exists, is relegated to separate positions in the social structure, usually marginal to the purposes of the organization or community. This guarantees that resources allocated to disaster processes will be minimal and organizational and community disaster priorities will always be secondary until disaster takes place or is imminent. If boundaries between the larger social, political, and economic planning and disaster planning were blurred, disaster planning would become routinized to an extent not possible thus far. This would assure greater sensitivity to mitigation measures and the high potential costs of ignoring catastrophic events.

4. Emphasize social rather than technical aspects of disasters. Although this point has been threaded throughout the above discussion, it needs to be made more explicit. Organizations and governments, in conjunction with the media and scientific community, have tended to emphasize the technology available to cope with disasters. This approach, the technological imperative, promotes the notion that science has reduced the severity of such events and therefore better science and more technology will enable us to deal with future events. For example, technology enables better adverse weather tracking, assures greater structural integrity of buildings during severe shaking from an earthquake, and provides more powerful fire fighting equipment. In our view, it is important to mitigation, preparedness, and recovery. However, whether such technology ever gets employed, how it is used, and whom it primarily benefits are essentially social questions. Thus, understanding how the social and political interact to create and enhance or control disasters is crucial to dealing effectively with them.

5. Consider the special needs of the poor and ethnic minorities. Because the response to, and recovery from, disasters differs by social class and ethnicity, it is essential to include these groups or categories of individuals in planning. As discussed previously, the poor are more vulnerable to, and have greater difficulty recovering from, disasters because they lack economic resources and are likely to reside in hazard-prone areas. Although their restoration needs are high, they have a greater problem securing aid (Nigg, 1995).

In 1987, Perry argued that existing theories of disaster behavior were limited because empirical studies provided little information about ethnic minorities. Although the amount of research is still small as we begin the millennium, an increasing number of studies are appearing dealing with the environmental aspects of gender, race, and ethnicity (Tierney, 1998). For example, results show that minority groups differ from majority group members in worldview, socioeconomic status, family organization and structure, participation in religious and other voluntary associations, political efficacy, and trust in social and political institutions. Therefore, group affiliation is likely to affect disaster preparedness and response. Understanding how people from various groups respond and prepare still requires substantial additional research to provide a more adequate empirical basis, which may well amend existing theory and create new frameworks.

CONCLUSION

We examined disasters in California in a social problems context, not because of the intrinsic harm to health or property that has occurred or that can be imagined. We consider disasters to be social problems because a significant part of the population today perceives these catastrophic events to be highly detrimental to the state's economic health and to their own well-being. Two decades ago, or perhaps even one decade ago, Californians had confidence in the growth and prosperity of their state. Today, that situation is reversed. Surely, disasters can't be blamed for this change, but on the other hand, natural, social, and technological disasters have occurred with such frequency recently that the public has become conscious of the fragility of the social systems within which people live and human vulnerability within these systems.

Along with this growing awareness is the perception that problems, even natural ones, do not just happen, that they are the product of humans grappling with events that are capable of social solutions. This chapter attempted to clarify the sociological understanding of disasters and the social processes involved in creating, maintaining, and controlling these events.

NOTE

1. This was a Ph.D. dissertation in sociology at Columbia University, which examined the social changes in the community following a massive ship explosion in a harbor of Halifax, Canada.

REFERENCES

Akimoto, R. (1987). Disaster and urban community. In R. Dynes, B. De Marchi, & C. Pelanda (Eds.), *Sociology of disasters: Contributions of sociology to disaster research* (pp. 153-179). Milan, Italy: Franco Angeli.

Barton, A. H. (1969). *Communities in disaster: A sociological analysis of collective stress situations.* Garden City, NY: Doubleday.

Bates, F., & Peacock, W. C. (1987). Disasters and social change. In R. Dynes, B. De Marchi, & C. Pelanda (Eds.), *Sociology of disasters: Contributions of sociology to disaster research* (pp. 291-330). Milan, Italy: Franco Angeli.

Britton, N. R. (1987). Toward a reconceptualization of disaster for the enhancement of social preparation. In R. Dynes, B. De Marchi, & C. Pelanda (Eds.), *Sociology of disasters: Contributions of sociology to disaster research* (pp. 31-55). Milan, Italy: Franco Angeli.

Bullard, R. D. (1993). Anatomy of environmental racism and the environmental justice movement. In R. R. Bullard (Ed.), *Confronting environmental racism: Voices from the grassroots* (pp. 15-39). Boston: South End Press.

Darley, J. M., & Gilbert, D. T. (1985). Social psychological aspects of environmental psychology. In G. Lindzey & E. Aronson (Eds.), *The handbook of social psychology* (3rd ed., pp. 956-959). New York: Random House.

The Disaster Center. (2000). Homepage [On-line]. Available: www.disastercenter.com.

Dynes, R. R. (1987). Introduction. In R. Dynes, B. De Marchi, & C. Pelanda (Eds.), *Sociology of disasters: Contributions of sociology to disaster research* (pp. 13-29). Milan, Italy: Franco Angeli.

Dynes, R. R. (1992). Social science research on earthquake hazard mitigation: Relevance for policy and practice. In *International symposium on building technology and earthquake hazard mitigation* (pp. 67-86). Buffalo, NY: University of Delaware Press.

Dynes, R. R, & Quarantelli, E. L. (1971). The absence of community conflict in the early phases of natural disaster. In C. G. Smith (Ed.), *Conflict resolution: Contributions of the behavioral sciences* (pp. 200-204). Notre Dame, IN: University of Notre Dame Press.

Dynes, R. R., Tierney, K. J., & Fritz, C. (1994). Foreword: The emergence and importance of social organization: The contributions of E. L. Quarantelli. In R. R. Dynes & K. J. Tierney (Eds.), *Disasters, collective behavior, and social organization* (pp. 9-17). Newark, NJ: University of Delaware Press.

Earthquake in India falls off the world news scale. (2001, February 7). *Toronto Star* (Edition 1).

Ferriss, S. (2001, January 19). Human error worsens impact of nature's wrath: Deforestation, erosion, and poor building standards in Central America heighten death, destruction of disasters. *The Atlanta Journal and Constitution* (Home Edition), p. B- 1.

Fitzpatrick, C., & Mileti, D. S. (1994). Public risk communication. In R. R. Dynes & K. J. Tierney (Eds.), *Disasters, collective behavior, and social organization* (pp. 71-84). Newark, NJ: University of Delaware Press.

Fritz, C. (1961). Disaster. In R. Merton & R. Nisbet (Eds.), *Contemporary social problems.* New York: Harper & Row.

Hal, C., Hurst, J., & Zamichow, N. (1993, November 5). Fire evacuees return, hoping to find homes instead of ashes. *Los Angeles Times* (Home Edition), p. A1.

Hecht, P. (1994, December 15). Surge knocks out power across eight western states. *Sacramento Bee*, p. B1.

Kaniasty, K., & Norris, F. H. (1995). In search of altruistic community: Patterns of social support mobilization following Hurricane Hugo. *American Journal of Community Psychology, 23*(4), 447-477.

Killian, L. M. (1952). The significance of multiple group memberships in disaster. *American Journal of Sociology, 57,* 309-314.

Lee, K. W. (1995, April 23). From dream to nightmare: The LA riots and the Korean American community. *Los Angeles Times,* Book Review Section, p. 2.

Lerten, B. (1984, October 2). Power outage hits 10 western states. *United Press International.*

Li, S.K.C., & Dahlburg, J.-T. (2001, January 27). Thousands die as a 7.9 quake strikes India. *Los Angeles Times* (Home Edition), p. A1.

Los Angeles Times. (1999, August 29). Home Edition, Foreign Desk, p. A1.

Lozano, C. (1994, January 30). Earthquake: The long road back. *Los Angeles Times,* p. A32.

Mileti, D. S., Drabek, T. E., & Haas, J. E. (1975). *Human systems in extreme environments: A sociological perspective.* Boulder: University of Colorado, Institute of Behavioral Science.

Mulligan, T. S., & Kristoff, K. M. (1994, March 7). A tale of two policies: How your quake ins. can save you or sink you. *Los Angeles Times,* p. D1.

Nigg, J. M. (1995). Disaster recovery as a social process. In *Wellington after the quake: The challenge of rebuilding* (pp. 81-92). Wellington, New Zealand: The Earthquake Commission.

Nordwall, S. P. (2001, February 15). Salvadoran quake's death toll hits 274. *USA Today,* p. 9A.

Palm, R. I., Hidgson, M. E., Blanchard, R. D., & Lyons, D. I. (1990). *Earthquake insurance in California: Environmental policy and individual decision making.* Boulder, CO: Westview.

Perry, R. W. (1987). Disaster preparedness and response among minority citizens. In R. Dynes, B. De Marchi, & C. Pelanda (Eds.), *Sociology of disasters: Contributions of sociology to disaster research* (pp. 135-151). Milan, Italy: Franco Angeli.

Platte, M. (1994, February 7). Heavy rain due; slide fears raised; weather: flash flood warnings are issued for Laguna Beach and other Southland areas that were hit by the fall wildfires. *Los Angeles Times,* p. B1.

Quarantelli, E. L. (1992). Disaster research. In E. Borgatta & M. Borgatta (Eds.), *Encyclopedia of sociology* (pp. 492-497). New York: Macmillan.

Quarantelli, E. L. (1993a). The environmental disasters of the future will be more and worse but the prospect is not hopeless. *Disaster Prevention and Management: An International Journal, 2*(1), 11-25.

Quarantelli, E. L. (1993b). The importance of thinking of disasters as social phenomena. *International Civil Defence Journal, 6,* 24-25.

Quarantelli, E. L. (1994a). *Emergency* (Working paper). Newark, NJ: University of Delaware Disaster Research Center.

Quarantelli, E. L. (1994b). Natural disasters. In R. Eblen & W. Eblen (Eds.), *Encyclopedia of the environment* (pp. 466-468). Boston: Houghton Mifflin.

Quarantelli, E. L. (1996). *The future is not the past repeated: Projecting disasters in the 21st century from current trends* (Working paper). Newark, NJ: University of Delaware Disaster Research Center.

Quarantelli, E. L. (1999a). *The disaster recovery process: What we know and do not know from research* (Working paper). Newark, NJ: University of Delaware Disaster Research Center.

Quarantelli, E. L. (1999b). *Disaster-related social behavior: Summary of 50 years of research findings* (Working paper). Newark, NJ: University of Delaware Disaster Research Center.

Quarantelli, E. L. (1999c). *Implications for programmes and policies from future disaster trends* (Working paper). Newark, NJ: University of Delaware Disaster Research Center.

Quarantelli, E. L., & Dynes, R. R. (1975). Response to social crises and disaster. In *Annual review of sociology* (Vol. 3, pp. 23-49). Palo Alto, CA: Annual Review Publications.

Quarantelli, E. L., & Dynes, R. R. (1976). Community conflict: Its absence and its presence in natural disasters. *Mass Emergencies, 1,* 138-152.

Reich, K. (1995, September 4). Challenges abound for state disaster chief. *Los Angeles Times,* p. A3.

Riley, M. (1999, November 1). When the earth moves. *Newsweek,* p. 43.

Rossi, P. H., Wright, J. D., & Weber-Burdin, E. (1982). *Natural hazards and public choice: The state and local politics of hazard mitigation.* New York: Academic Press.

Stallings, R. A. (1988). Conflict in natural disasters: A codification of consensus and conflict theories. *Social Science Quarterly, 69,* 569-586.

Stallings, R. A. (1995). *Promoting risk: Constructing the earthquake threat.* New York: Walter de Gruyter.

Supporting El Salvador [Editorial]. (2001, February 23). *The Washington Post*, p. A22.

Susman, P., O'Keefe, P., & Wisner, B. (1983). Global disasters, a radical interpretation. In K. Hewitt (Ed.), *Interpretations of calamity*. Boston: Allen & Unwin.

Tierney, K. J. (1998). *The field turns fifty: Social change and the practice of disaster field work* (working paper). Newark, NJ: University of Delaware Disaster Research Center.

Tierney, K. J. (1999). *Toward a critical sociology of risk* (Working paper). Newark, NJ: University of Delaware Disaster Research Center.

Webb, G. R. (1999). *Individual and organizational response to natural disasters and other crisis events: The continuing value of the DRC typology* (Working paper). Newark, NJ: University of Delaware Disaster Research Center.

Wenger, D. E. (1987). Collective behavior and disaster research. In R. Dynes, B. De Marchi, & C. Pelanda (Eds.), *Sociology of disasters: Contributions of sociology to disaster research* (pp. 213-237). Milan, Italy: Franco Angeli.

EPILOGUE
Deregulating Energy in California

James A. Glynn

In 1953, Lewis Strauss, chairman of the U.S. Atomic Energy Commission, made a famous prediction. He said that, in the future, electric power would be "too cheap to meter."

Randall E. Stross

When the first edition of *California's Social Problems* was planned and the authors had agreed to participate, Californians complained about the cost of energy about as much as they whined about the quality of their drinking water or another price hike by the U.S. Postal Service. The year was 1996. Between the date that contracts were signed and the book was actually published, the state of California deregulated its energy industry. A couple of years later, Ralph Nader called it the "largest corporate ripoff in American business history" (Wasserman, 1998).

The concept for this edition of the textbook was formalized in April 2000, and a couple of months later, San Diego experienced the first of the blackouts that have come to plague California's commerce and homeowners. That summer, power bills skyrocketed by as much as 300% for many of San Diego's residential customers (Asmus, 2001). John Greenwald (2001) summarizes the situation by telling us that "California dismantled its private power-generating industry without securing adequate power supplies" (p. 39). In a nutshell, he's right, but there is more to the story.

THE HISTORY OF THE PROBLEM

Two decades ago, California was at the cutting edge of energy production. Under then-Governor Jerry Brown, who is often referred to as Governor Moonbeam because of his unconventional proposed solutions to many of the state's social problems, the Golden State began to build small hydroelectric plants and to supplement their output with solar, geothermal, and biomass operations. Even the power of the wind was harvested by huge windmills that are conspicuous as one crosses the Tehachapi Mountains between Bakersfield and the Mojave Desert. As Peter Asmus (2001) points out, "an entire domestic renewable energy industry was spawned in California" (p. 18). Under Governor Brown, more than 15,000 windmills were built, and California once produced 90% of the world's wind power. The 450 windmills along the California/Oregon border, alone, have sufficient capacity to power 70,000 homes (Wasserman, 2001, p. 14).

But big contractors, with plans for large, centralized, and traditional power plants, reentered the picture when Brown's term of office was over. Backed by the California Manufacturers and Technology Association (CMTA), they built huge facilities that were supposed to supply enough electricity for many decades to come (Nemec, 2001). But then the state's Silicon Valley began to boom, and the economy expanded by 34% (Greenwald, 2001, p. 38). As dot-com companies proliferated, they required more and more access to the Internet. Web sites, however, depend on "server farms" to support their energy demands. These server farms are concentrated in Silicon Valley. Although they are housed in relatively inconspicuous buildings, they consume about the same amount of power that is necessary to run 10,000 homes. "Exodus Communications, the company that takes care of eBay and Yahoo!, among others, has six data centers . . . and will soon draw an estimated 25 percent of the area's power," according to Randall E. Stross (2001, p. 33). On top of that, population continued to increase (see Chapter 10), and more households acquired even more plug-in devices, from personal computers to microwave ovens to Surround-Sound home entertainment centers, focused on big-screen television sets and stereophonic systems.

Clearly, something had to be done to keep electricity affordable yet avoid the costs of building new power plants. (In fact, not a single new plant has been constructed in California in more than a decade.) So, a plan was hammered out in the boardrooms of the state's Big Three utility companies: Pacific Gas and Electric (PG&E), Southern California Edison (SCE) (now known as EdisonInternational), and San Diego Gas and Electric (SDGE). The scheme, formulated as Assembly Bill 1890, would allow the regulation of distribution lines, separate the business of generating power from the service of supplying it to the customer, and spin off much of the generating capacity to the utilities' parent companies, such as Duke Energy,

Dynegy Corporation, Reliant Energy, and The Williams Company. Then, because PG&E, SCE, and SDGE would simply be service distributors, competition would increase and prices would drop (Wasserman, 2001).

THE PROBLEM IN CALIFORNIA TODAY

AB 1890 sounded good when it was presented to the public in 1996 because, under the law, the state would not allow the intermediary companies (formerly the Big Three) to enter into long-term power-purchasing agreements, and the companies had to lock in the rate that they charged to customers until 2002. However, according to John Greenwald (2001), "the wholesale price of energy jumped from less than 5 cents per kW-h (kilowatt hour) in January 2000 to nearly 40 cents per kW-h (in December of that same year)" (p. 42). Moreover, about 40% of the energy is produced in facilities that were built more than 30 years ago and are prone to constant maintenance problems. Meanwhile, Duke Energy, which lost $189 million in the fourth quarter of 1999, amassed $284 million in the fourth quarter of 2000 (Greenwald, 2001, p. 42).

Knowing that power suppliers are making hundreds of millions of dollars while businesses and homes are tightening their belts as never before has captured the attention of Californians, according to the Public Policy Institute of California (PPIC). A survey conducted in mid-January 2001 showed that 84% of Californians are paying close attention to the cost, supply, and demand for electricity, compared to 60% who said they were following such news in October 2000. In fact, Californians mentioned electricity and deregulation (25%) almost as often as education (26%) as the No. 1 issue facing the governor and the legislature. No other issue was mentioned by more than 4% of the respondents (PPIC, 2001).

"Californians are deeply worried about the implications of this crisis for the state economy and their own pocketbooks," according to survey director Mark Baldassare (PPIC, 2001, p. 1). In his analysis, people are still somewhat optimistic, but they think that the state will suffer serious setbacks unless the problem is solved soon. According to Greenwald (2001, p. 44), the crisis has already caused layoffs and shutdowns, particularly in the northern part of the state, the area hit hardest by the effects of deregulation. Chris Taylor (2001, p. 44) reports that Carl Guardino of the Silicon Valley Manufacturing Group, which represents the chief executive officers of companies such as Apple, eBay, and Intel, is already looking for places to relocate if California cannot supply the energy required by these companies. Likewise, Greenwald (2001, p. 44) says that a Miller brewery has shifted its production to Texas because of similar uncertainties.

In December 2000, a shortage of rain and snow in the Pacific Northwest deprived hydroelectric plants of the water that they need to generate electricity. To Californians, that meant the loss of 3,000 mK/h. Then, another 11,000 mK/h were lost to routine maintenance and unexpected shutdowns. The shutdowns were caused by a gridlock—similar to a Los-Angles freeway jam—on Path 15, a series of power supply lines that run through the Central Valley. On December 7, 2000, the first Stage 3 alert was issued, meaning that emergency power had dropped below 1.5%. Thereafter, many Californians experienced a series of *rolling blackouts,* when certain grids were shut down to preserve emergency energy for operations such as hospitals, prisons, and police departments. Power was then purchased at exorbitant prices on the spot market, but another Stage 3 alert was issued on January 16, 2001, and 32 days passed before regulators eased it back to Stage 2. Yet, most daily routines continued.

The PPIC (2001, p. 2) says that, overall, residents of the state place the blame for the current situation on the process of deregulation (47%) rather than on the power companies (25%). About 10% blame consumers (themselves). Interestingly, only 9% blame the legislature, which, of course, passed AB 1890. Amazingly, 10% blame the current governor, who had nothing to do with bringing about the situation; it was a former governor who jumped at deregulation as a means of turning the problem over to private companies and signed the assembly bill into law. It is intriguing to note that only one quarter of those surveyed place blame on the power companies, which had amassed a debt of over $12 billion by January 2001. By late February, when the governor proposed buying 26,000 miles of transmission lines from PG&E and SCE, the estimate of the state's cost to bail out the power companies had jumped to about $20 billion (Thompson, 2001). Both PG&E and SCE could have been forced into bankruptcy by either bankers or power suppliers (Greenwald, 2001, p. 42). And, of course, the consumer will eventually pay off the debt in some way. Customers can expect to pay at least 19% more than before the crisis peaked. The current 9% emergency rate is expected to become permanent, and the 1996 law allows the companies to increase rates by 10% beginning in March 2002 (Thompson, 2001).

Moreover, customers have been billed an additional $1.6 billion since 1968 by the power companies to pay for placing electricity wires underground. Aside from the aesthetics, this plan was seen as a safety feature because sparks, which may start fires, are produced when swinging power lines collide during storms. Because of the current crisis, those plans have been shelved (Associated Press, 2001).

CALIFORNIA'S PROBLEM COMPARED WITH THE UNITED STATES

Among the 50 states, only 8 are currently taking no action that could result in the deregulation of their utility companies: Georgia, Alabama, Tennessee, Kansas, Nebraska, South Dakota, Idaho, and Hawaii. The restructuring of energy delivery is being studied in about half of the states. Sixteen others are involved in legislative investigation. Legislation is pending in South Carolina, and New York has already enacted a deregulation decree (Eisenberg, 2001). Although this does not mean that the rest of the United States could end up with the crisis that is currently facing California, the critics of such energy policy believe that the verdict is in: Deregulation does not always work.

Proponents say that deregulation of the airline industry brought ticket prices down, and deregulation of the telephone company made long-distance phone calls less expensive. But Eisenberg (2001) makes the point that the "Balkanized transmission grid" (p. 45) that straddles the 48 contiguous states is too old and worn to handle so much long-distance travel for a commodity that cannot be stored. In other words, the country must have an infrastructure in place before it is ready to transport energy over huge geographic regions. Of course, airlines have been criticized for neglecting smaller cities and having a poor record of leaving/arriving on time. And many analysts claim that long-distance telephone rates were overpriced at the start.

One success story can be found in Pennsylvania, where about 10% of the state's population, choosing among a variety of producers, have saved about $3 billion on their electric bills. But, unlike California, Pennsylvania already had surplus energy supplies. By contrast, California's Independent System Operator (ISO) has had to import 6,000 megawatts a day from sources outside the state (Greenwald, 2001, p. 42) So, having seen the problems experienced by Californians, states such as West Virginia, Oklahoma, Arkansas, Nevada, and New Mexico have already scrapped their plans for deregulation (Eisenberg, 2001, p. 46).

Observers say that one of the biggest mistakes made in California was requiring the Big Three to sell their generating plants to become distributors. Texas, like Pennsylvania, allowed companies to keep their power generators, and Texas, unlike any other state, operates its own electricity grid, thereby avoiding the bottlenecks that have tormented the Golden State. Moreover, unlike California, Texas has encouraged power-plant construction, with 22 new plants coming on line since 1995 and 15 more being built over the next couple of years (Eisenberg, 2001, p. 47). Massachusetts, which promised its consumers a 15% rate decrease, has actually been forced to raise customers' bills by as much as 50%. A similar story can be told about New York, where residents of Westchester County and New York City have experienced 30% increases in their energy bills (Eisenberg, 2001, p. 47).

CALIFORNIA'S PROBLEM IN GLOBAL CONTEXT

For thousands of years, human beings have been harnessing energy, and for most of that time, their reliance has been on burning wood. The practice can be dated back to the dwellers of the Escale cave in southern France (Dunn, 2001), and billions of people still rely on this technology. As late as 1850, 90% of the world's energy output was still based on wood. However, by the early 20th century, coal replaced wood as the major generator of energy, and that remained the case until the 1960s, when petroleum replaced coal. Then, in 1999, natural gas use surpassed that of coal. Respectively, oil, natural gas, and coal supply 32%, 22%, and 21% of the world's energy (Dunn, 2001, p. 88).

However, that energy is not evenly distributed. In 1998, the United States was by far the greatest consumer of energy, using 94.57 quadrillion British thermal units (BTU) per year. The closest competitor was China, at 33.93 quadrillion BTU. However, the United States was also the major producer of energy (72.55 quadrillion BTU), with Russia a distant second (41.04 quadrillion BTU). The important point to be made is that there is a significant difference, amounting to more than 22 quadrillion BTU, between what the United States produces and what it consumes (*World Almanac*, 2001, p. 173). This means that a significant proportion of energy must be imported, and usually at higher prices than domestically produced electricity.

Within the United States, the distribution of energy consumption is likewise uneven. California's annual demand for 7,275.5 trillion BTU is second only to Texas (11,396.1 trillion BTU). The other states require so much less energy that it would take Ohio, New York, and Louisiana (the No. 3, 4, and 5 consumers) combined to oust Texas from first place (*World Almanac*, 2001, p. 174). But, whereas Texas is able to meet its own needs, California has had to seek energy from other states, transporting it across the shaky national grid previously discussed.

In light of the fact that fossil fuels (coal, oil, and natural gas) are finite quantities, many countries of the world have turned to nuclear power plants. Although the United States has far more nuclear reactors (104) than France (59), its closest competitor, the United States relies far less on nuclear energy (*World Almanac*, 2001, p. 176). France gets 75% of its energy from nuclear sources, and the other top 10 countries are in northern Europe, eastern Europe, or the Far East. In order, they are Lithuania (73.1%), Belgium (57.7%), Bulgaria (47.1%), Slovak Republic (47%), Sweden (46.8%), Ukraine (43.8%), South Korea (42.8%), Hungary (38.3%), and Armenia (36.4%) (*World Almanac*, 2001, p. 175). The reluctance of the United States to turn to nuclear energy reflects the attitude of Californians, who have been closing down nuclear reactors. The consequence is that the Golden State may be forced to import as much as 25% of its energy requirements from both other states and the rest of the world (Asmus, 2001, p. 19), and the cost is likely to be enormous.

ALTERNATIVE EXPLANATIONS FOR THE PROBLEM

More than 50 years ago, sociologist Robert Merton (1949) drew a distinction between manifest functions and latent functions. In sociological terms, functions are the consequences of social structures. Manifest functions are the intended outcomes, for example, learning to read and write in school. Latent functions are unintended consequences, for example, the development of youth subcultures within a school that may be at variance with the goals of education. Steven M. Gillon (2001) points out that Merton "classified ignorance, or the failure to obtain sufficient information about possible outcomes, as a major contributor to unintended consequences" (p. 54). That observation can be supplemented by a comment from columnist Dan Walters (2001a): "Elected officeholders put short-term expediency above the public interest" (p. A9).

This structural-functional approach may explain why the Big Three were lured into backing AB 1890, and then-Governor Pete Wilson, a potential presidential candidate at the time, eagerly signed it. As Merton said, "the imperious immediacy of interest" may cause people to overlook the unintended consequences of their actions. Yet, many people outside the corporate and political power structures took a more cautious approach. In 1998, consumer and environmental groups gathered 700,000 signatures in less than 5 months to put an initiative on the California ballot to repeal deregulation. But, the utility companies spent $40 million to oppose the measure, and the initiative went down in defeat (Wasserman, 2001, p. 12).

Conflict theorists might see California's problems stemming from a coalition of powerful organizations that have seen an opportunity to exploit a relatively powerless population of consumers, who have little choice but to pay the bills. Richard Nemec (2001, pp. 11-12) points out that the major players in trying to put the current crisis at rest are (a) state agencies, such as the California Public Utilities Commission, California Energy Commission, California ISO, California Power Exchange, and the California Electricity Oversight Board; (b) private sector generators, such as AES Corporation, Duke Energy, Dynegy Corporation, NRG, Reliant Energy, Southern Company, and The Williams Company; (c) the Big Three, that is, PG&E, SCE, and SDGE; (d) the Federal Energy Regulatory Commission; and (e) consumer and environmental groups, along with organized labor, including The Utility Reform Network, Utility Consumer Action Network, Foundation for Taxpayer and Consumer Action Network, Natural Resources Defense Council, Sierra Club, and statewide utility employee unions. These, according to Nemec, are the same groups that made up the coalition that pushed AB 1890 through a political process that included a Republican governor and a Democrat-controlled state legislature.

Supporting conflict theory is the fact that, in June 2000, when the temperature in usually cool San Francisco hit 103 degrees, power supplies declined to a dangerously low level, and power was cut off to more than 100,000 houses. However, when the weather moderated in July, electricity costs spiked to an all-time high (Egan & Verhovek, 2001), even though consumption decreased. California almost immediately ordered a price freeze on the utility companies. In his column, Dan Walters summed up the ripoff. He pointed out that the legislature in 1996 financed bonds that would be repaid by customers over the next 10 years. However, the amount of the bond debt would only guarantee a price freeze for a few more years. "It was like buying an automobile with a useful life of five years with a 10-year loan, and the politicians want to do it again" (Walters, 2001b, p. A11).

As usual, California will try to solve the energy crisis with a surfeit of legislation. According to Noel Brinkerhoff (2001), the number of new bills addressing the problems of deregulation will be surpassed only by the number of jokes about how many Californians it takes to screw in a light bulb. (The ultimate answer is: All of them.) And, although legislation is undoubtedly called for, other considerations need to be addressed. Columnist Dan Walters points out that, at the same time that lawmakers must wrestle with the energy crisis, they are saddled with problems concerning the water supply, housing, college education, airport capacity, deteriorating highways, and health care issues. And on top of that, the state is committing hundreds of millions of dollars to build a 10th campus of the University of California in Merced (Walters, 2001a).

FUTURE PROSPECTS

Peter Asmus (2001) points out that many politicians who were swayed by the 1996 AB 1890 were unaware that "prices go up and down in response to supply and demand" (p. 18). California's large, centralized power plants were built during the 1970s. They have reached their capacity, whereas electricity consumption in Silicon Valley is growing by 5% per year. In a few years, demand for electricity will double our current requirements. Furthermore, virtually all new power plants will burn natural gas, a commodity whose price has skyrocketed in the past year. In 2000, Stephanie Anderson Forest (2000, p. 150) pointed out that the demand for natural gas had increased by 4.3%, whereas production increased by only 1%. Moreover, she predicted that in 2001, demand would increase by another 3.2% and production would hold at the 2000 level. Without new sources of fuel, the near panic in California will induce legislators to remove environmental requirements from existing power facilities, likely increasing air, water, and soil pollution.

The power emergency is currently costing the state an additional $45 million a day to keep the system running, as legislators search for a long-term remedy

(Maxwell, 2001). In 2002, when the rate cut expires, the price of energy in California will likely increase by at least another 10%. Unless solutions are found, the massive rolling blackouts in cities from central California to the Oregon border, a distance of about 500 miles, which affected 675,000 homes and businesses, will likely reoccur. This scenario will cost billions of dollars in manufacturing productivity and will almost surely threaten the state's two largest utilities with bankruptcy. Because California accounts for about one eighth of the nation's gross domestic product, such an occurrence could "tip the U.S. economy into a downturn" (Whitman, 2001, p. 27).

SOLUTIONS/SUGGESTIONS

As previously stated, no power plants have been built in California for the past decade. However, in 2001, Governor Gray Davis took action to license nine new power plants (see Figure Epi.1), five of which are currently under construction (Whitman, 2001, p. 29). Unfortunately, none will be completed by summer, when energy demand will again peak and the state will likely face another round of rolling blackouts. Furthermore, there is a problem in siting the plants because of poor air quality in the areas where the power is most needed (Asmus, 2001). Sometimes, these difficulties can be overcome by *offsets*, payments made by the power companies to other industries to reduce their pollution.

Furthermore, bottlenecks in the California grid system must be addressed. If the state takes control of Path 15 (the lines that run from Bakersfield in the southern Central Valley to Los Banos, north of Fresno), it may be able to solve some of the transmission problems. Also, there is a plan to fix three 500,000-volt lines that run from Los Angeles to Bakersfield, three 500,000-volt lines between Oregon and Sacramento, and a transformer in the San Francisco Bay Area (Bridges, 2001). Still, even if everything goes right, these steps will not be sufficient to supply power for the expected growth in the state's population or for industrial expansion. Dennis Randall, senior director of real estate at Opus U.S. Corp., which built more than 29 million square feet of commercial space last year, said, "If you can't rely on basic infrastructure, people will start leaving the area" (cited in Ginsburg, 2001, p. 106).

Opus and similar real estate development companies are taking steps to protect themselves from the vagaries of the overloaded grid. The new technique, praised by energy experts as well as environmentalists, is called distributed power (DP). Essentially, this is a system that allows small energy producers to send power that they may not need at the moment onto the grid. Traditionally, the grid allowed one-way transmission, but DP redesigns the grid from the ground up. Small plants, which might even include diesel engines, would be closer to the consumer, so much of the

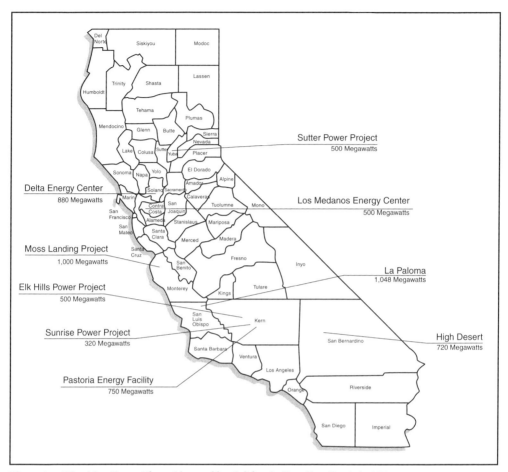

Figure 1. Nine New Power Plants Licensed by California Gov. Gray Davis in 2001
SOURCE: California Energy Commission.

grid would be freed for other transmissions. Janet Ginsburg says that "as much as one-fifth of all electric generation in the U.S. by 2010" (Ginsburg, 2001, p. 106) could be produced by DP.

DP would rely on central power stations, as has been the tradition. However, all of the following would be contributors to the state's energy demands: wind turbines on farms, solar panels on top of gas stations and other buildings, diesel generators and flywheel storage that could power "server farms" for Internet companies, microturbines in places like hospitals and police stations, homes with solar shingles, homes with fuel cells, and waste methane from garbage dumps and landfills. Ginsburg makes the following analogy: Many experts said that personal computers would replace mainframes. In a similar manner, California may always rely on large, centralized power stations, but much of tomorrow's energy may be supplied by microsystems (Ginsburg, 2001, p. 107). This would also help resolve a problem that is voiced by Dan Berman: "In the long run, public ownership is central to any real solution to the problems of the electric-utility grid" (cited in Wasserman, 2001, p. 14).

REFERENCES

Asmus, P. (2001, January). California's new energy legacy? A desperate innovator? *California Journal*, pp. 17-21.
Associated Press. (2001, February 19). Two utilities drop project to put lines underground. *Fresno Bee*, p. A11.
Bridges, A. (2001, February 18). State could fix system, at a price. *Fresno Bee*, p. A19.
Brinkerhoff, N. (2001, January). Deregulation under the dome: It's all in the delivery. *California Journal*, pp. 10-11.
Dunn, S. (2001). Decarbonizing the energy economy. In L. Stark (Ed.), *State of the world 2001* (pp. 83-102). New York: Norton.
Egan, T., & Verhovek, S. H. (2001, February 12). Companies profited from buyer's panic. *Fresno Bee*, pp. A1, A6.
Eisenberg, D. (2001, January 29). Which state is next? *Time*, pp. 45-48.
Forest, S. A. (2000, September 18). There is not enough gas around. *Business Week*, pp. 149-150.
Gillon, S. M. (2001, March-April). Unintended consequences: Why our plans don't go according to plan. *The Futurist*, pp. 49-55.
Ginsburg, J. (2001, February 26). Reinventing the power grid. *Business Week*, pp. 106-107.
Greenwald, J. (2001, January 29). The new energy crunch. *Time*, pp. 37-44.
Maxwell, L. A. (2001, February 14). State purchase of lines nearer. *Fresno Bee*, pp. A1, A12.
Merton, R. K. (1949). *Social theory and social structure*. Glencoe, IL: Free Press.
Nemec, P. (2001, January). When the lights went out. *California Journal*, pp. 12-16.
Public Policy Institute of California. (2001, January 18). *Electricity issue registers with Californians* [On-line]. Available: http://www.ppic.org/publications/CalSurvey16.press.html.
Stross, R. E. (2001, January 29). The price isn't right. *U.S. News and World Report*, p. 33.
Taylor, C. (2001, January 29). It's grim and dim for the dotcoms. *Time*, p. 44.
Thompson, D. (2001, February 19). Cost likely to top $20b. *Fresno Bee*, pp. A1, A11.
Walters, D. (2001a, February 12). Power crisis just the first growth woe. *Fresno Bee*, p. A9.
Walters, D. (2001b, February 14). Shell game leaves state in the dark. *Fresno Bee*, p. A11.
Wasserman, H. (1998, March 16). The last energy war. *The Nation*, pp. 16-20.
Wasserman, H. (2001, February 12). California's deregulation disaster. *The Nation*, pp. 11-14.
Whitman, D. (2001, January 29). California unplugged. *U.S. News and World Report*, pp. 26-30.
World Almanac and Book of Facts 2001. (2001). Mahwah, NJ: World Almanac Books.

GLOSSARY AND INDEX

Abortion, 161, 165
Academic freedom, 242, 251, 257
Academic headhunting: Attempts by people and agencies to lure teachers from other districts to teach in their schools, 300
Acid rain: Rain that deposits significant levels of nitrogen on the earth because of polluted air, 281
Acquired Immune Deficiency Syndrome. *See* AIDS epidemic
ACT-UP (AIDS Coalition to Unleash Power), 98
Addiction. *See* Alcohol abuse; Illegal drugs; Substance abuse
Administrative state, 42
Aerospace industry, 276, 290, 331
Affirmative action, 11, 76
African American community, 10, 11, 12
  AIDS in, 120, 132, 133 (figure)
  arrest/incarceration and, 54-56, 55 (table), 74
  drug trafficking and, 48
  feminization of poverty and, 154-156, 155 (figure)
  firearm deaths, 58
  gang membership and, 10, 45-46, 52
  hypersegregation of, 69
  intelligence and, 75-76
  nonmarital birthrates, 159, 160 (figure)
  political organizations, 46
  population levels, 291, 292 (figure), 295
  slavery and, 65
  teen parenting in, 153(figure), 158
  triple discrimination in, 167
  See also Coalition politics; Race/ethnic relations
Agriculture, 287, 289, 290, 291
AIDS epidemic, 3, 86, 88-89, 97
  anti-gay sentiment and, 89, 100, 130-131
  deviant subcultures and, 129-132, 135
  disease course, 120-122, 121-122 (figures)
  disease description, 117-119
  economic/human costs, 124-125
  educational solutions to, 134, 135
  global context of, 127-129, 168-169
  history of, 119-120
  infection rates, 122-123, 123 (figure)
  means of infection, 126-127, 127 (figure)
  national trends in, 125-127, 126-127 (figures)
  population levels and, 128-129
  public health response, 134, 135
  San Francisco bathhouses and, 98
  socioeconomic status and, 125, 128, 129, 135
  strategies for solution, 134-135
  treatment costs, 123-124
  vaccine research, 134-135
  youth infection, 125, 132, 133, 134, 169
Aid to Families with Dependent Adults policy, 30
Aid to Families with Dependent Children (AFDC): An income assistance program available to poor families as an entitlement

under the welfare system that existed prior to welfare reform, 27, 72
  caseload reduction and, 208, 209 (figure)
  poor families and, 190
  welfare reform and, 143, 162, 204
Airborne particulate matter: Fine particles of dust and other substances that are carried by the wind and can cause lung damage, 268, 280
Air pollution, 2, 271, 279-280
Air quality: The condition of the air in a region, 280
Alcohol abuse, 8, 9
  disability benefits and, 29-30
  homelessness and, 29
  public drunkenness, 7, 16
Alien Land Law (1913), 331
Amendment 2, Colorado, 103
American Association for Higher Education (AAHE), 245
American Association of University Professors (AAUP), 246, 251, 255, 257, 258, 259
American Civil Liberties Union, 23
American Council on Education (ACE), 245
American Sociological Association, 251, 256, 258
Anaheim, 76
Anglo American settlements, 65
Anomie: The normlessness that is produced by rapidly shifting moral values. An anomic person has few guides to what is socially acceptable behavior. According to sociologist Robert Merton, anomie is a condition that occurs when personal goals cannot be achieved by available means, 47, 69
Anti-community: Urban conditions that are detrimental to a sense of community and to the quality of life, 278
Apartheid, 65
Aquiculture: Commercial breeding of fish for food, 290
Aquifers: Natural underground areas where water can be stored, 281
Arrest rates, 54, 55 (table), 73-75
Asian community:
  Chinese, 66, 78, 267, 275, 283, 284, 288, 332
  immigrants, 11, 66, 285
  Japanese, 66, 184, 283, 331
  Koreans, 11, 12, 66, 70, 71
  nonmarital birthrates, 159

  population growth, 284-285, 285 (figure), 291, 292 (figure), 295, 331
  poverty and, 189
  teen birthrate in, 153 (figure)
  Vietnamese, 284
Automobiles:
  assembly factories, 290
  design centers, 286
  environmental impact of, 2, 268, 278, 280

Basque population, 284
Battelle Human Research Institute, 332
Bilingual education: A system of providing students with education both in their own language and in English, 11, 284
  See also Proposition 224
Bioamplification: The process by which poisons become more toxic as they are passed up food chains, 268, 282
Biodiversity: The variety of plant and animal life that inhabits an area, 268, 282
Birthrate: Number of annual births per thousand population, 149
Black Panther Party, 46
Bloods, 46, 52
Bowers v. Hardwick (1986), 98
Boxer, B., 91
Boy Scouts of America v. Dale, 104, 105, 111
Bracero program: Created by Congress in 1932 to recruit Mexican laborers to fill in for the shortage of workers in the United States created by World War II, 332
Bradley, T., 77, 78
Brain drain, 5
Briggs, J., 86, 97
Brown, E. G., 237
Brown, J., 97, 354
Bryant, A., 97
Business interests: The stress on concerns of the business community, such as profit making and managerial control:
  deregulation and, 184
  distance learning and, 240-241
  higher education and, 147-148, 238-239, 244-246
  land use and, 268
  occupational structures, 185-186, 186 (table), 188 (figure)

Cal Grant legislation, 236, 237
California, 1, 6
  poverty and, 15-16
  race relations in, 11, 12
  racial diversity in, 11, 52
  smart growth plan, 269
  social problems in, 1-2
  state agencies, policing role of, 270
California Child Care Resource and Referral Network, 224
California Civil Rights Initiative, 270, 331
California Community College (CCC) system. *See* Higher education
California Department of Health Services (DHS), 119, 120
California Department of Housing and Community Development, 18, 26
California Faculty Association (CFA), 251, 259
California Judicial Council, 74
California Manufacturers and Technology Association (CMTA), 354
California Office of AIDS, 120, 122
California Postsecondary Education Commission (CPEC), 5, 247
California Self-Sufficiency Standard, 189
California State University (USC) system. *See* Higher education
California Teachers Association, 257
California Work Opportunity and Responsibility to Kids Act. *See* CalWORKs
California Youth Authority (CYA): The maximum security juvenile correction facility in California, widely considered the largest juvenile prison system in the country, 42
CalWORKs: Welfare-to-work program created in California under the California Work Opportunity and Responsibility to Kids Act, the state's welfare reform law, 143-144, 205-207
  allocation targeting/spending, 216-217
  CW7 form, 213
  educational triage and, 220, 219 (table)
  housing issues, 221-223
  human capital and, 145, 196, 213-214
  income increases, 220-221
  outcomes evaluation, 215-216
  See also Child care; Welfare reform
Catastrophe: An event that is qualitatively more severs, disruptive, and overwhelming to a community than a disaster. In such an event, the facilities and operational bases of almost all emergency organizations are generally adversely affected. A community affected by a catastrophe would likely be overwhelmed by the response needed to deal with its aftermath, 270, 331
Centers for Disease Control (CDC), 117, 120, 131, 134
Central Valley, 280, 282, 283, 284, 290, 291, 293
Chicago Bar Association, 41
Child abuse, 3, 141
  See also Sexual abuse
Childbearing. *See* Families; Nonmarital childbearing; Teen pregnancy
Child care, 145, 212-215, 215 (table), 223-225
  cost/availability of, 224-225
  solutions to, 226
  three-stage system of, 225-226
Child Care and Development Fund, 224
Children:
  homelessness and, 7, 8, 223
  infant mortality rate (IMR), 168
  one-parent families and, 166-167
  poverty of, 15, 73, 163, 164 (figure)
  welfare reform and, 145
  See also Teen pregnancy
Child support payment, 156, 162, 171, 205
Chinese Americans, 66, 78, 267, 275, 283, 284, 288, 332
Chinese Exclusion Act: Congressional decision to exclude all Chinese from immigration privileges and naturalization, 66, 288
Christopher, G., 94
Chronically poor, 191 (table), 192
Claims makers: Individuals or groups who attempt to publicly define situations as harmful and to promote action to change them, 331
Clean Air Trust, 280
Clinton, W. J., 104, 112, 161, 258
Clitoridectomy: Surgical procedure to remove the clitoris, 168
Coalition politics: Strategy of political cooperation in which diverse groups, such as different races, work together in an effort to bring about social change for the benefit of the greater good, rather than one specific interest, 12, 77-79, 101, 109-110
Cohabitation: Situation when unmarried couples live together without being married (usually refers to heterosexual relationships), 165, 166

Collective stress: A situation that occurs when many members of a social system fail to receive expected conditions of life from the system, such as safety, protection from attack, food, shelter, and income, 331

Colorado, Amendment 2, 87, 103

Committee to Outlaw Entrapment, 93

Communities:
  anti-community conditions, 278
  moral decay in, 76-77
  strengthening of, 58-59

Community Resources Against Street Hoodlums (CRASH), 56-57

Community United Against Violence (CUAV), 101

Computer industries. *See* High-tech industries

Confinement. *See* Incarceration

Conflict theorists: Intellectuals who believe that society is held together by the exercise of power, 268, 290

Conflict theory, 20, 87, 89, 109-110
  development/growth and, 290
  disasters, reaction to, 271-272, 335-337
  energy deregulation and, 359-360

Consensus theory, 271, 334-335

Consolidated Metropolitan Statistical Area (CMSA): Merged stretches of metropolitan areas and their suburbs, 278

Cost-of-living squeeze: Loss of purchasing power, 167

Council on Families in America, 150

Council on Religion and the Homosexual, 95

Courts. *See* Judicial system; Juvenile justice system

Crime:
  juvenile crime rates, 9, 36, 37-39 (tables)
  rehabilitation and, 41
  treatment-oriented prisons, 42
  See also Differential justice

Criminalization: Using the law to address a social problem, such as enacting legal statutes that prohibit behaviors or more rigorously enforcing existing legal statutes, 7, 16
  homelessness and, 23
  injection equipment possession, 134

Criminal Justice Statistics Center, California Department of Justice, 101

Crips, 46, 52

Cropland: Fertile land that can be used to grow food, 268, 269, 282

Culture:
  immigrant, 331-332
  inequality and, 178-179

Cyclically poor, 191-192, 191 (table)

Daughters of Bilitis, 85, 93

Davis, G., 86, 146, 236, 361

Defense industry, 276

Defense of Marriage Act, 104

Deforestation: Denuding of forests, principally by a method called clear-cutting, 268, 290

Demographics:
  gang membership and, 52
  homeless population, 18-20
  population changes, 268-269

Department of Housing and Urban Development (HUD), 24

Deukmejian, G., 98

Development theories:
  conflict theory, 290
  environmental racism, 291
  ethnic antagonism, 288-289
  sociocultural evolutionary theory, 289-290, 289 (figure)
  See also Smart growth

Deviant subcultures: Subcultures that do not conform with one, some, or many existing or traditional social norms. These subcultures are often regarded as relatively defective or pathological when they challenge power and privilege, 129
  See also Subcultures

Differential justice: Situation in which lower status groups face patterns of prejudice and discrimination in the legal and judicial system resulting in a higher arrest rate, incarceration rate, and execution rate, 73-75

Disabilities, 9, 29-30

Disaster: Sudden event of a catastrophic nature that affects a major portion of a human community or society. Disasters tend to overpower the capacity of the community to quickly absorb and recover from the loss. Generally, an integrated multiorganizational response from within the community and across communities is required, 270, 331-332
  California experience of, 332-332, 337-339

claim makers and, 331
collective stress situation, 332-331
communication and, 339-341
conflict theory and, 271-272
consensus theory and, 334-335
coping strategies, 346-348
developing nations, 333
global perspective, 271
management strategies, 272
media messages, 272, 341, 342-343
modernization and, 345-346
myths about, 343
natural/technological, 331-331, 331
planning for, 272, 339
rescue/relief organizations, 272
research history, 331-332
response to, 271, 340-341
risk warning, 340-341
role/group organization in, 343-345, 344 (figure)
social constructionist view, 337
social event of, 271, 331-332, 349
sociological perspectives and, 334-337
symbolic interactionism and, 272
systems complexity and, 332-333
Disaster Research Center, 271, 331, 332
Discrimination:
cultural values and, 178-179
sexual orientation and, 87, 91, 97, 99
Disproportionate minority confinement rate requirement: Because of concerns that African Americans and Latinos are disproportionately included among the confined population of juveniles, the federal government has monitored minority rates of confinement for each state and has required each state to assess the causes for any disproportionality in these rates, 52, 53-54 (table), 54
Distance learning: The use of computers and the Internet to transmit higher education from a central location to many students at physical distances from the academic point of origin, 5, 147, 238, 239-241
Diversity:
racial, California, 11, 52, 67-69, 68 (figure), 268
sexual orientation, 87, 111-112
Divorce rate: Number of annual divorces per 1,000 population, 139, 149, 151, 158, 165

Domestic partners: Two people of the same or opposite sex who are in a relationship and are not legally married but may have some rights as a couple, depending on the policies of a company, city, county, or state, 86, 99, 166
Doubling time: The number of years that it takes for a population to double in size, 168
Due process revolution: Beginning in the late 1960s, the U. S. Supreme Court decided a series of landmark cases that dramatically changed the character and procedures of the juvenile justice system. The most important of these cases, the 1967 Supreme Court ruling *In re Gault*, mandated that juveniles must be afforded the right to due process, ensuring that basic constitutional rights must be afforded to juveniles. This meant that juveniles now have the right to notice and counsel, to question witnesses, and to protection against self-incrimination, 42-43
Durkheim, E., 47

Earned income tax credits (EITCs), 220
Earthquakes. *See* Disaster
Economic issues:
AIDS epidemic, 169
earned income tax credits (EITCs), 220
families, economic restructuring and, 167
growth benefits, inequalities in, 26
housing availability, 21-22
inequality and, 177-178
poverty/homelessness and, 26-29
self-sufficiency, 144-145, 190, 196, 198 (figure)
structural transformation, 10, 12, 25-26, 47, 48, 70
taxation policy, 142-143, 178, 183-184, 186-187
underclass, minority communities, 12, 72-73, 75
vouchers, cash substitutes, 27
See also Human capital; Inequality; Welfare reform
Economic/social dislocation, 10
gang membership and, 48
industrial labor markets and, 204
Ecosystem: A self-sustaining biological environment, 282
Education:
AIDS education, 134, 135

business model of, 147-148, 238-239, 244-246
corrections spending and, 74
critical-hours programs, 58
English as a Second Language (ESL) classes, 270, 331-332
gay/lesbian teachers/students, 86, 97, 105
human capital requirements and, 193-196, 194 (figure), 195 (table), 197-198 (figures)
inequality and, 178
The Master Plan for Higher Education, 4, 146-148, 235, 258, 260-261
newhereditarians and, 75-76
pregnancy prevention, 141
self-initiated programs (SIPs), 206, 207, 216, 220, 219 (table)
student population levels, 4-5
welfare and, 28
See also Bilingual education; Higher education
Elderly population, 298-299
Electoral initiatives, 2, 269
Electronics industry, 276, 290
Emergency: A significant natural or technological event that disrupts a community and requires a rapid response. A community can usually use its traditional response organizations because the event does not overwhelm the community.
Emergent norms: Norms that emerge as the result of a disaster. They tend to be spontaneous and less rule-bound than those that operate within the context of existing social structures, 335, 337
Employment:
  foreign-trained professionals, 5, 284, 287
  job stimulation, 221
  language barriers to, 270, 331, 332
  poverty/minimum wage and, 25-26, 220-221
  racism and, 65, 66, 67
  state planning for, 269, 300-301
  structural transformation in, 10, 12, 25-26, 47, 48, 70
  welfare recipients and, 28, 72-73, 143, 162-163, 204
  See also High-tech industries; Human capital; Labor market; Unemployment
Energy deregulation, 353
  conflict theory and, 359-360
  distributed power (DP), 361-363
  energy crisis, 6, 332, 355-356
  global context, 357-358
  history of, 354-355
  national context, 356-357
  power-purchasing agreements, 355
  rolling blackouts, 38, 360-361
  solutions, 361-363, 362 (figure)
  structural-functional approach, 359
  wind power, 354
England:
  English Poor Laws, 27
  institutionalized racism, 74
English as a Second Language (ESL) classes, 270, 331-332
Entitlement programs. *See* Welfare; Welfare reform
Environmental concerns, 5, 268
  acid rain, 281
  air pollution/smog, 1, 268, 271, 279-280
  automobile use and, 278, 280
  land use conversion, 282
  open space preservation, 300
  species destruction, 282-283, 283 (figure)
  urban sprawl, 268
  water availability, 281-282
  water marketing, 268, 282
  water pollution, 280
Environmental racism, 291
Episodic poor, 190-191, 191 (table)
Equal Protection Clause, Fourteenth Amendment, 24, 103
Ethnicity. *See* Race/ethnicity; Race/ethnic relations
Exclusion: The process in which lower paid workers are kept out of competition with higher-paid workers.

Faculty Coalition for Public Higher Education, 241, 248, 251
Families:
  demographics of, 4, 154
  divorce rates, 139, 149, 151, 158, 165
  economic restructuring of, 167, 171
  father's role in, 141, 161-162, 167, 171
  global perspective on, 141, 168-169, 170 (figure)
  homelessness and, 8, 19, 221-222
  housing assistance, 222-223
  national trends, 158-164, 160 (figure), 164 (figure)
  nonmarital childbearing, 139, 141, 149, 151-152, 152 (figure), 158-159, 160 (figure), 164 (figure)

Glossary and Index 371

one-parent families, 166-167
poverty and, 73, 139-140, 154-156, 155 (figure), 157 (figure)
teen pregnancy, 140, 141, 152-154, 153 (figure), 156-158
See also Inequality; Marriage/civil unions; Welfare; Welfare reform
Federal government:
 administrative state and, 42
 affordable housing and, 21
 anti-gang law, 57-58
 due process revolution, 42-43
 homelessness and, 24-25
 homosexuality, court rulings, 94-95
 taxation policy, 142-143
 welfare system, 27-28
Feminization of poverty: Tendency of low-income families to be headed by single women, 139, 150, 154-156, 155 (figure), 157 (figure)
Filipinos, 66, 284
Fires. *See* Disaster
Food stamps, 214, 215 (table)
Foreign-trained professionals, 5, 284, 287
Fourteenth Amendment, 24, 103
Functionalist perspective, 20

Gangs. *See* Urban gangs
Garcetti, G., 43
Gay and Lesbian Alliance Against Defamation (GLAAD), 98
Gay/lesbian community:
 AIDS epidemic and, 3, 86
 anti-gay backlash, 96-97, 98
 Black Cat Cafe, 92-93, 94
 coalition politics in, 101, 109-110
 conflict theory and, 109-110, 111
 discrimination against, 87, 91, 97, 99, 104-105
 domestic partnership, 86, 99, 166
 global context and, 87, 106-108
 hate crimes/violence and, 86, 87, 96, 98-99, 101, 102 (figure), 103-104, 105
 law enforcement activity and, 93, 94, 95
 legislative initiatives, 85, 86-87, 88, 94, 97, 98, 99, 100 (figure)
 marriage/civil unions within, 87, 91, 104
 media representation of, 98, 108, 113
 military service and, 93-94
 national political climate and, 103-106

oppression/resistance, 87-88, 112-114
organization of, 85-86, 93, 100
political action, 95-96, 97, 98, 112-114
religious/conservative leadership and, 98, 99, 100
special rights for, 103, 104
Tavern Guild, 95
visibility of, 111-112, 113-114
youth outreach/clubs, 104-105, 111
See also AIDS epidemic
Gay and Lesbian Community Services Center, 95-96
Genocide, 65
Global brain drain, 5
Gore, A., 258
Government. *See* Federal government
Graduate Record Examination (GRE), 242
Greaser Laws, 332
Great Depression: Era of the 1930s when unemployment was extremely high, 5, 27
 employment initiative, 171
 immigrant deportations, 44-45
 repatriation policies, 267, 269
Great Valley Center: A think tank for anticipating and solving the problems of growth in the San Joaquin Valley, 269, 299
Growth:
 automobile and, 278, 280
 urban sprawl, 278, 279 (figure)
 See also Development theories; Population growth
Gun control, 58-59, 60 (figure)

H-1B visas: Work permits issued to preferred categories of professionals, 287
Hate crimes: Violence, threats of physical harm, verbal abuse, and other types of aggression directed against individuals because of their actual or perceived sexual orientation, race and ethnicity, gender, or religion, 98, 100-101
 gay/lesbian targets of, 86, 87, 98-99, 105
 legislative activity and, 103-104
 statistical records of, 101, 102 (figure)
Hate Crimes Prevention Act (2000), 105
Head Start, 75
Health insurance, 29
Health/welfare services, 17

Heterosexism: The set of institutional practices and assumptions that privilege heterosexuality, 92, 109
Higher education: The tertiary level of formal education, beginning after graduation from high school, with typical degrees awarded being bachelor's degree, master's degree, medical degree, law degree, and doctor of philosophy degree:
  accountability measures, 242, 247-249
  administrative control of, 243, 252-253
  affordability model, 237
  business models of, 147-148, 238-239, 242, 244-246
  Cal Grant legislation, 236, 237
  commodification of, 242, 249-250
  department elimination, 243, 252-253
  distance learning and, 5, 147, 238, 239-241
  faculty hiring, 243, 252
  financial commitment to, 235-238, 256-257
  global context of, 253-254
  intellectual property rights in, 242, 250, 257
  learning-centered, 242, 246-247, 249
  legislative protections of, 255
  The Master Plan for Higher Education, 4, 146-148, 235, 258, 260-261
  media coverage of, 257, 260
  part-time instructors, 242, 250-251
  political action and, 257-260
  Proposition 13 and, 236
  solutions strategies, 255-260
  strategic alliances, education/business, 242, 249
  strategic planning trends, 238, 241-244, 254-255
  students in, 259, 260
  tenure and, 239, 242, 248, 251
  threats to, 146-147
High-tech industries, 287
HIV. *See* AIDS epidemic
Hmong population, 284
Homeless: Narrowly defined, those people who are without suitable, conventional housing, or broadly defined, all people who are inadequately housed or at high risk for homelessness, 7-8, 15-16
  characteristics of, 18-20
  numbers of, 16-18
  property rights and, 24
  visibility of, 16
Homeless advocacy groups, 8, 16

  homelessness rates, 17-18
  legal activity of, 23
Homelessness, 7, 22, 30-31
  definition of, 17
  disabilities/illness and, 9, 29-30
  economic solutions to, 8-9, 26-29
  factors in, 20-22
  families and, 8, 19, 221-222
  housing and, 8, 24-26
  legal approaches to, 8, 23-24
  measurements of, 7-8, 16-18
  personal failure and, 8, 21
  See also Poverty
Homicide rates, 59, 60(figure)
Homophobia: The irrational fear and hatred of homosexuals, 91
Housing:
  affordable, 8, 16, 21-22, 26
  government involvement in, 222, 298
  minimum wage and, 25-26
  shortage of, 145
  voucher system and, 21, 30, 223
  welfare reform and, 214, 215 (table), 221-223
Housing California, 25, 26
HUD. *See* Department of Housing and Urban Development (HUD)
Human capital:
  administrative-support/clerical positions, 196
  educational levels, 194-196, 194 (figure), 195 (table)
  educational triage, 220, 219 (table)
  job growth, 196-198, 197-198 (figures)
  job matching and, 178, 193
  precision production, 196
  service occupations, low level, 195, 270
  unskilled/semiskilled labor, 195-196, 270
  welfare reform and, 145, 213-214
Human immunodeficiency virus (HIV): Condition that may lead to AIDS, 117, 168
  See also AIDS epidemic

Illegal drugs, 10
  AIDS transmission and, 88, 89, 126, 127, 134
  distribution of, gangs and, 47, 48 (figure)
Illegal immigrants: People who are in the United States without proper documentation, 5, 285, 286 (table), 291

Illegal Immigration Reform and Immigrant Responsibility Act (IIRIRA), 27
Immigrant populations, 3, 5
  countries of origin, 269, 270
  family patterns among, 154
  job flight and, 270
  poverty and, 73
  See also Illegal immigrants; Nativism; Net migration; Proposition 187
Immigration: Movement of people into a geographic area from some external location, 5, 273-274
Immigration Reform and Control Act (IRCA), 331
Immigration and naturalization Service (INS), 331
Incarceration:
  arrest rates, 54-56, 55 (table)
  differential justice and, 73-75
  minority group membership and, 52, 53-54 (table), 54, 74
Income determinant: This process determines the manner in which income is distributed unequally, that is, why some get more than others and how great the magnitude of that inequality is, 177
Income levels:
  California, 15
  gang formation and, 50-51
  housing availability and, 8, 16, 21, 25-26
  minimum wage, 25
  poverty and, 220-221
  single mothers, 156, 157 (figure)
Indigenous populations, 3, 11, 64-65, 267, 275
Inequality, 141-142, 198-199
  human capital requirements, 194-196, 194 (figure), 195 (table), 197-198 (figures)
  income and, 180-182, 180-182 (figures)
  income determinants, 177-178
  labor market and, 178, 184-185, 193
  manufacturing sector and, 185
  middle class, 184-188, 186 (table), 188 (figure)
  occupational structure, shift in, 185-186, 186 (table), 188 (figure)
  poverty, categorization of, 189-192, 191 (table)
  social factors in, 178-179
  social stratification system, 177
  taxation policy and, 142-143, 178, 183-184, 186-187
  upper class, 183-184
  wealth and, 183

Infant Mortality Rate (IMR): Number of live births per 1,000 who do not live through the first year, 168
In re Gault, 43
Integration, 76-77
Interactionist perspective, 20
International Association of Lesbian and Gay Pride, 108
International Gay and Lesbian Human Rights Commission (IGLHRC), 107
International Lesbian and Gay Association (ILGA), 107
International Sociological Association, 332
Internet resources:
  distance learning, 239-240, 246
  gay/lesbian networks, 108, 111
  student lobbying efforts, 259
Interpride Conference, 107
Irish Americans, 51, 52
Italian Americans, 51, 66

Jankowski, M. S., 48, 49, 51
Japanese Americans, 66, 184, 283, 331
Jet Propulsion Laboratory, 276
Jewish Americans, 70
Job queue: Process through which jobs in a specific occupational category in a specified geographical area are matched with the number of individuals seeking such jobs, 178, 193
Joint United Nations Programme on HIV/AIDS, 127-128, 129
Judicial system:
  anti-gay crimes/violence, 113
  incarceration rates, 74
  minorities in, 74-75
  See also Juvenile justice system
Juvenile delinquency, 9
  Chicago principles of rehabilitation, 41
  communities, strengthening of, 58-59
  curfews/critical-hours programs, 58
  decision-making process in, 58
  declining rate of, 36, 37-39 (tables)
  ethnicity and, 51-5
  gun control and, 58
  homicide/suicide rates, 59, 60 (figure)
  poverty and, 59
  punitive response to, 37, 40-41 (table), 41
  super-predator, 36

threat of, 35-36
See also Urban gangs
Juvenile Justice and Delinquency Prevention Act, 52
Juvenile justice system:
  adult court processing/sentencing, 35, 37, 40-41 (table), 41
  arrest rates, 54, 55 (table)
  California Youth Authority (CYA), 42
  community-based correctional facilities, 9-10
  due process in, 42-43
  minority group membership and, 52-56, 53-55 (tables)
  punishment/rehabilitation and, 9, 37, 41, 43, 45, 61
  triage model of juvenile justice, 43
  See also Disproportionate minority confinement rate requirement

Kent v. United States, 43
King, R., 12, 69
Kiplinger Editors Group, 284, 286, 287, 292
Knight, W., 91
Know-Nothing political party, 331
Koreans, 11, 12, 66, 284
  African Americans and, 70
  economic loss among, 71
  Latino community and, 71
  poverty and, 189

Labor market:
  female workers, 184-185, 187, 196
  inequality and, 178
  job queues, 178, 193
  manufacturing sector, 185, 204, 268, 270
  minorities/immigrants and, 270
  occupational structure, shift in, 185-186, 186 (table), 188 (figure)
  part-time/temporary workers, 184
  unions and, 185, 187
  See also Employment; Human capital
Lake Tahoe, 280-281
Land use, 269
  marsh conversion, 282
  state planning of, 297-298
Language barriers, 270, 331, 332
LaRouche initiative, 98

Latinos, 12
  AIDS infection and, 120, 133 (figure)
  arrest/incarceration and, 54-56, 55 (table), 74
  drug trafficking and, 48
  feminization of poverty and, 154-156, 155 (figure)
  gang membership, 44-45, 52
  population growth, 284, 291, 292 (figure), 295
  teen birthrate among, 153-154, 153 (figure)
  youth groups, 10, 44
  Zoot Suit riots, 10, 45, 70
  See also Race/ethnic relations
Lau v. Nichols, 284
Law enforcement:
  entrapment and, 93, 95
  gang suppression, 56-57, 58
  gay/lesbian arrests, 86, 92, 93, 94, 95
  mistreatment by, 69-70
  state agencies, policing role of, 270
Legal Aid Foundation of Los Angeles, 213
Legal issues:
  anti-/pro-homosexual legislation, 85, 86-87, 88, 94, 97, 98, 99, 100 (figure), 104-105
  civil unions, same-sex partners, 87, 104
  Colorado, Amendment 2, 103-104
  disenfranchisement, 67
  due process revolution, 42-43
  homelessness and, 23-24
  juvenile crime and, 36-41, 40-41 (table)
  law enforcement/legislative response, 56-58
  Oregon, Ballot Measure 9, 103, 104, 105
  public nuisance statutes, 7, 16, 23
  sodomy laws, 103, 112
  See also Differential justice; Gay/lesbian community; Higher education; Judicial system; Juvenile justice system; Race/ethnic relations
Legislation. *See* Legal issues
Lesbian community. *See* Gay/lesbian community
Little Saigon: Area of Orange County that is heavily populated by immigrants from Vietnam, 284
Lockheed, Martin-Marietta, 276
Long Beach, 93-94, 295, 297 (figure)
Longitudinal study: Study that follows subjects over a long period of time, 167
Los Angeles:
  African American gangs, 46
  African American middle class, 65

gang membership in, 46, 47, 52
gay/lesbian community in, 94, 95-96, 107-108
growth projections for, 295, 297 (figure)
hypersegregation in, 69-70, 76
integration in, 76
Latino gangs, 44-45
law enforcement actions, 56-57, 58
megalopolis of, 268
race relations in, 11-12
Watts riot, 10, 12, 45-46
Zoot Suit riots, 10, 45
See also AIDS epidemic
Low-cost housing. *See* Housing
Lungren, D., 103

Maintenance of effort funding (MOE), 205, 217, 223
Maltreatment. *See* Child abuse
Manpower Demonstration Research Corporation, 213
Manufacturing, 185, 204, 268, 270, 290, 300, 332, 331
Maquiladoras: Foreign-owned factories located in Mexico, 268, 287, 290
Marriage/civil unions, 87, 91, 104
 cohabitation, 165, 166
 domestic partnership, 86, 99, 166
 incidence of, 140-141, 150
 poverty and, 6, 157 (figure)
 sexual revolution and, 165, 171
 teenagers and, 152-153
 transition to, 165-166
 See also Divorce rate; Families
Mass media. *See* Media coverage
The Master Plan for Higher Education, 4, 146-148, 235, 258, 260-261
Mattachine Society, 85, 91, 93
McCarthy, J., 93
McKinney Act. *See* Stewart B. McKinney Homeless Assistance Act
Media coverage, 16
 disasters, 272, 341, 342-343
 Gay and Lesbian Alliance Against Defamation (GLAAD) and, 98
 gay/lesbian images/communications, 108, 113
 gayola scandal, 95
 higher education, trends in, 257, 260
 illegal drug distribution, 47

juvenile crime threat, 36
moral panics and, 45
Median age at first marriage: Age at which half of the population are younger when they get married and half are older, 152
Medicaid: A federal medical assistance program for those in need in the United States. California's version of this program is known as Medi-Cal:
 addiction/disability and, 29-30
 AIDS epidemic and, 124
 welfare-to-work programs and, 214, 215 (table)
MediCal: *See* Medicaid.
Megalopolis: Continuous stretch of cities that covers most of the California coastal area, 278
Mental health treatment, 9, 22, 29, 30
Mental illness: Any psychiatric disorder of the mind that causes untypical thoughts, emotions, or behavior, 9, 29-30
Merton, R., 47, 359
Metropolitan Community Church, 95
Metropolitan Statistical Area (MSA): Large city and its suburban areas, 278
Mexican Americans:
 California stratification system and, 66-67
 environmental racism and, 291
 nativistic sentiment and, 269-270, 332, 331-332
 political clout of, 71
 poverty and, 189
 Zoot Suit riots, 10, 45, 70, 331
 See also Latinos; Nativism
Migration: The physical movement of people into or out of a geographic region, 269, 275, 298-299
Military bases, 93-94, 290, 300
Milk, Harvey, 86, 92, 96, 97
Minimum wage: A rate of pay determined either by a collective bargaining agreement or by government statute as the lowest wage payable to specified categories of employees, 25
 housing issues and, 25-26
 poverty level and, 220-221
Mining, 275, 289, 290
Minority groups, 3
 AIDS infection in, 88, 89, 119, 125, 132-133, 133 (figure)
 discrimination against, 5
 economic underclass, 72-73, 75
 environmental pollution/degradation and, 291
 health care/education and, 12

job flight and, 270
juvenile delinquents and, 10
racism and, 65-67
underclass status, 12, 72-73, 75
See also Coalition politics; Differential justice
Mojave Desert, 268, 280, 281 (figure)
Monterey Park, 78
Moral panic: A state in which public reaction to a problem is out of proportion to the actual threat offered. The panic is exacerbated when "experts" perceive the threat in crisis proportions, regardless of the actual threat, and when the media representations universally stress "sudden and dramatic increases" in the numbers involved or the events themselves. The term *moral panic* was created by British sociologist Stanley Cohen in his book, *Folk Devils and Moral Panics*, to identify a form of collective behavior characterized by widely circulating rumors disseminated by the mass media, which exaggerate the threat posed by some "newly identified" type of moral deviance, 10, 44, 45
Moscone, G., 86, 97
Motion picture industry, 267-268, 276, 300
Municipal Elections Committee of Los Angeles (MECLA), 97

Nabozny v. Podlesny, 105
National Academy of Sciences, 331
National Alliance to End Homelessness, 25
National Association of State Alcohol and Drug Abuse Directors, 29
National Campaign to Reduce Teenage Pregnancy, 161
National Coalition for the Homeless, 17, 25, 26, 27, 29, 31
National Health Care for the Homeless Council, 30
National Institute of Mental Health, 258
National Labor Relations Act, 184
National Law Center on Homelessness and Poverty, 18, 19
National Law Enforcement Institute, 46, 52
National Low Income Housing Coalition, 25
National Opinion Research Center (NORC), 331
National Resource Center on Homelessness and Mental Illness, 29
National Science Foundation, 258

National Youth Gang Survey, 47
Native Americans, 64-65, 189, 267, 275
Nativism: An ideological set of beliefs that depict outsiders, primarily immigrants, as a threat to U. S. society, 5, 269, 331-332, 331
California, immigrant population and, 332-331
context of, 331-331
employment and, 331-332
history of, 331-331
language barriers, 270, 331, 332
Mexican immigrants and, 331-332, 332, 331-332
Proposition 187 and, 332-331
Proposition 224 and, 332, 331
social factors in, 332-331
worker segmentation, 331-331
Nativist mind-set: The perception of immigrants as a threat to Anglo society, 11, 66, 331
Natural disasters, 5, 270, 271, 331, 331, 332
See also Disaster
Natural increase: The difference between the number of births and the number of deaths in a population, 284
Net migration: The difference between the number of immigrants and the number of emigrants, 285
New Deal, 27
Newhereditarians: Advocates who reverted to older theories claiming a biological basis for discernible differences among races, such as intellect, in contrast to the prevalent social science perspective, which sees race as a social construct, 12, 75-76
No-fault divorce, 151
Nonmarital childbearing, 139, 141, 149, 151-152, 152 (figure), 158-159, 160 (figure), 164 (figure)
Norms, 3, 6
North American Free Trade Agreement (NAFTA): Pact involving Canada, the United States, and Mexico, which allows for free trade across the borders, 268, 287

Oakland, 76, 77, 94
Oak Ridge National Laboratory, 332
Oblinger, D. G., 244-247, 249-253
Office of Emergency Services, 331
One, Inc., 93, 94
Operation Wetback: A program created by Congress in 1954 to apprehend Mexican laborers in

Glossary and Index 377

the United States who had not entered the country as legal seasonal workers, 332, 332
Opportunity costs: Chances of becoming a pregnant teen if one sees little opportunity for the future, 156
Oregon, Ballot Measure 9, 87, 103, 104
Out-of-wedlock births: Births to unmarried women, 139, 141, 149, 151-152, 152 (figure), 158-159, 160 (figure), 164 (figure), 169, 170 (figure)
Ozone concentration: Ozone content of the air measured in parts per hundred million (pphm), 268, 280

Pacific Rim: Trade zone of lands that ring the Pacific Ocean, 287
Palo Alto, 99
Per capita income: Amount of income per person in a household, 168, 287
Period prevalence counts, 17
Perry Pre-School Program, 75
Personal Responsibility and Work Opportunity Reconciliation Act (PRWORA), 27, 72, 143, 204
Personal Rights in Defense and Education (PRIDE), 92
Pesticides, 291
Photochemical smog: Air pollution that arises from the mixture of carbon monoxide and sunlight, 268, 278, 280
Physical disabilities, 9
Point-in-time counts, 17
Police. *See* Law enforcement
Politics:
  coalitions in, 12, 77-79, 101, 109-110
  disenfranchisement, 67, 78
  gay political action, 95-96, 97, 98
  homelessness and, 17-18
  Mexican Americans in, 71
  power redistribution, 79
  See also Higher education; Legal issues
Pollution. *See* Air pollution; Environmental concerns; Water pollution
Population density: Number of people per square mile, 278
Population diversity, 268-269
  elderly population, 298-299
  foreign-born residents, 270
  indigenous people, 3, 11, 64-65, 267, 275

multi-national pool, 283-285, 285 (figure)
  redistribution plan, 269
Population growth, 5, 168, 169, 267-268
  colonization, 274-275
  demographic changes, 269, 284-286, 285 (figure), 291, 292 (figure), 295
  gold rush, 275
  immigrants/minorities and, 5, 285, 286 (table)
  motion picture industry and, 267-268, 276
  new-technology industries, 276
  projected growth, 292-296, 293-296 (figures)
  Spanish settlement, 64-65, 267, 274-275
  waves of, 273-276, 274 (figure), 277 (figure)
  See also Development theories; Smart growth
Population Reference Bureau, 292
Poverty: People whose resources fall relatively significantly below those of the larger community are generally said to be living in poverty. *Absolute poverty* describes a lack of access to the minimum basic needs required to sustain life (such as food, shelter, clothing, and clean drinkable water):
  AIDS incidence and, 89, 128
  California ranking, 15-16, 72, 73
  categories of, 190-192, 191 (table)
  deserving/undeserving poor, 22, 27
  housed vs. homeless poor, 8, 20, 21
  immigrants and, 73
  juvenile delinquency and, 50-51, 59
  marginal middle class and, 187
  self-sufficiency and, 144-145
  single mothers and, 156, 157 (figure)
  See also Feminization of poverty; Homelessness; Inequality; Unemployment; Welfare; Welfare reform
Power-conflict paradigm. *See* Conflict theory
Pregnancy. *See* Families; Nonmarital childbearing; Teen pregnancy
Presidios: Military outposts during the colonial period, 275
Primary economic industry: Gathering raw materials, 289-290
Primary Metropolitan Statistical Area (PMSA): Metropolitan Statistical Areas within a Consolidated Metropolitan Statistical Area, 278
Proposition 6, 86, 97
Proposition 13, 4, 236
Proposition 21, 9, 35, 36, 37, 58

Proposition 22, 91, 105, 111
Proposition 64, 98
Proposition 69, 98
Proposition 98, 237
Proposition 187: An initiative approved by California voters in November 1994, which denied social welfare and educational services to immigrants and their families, 269, 270, 291, 331-331, 332, 331, 332-332
Proposition 207, 270, 332
Proposition 224: An initiative approved by California voters in June 1998, which eliminated bilingual instructional services in California schools, 11, 269, 270, 284, 331-332, 331, 332-331, 332-332
Public nuisance statutes, 7, 16, 23
Public opinion, 16
Public Policy Institute of California (PPIC), 355, 356
Pueblos: Towns that grew up around presidios, 275
Pull factors: Conditions that make people want to immigrate., 276, 300, 332

Race/ethnicity, 3
   biological view of, 75-76
   diversity in California, 11, 67-69, 68 (figure)
   gang membership and, 10, 49 (table), 51-52
   juvenile justice system and, 52-56, 53-55 (tables)
   segregation and, 12
   white ethnic group identity, 51-52
   See also Newhereditarians
Race/ethnic relations:
   African Americans, 65, 69-70
   Anglo American settlements, 65
   antagonism, 288-289
   antagonism in, 288-289
   apartheid and, 65
   Asian immigrants, 66
   California/national contexts and, 72-73
   coalition politics and, 77-79
   economic downturn and, 70
   global context of, 73-75
   history of, 63-67
   hypersegregation and, 69
   indigenous populations, 3, 11, 64-65, 267, 275
   Mexicans, 66-67, 69, 70
   nativist mind-set and, 11, 66, 331

   newhereditarian attitude and, 75-76
   racism and, 64-67
   segregation/integration, trends in, 76-77
   social change and, 69-70
   Spanish settlements, 64-65
   urban riots, 69-71
Racism: Ideology supporting a system of stratification in which certain races are defined as inherently inferior. These beliefs justify patterns of prejudice and discrimination toward these minority groups, 64
Railroads, 275-276
Rational choice: Theory that views crime as a function of a decision-making process in which the potential offender weighs the potential costs and benefits of an illegal act, 48-49, 50
Reagan, R., 91, 97, 98, 184, 236
Reconquista: Notion that Southern California is being reclaimed by the Latino population, 268, 288
Regional planning: Decision-making processes that are shared among city and county administrators, 269, 297, 299-300
Religious fundamentalism, 98
Relocation camps, 283
Renewable energy industry, 354
Repatriation, 5, 269, 332, 331
Residential instability. *See* Homelessness
Residential treatment/rehabilitation programs, 9, 10
Rich/poor divide. See Two-tiered system of stratification
Riots, 10, 12, 45-46, 69-71
Risk promoters: Claims makers or individuals who take some potential hazard or disaster and advocate the threat or riskiness involved in it to various publics, 331, 337
Rochester Youth Development Study, 50
Roe v. Wade, 165
Romer v. Evans, 103

Sacramento, 101
Sanders, W., 50, 51
San Diego:
   blackouts in, 353
   gangs in, 46, 47-48, 51, 52
   gay/lesbian community in, 93, 94, 96

San Diego Association of Governments (SANDAG), 47
San Francisco:
  gangs in, 51, 52
  gay subculture in, 85-86, 94-95, 96
  growth projections for, 293, 297 (figure)
  megalopolis of, 268
  segregation in, 76
  transitional job programs, 207
  See also AIDS epidemic
San Gabriel Valley, 78-79
San Joaquin Valley, 269, 280, 282, 299
San Jose, 76
Santa Ana, 23
Santa Paula, 67
Scapegoats: Targets of nativist attacks, 332
Scholastic Aptitude Test (SAT), 242
Secondary economic industry: Manufacturing, 289 (figure), 290
Section 8 housing, 22, 222
Segregation, 76-77
Self-initiated educational programs (SIPs), 206, 207, 216, 220, 219 (table)
Self-sufficiency, 144-145, 190, 196, 198 (figure)
Service industries, 268, 289 (figure), 290, 332, 331
Sexual abuse, 141, 162
Sexual orientation. *See* Gay/lesbian community
Sexual revolution: Accepting attitude toward sex outside marriage, 165
Shelter Partnership, 18
Shelters, 16, 17
  Aid to Families with Dependent Adults, 30
  short-term relief and, 25
Shilts, R., 89, 130, 131, 132
Shockley, W., 276
Silicon Valley: Area between San Jose and San Francisco that is known for its high-tech industry, 268, 276, 298, 360
Silicon Valley Manufacturing Group, 355
Single-mother families, 4, 139
  children, impact on, 166-167
  income levels, 156, 157 (figure)
  See also Feminization of poverty
Slavery, 65
Smart growth: A planning strategy to integrate housing, work, and shopping in a way that avoids urban sprawl, 269, 296-297
  employment, 300-301
  land use, 297-298

  migration, 298-299
  regional planning, 299-300
Social change perspective, 150
Social problems perspective, 150, 179, 331
Social sciences:
  homelessness, perspectives on, 20
  racial categorization, 76
  social engineering and crime, 59
  treatment-oriented prisons and, 42
Social Security Act (SSA), 27
Social Security Disability Insurance (SSDI), 29
Social welfare, 7, 143, 204
Society for Individual Rights, 95
Sociocultural evolution theory: A concept that says social structures tend to become more complex over time, 289-290, 289 (figure)
Socioeconomic status: An individual's position in society relative to others, frequently measured in terms of social category, relative wealth and/or income, educational attainment, and/or occupation. With regard to issues of health and disease, higher status generally determines relatively better health and access to medical care, 125, 128
Sojourners: Usually single men who immigrate temporarily with the intention of making a "fortune" and then returning home, for example, the Chinese in 19th century California, 275, 288
Spanish settlements, 64-65, 267, 274-275, 274 (figure)
Species destruction, 282-283, 283 (figure)
Split-labor market: A workplace where two or more identifiable groups do the same or comparable work, but for different wages, 288-289
Stanford University, 276
Stewart B. McKinney Homeless Assistance Act, 24-25, 29
Strategic plans: A term borrowed from business referring to broad-scale projects for organizations often entailing anticipated large changes, 238
Stratification system: Through this process, a society distributes social goods, including such things as wealth, status, and power. It consists of social hierarchies of individuals in terms of the greater and lesser degrees of social goods that they receive, and stratification determinants,

which are the processes through which some get more and others get less, 177
See also Two-tiered system of stratification
The Street Terrorism Enforcement Prevention (SSTEP) Act (1988), 57
Student Declaration of Educational Rights, 259
Subcultures: Social life worlds, part of a larger world, with their own meaning systems and values., 129, 130
See also Deviant subcultures
Substance abuse: Excessive use of addictive substances, especially narcotic drugs and alcohol, as a result of a craving or strong urge for the substance caused by genetic, psychological, and social factors, 9, 29
  disability benefits and, 29-30
  homelessness and, 16, 19
  See also Alcohol abuse; Illegal drugs
Suicide rates, 59, 60 (figure)
Supplemental Security Income (SSI), 27, 29, 30, 214
Symbolic interactionist theory, 87, 110, 272, 337

Tavern Guild, 95
Taxation policy, 142-143, 178, 183-184, 186-187
Technological developments, 276
Teen pregnancy, 140, 141, 149, 152-154, 153 (figure)
  abortion and, 161
  difficulties of, 156-157
  family support system, 141, 157-158, 172
  fathers in, 161-162
  marriage and, 152-153
  national trend in, 159, 161
  pregnancy prevention, 171-172
  sexual abuse and, 162
  social class and, 156
Temporary Assistance to Needy Families (TANF): Assistance program that replaced AFDC under welfare reform. It requires the performance of work activities to receive assistance and places a lifetime limit on the number of weeks a family can receive it, 27-28, 72-73, 143, 162-163, 204
  caseload reduction, 208-211, 209-210 (figures)
  child welfare considerations, 145
  housing assistance, 223
  teen parents and, 141, 172

Tenure: The achievement of permanent faculty status awarded after several years of graduate school training and then 6 years of intensive faculty and administrative evaluations of academic performance in terms of teaching effectiveness, professional growth, and service to the university and the larger community, 239, 242, 248, 251
Tertiary economic industry: Providing services, 268, 289 (figure), 290, 332
Textile industries, 287
Theory of ethnic antagonism: Edna Bonacich's theory, which holds that a split-labor market can lead to antagonism by the higher paid workers toward the lower paid workers, 288-289
Three strikes rule, 74
Tobacco use, 3
Toxic waste sites, 271, 331
Traditional Values Coalition, 91, 100
Treaty of Guadalupe Hidalgo: Pact that ended war with Mexico, which ceded most of the west to the United States, 67, 275
Triage model of juvenile justice: The current process of decision making by the prosecutor's office regarding whether or not to send a juvenile to adult court for processing or to charge the minor as a juvenile. The decision is often swayed by the offender's history. Those with previous serious offenses are deemed unable to be helped by the juvenile justice system, and the triage model demands that they be processed in adult court, 43
Triple discrimination: Inequality based on age, race, and gender, 167
Two-tiered system of stratification: Process by which the rich and the poor are increasingly separated, 4, 5, 269, 290, 298
Tylenol scare, 89, 130-131

Underclass status, 12, 72-73, 75
Undocumented workers. *See* Illegal immigrants
Unemployment:
  California rates of, 15, 72
  gang membership rates and, 49
  homelessness and, 20-21, 25
  immigrants and, 332
  job stimulation, 221

University of California (UC) system. *See* Higher education
Urban gangs, 9
  African American gangs, 45-46
  Chicago anti-gang ordinance, 57-58
  criminality and, 50
  despair and, 49-50
  economic/social dislocation and, 48-49
  emergence of, 10
  ethnicity and, 10, 51, 52
  illegal drug distribution and, 47, 48 (figure)
  inter-gang conflict, 52
  Latino gangs, 44-45
  law enforcement/legislative response, 56-58
  low-income lifestyles and, 50-51
  membership levels, 46-47
  moral panic and, 44, 45
  motivation to join, 48-51, 49 (table)
  rational choice view and, 48-49, 50
  tribal group process and, 50
  See also Juvenile delinquency
Urban renewal, 21, 22
Urban sprawl: City populations that spill over into countryside and agricultural areas, 268, 269, 278
Urban underclass, 12, 72-73, 75
U.S. Organization, 46

Vagrancy, 16
Vehicle emissions, 2, 268
  acid rain and, 281
  See also Photochemical smog
Vermont, 87, 104, 111
Verville, A. -L., 244-247, 249-253
Voter initiatives, 2, 269
Vouchers:
  cash substitute, 27
  housing, 21, 26, 30, 145, 223
  substance abusers and, 29-30

War on Drugs, 74
Water exchange: The siphoning of salt water into fresh water areas as fresh water sources are depleted, 268, 282
Water marketing: Selling surplus water, 268, 282

Water pollution: Impurity of water sources because of the use of fertilizers, pesticides, agricultural practices, or other means, 280
Water wars: Nearly constant conflict over the best use of the state's water supply, 268, 298
Watts riot, 10, 12, 45-46, 70
Wealth, 183
Welfare: In the United States, this term usually refers to government-funded programs designed to provide economic support, goods, and services to unemployed or underemployed people, although the term can be used to refer to public and private support for people in need, 27
  childhood poverty, 163, 164 (figure)
  employment and, 28
  homeless populations and, 28-29
  maintenance of effort (MOE) funding, 205, 217, 223
Welfare: In the United States, this term usually refers to government-funded programs designed to provide economic support, goods, and services to unemployed or underemployed people, although the term can be used to refer to public and private support for people in need.
Welfare reform: Program that ended welfare as an entitlement to individuals in need and replaced it with limited assistance with a specific time limit, required that program participants put in a certain number of hours of work-related activities per week to get it, and gave control over programs to state governments, 27-28, 72-73, 203
  caseload reduction, 208-211, 209-210 (figures)
  child care assistance, 214, 215 (table), 215
  evaluation of, California, 144, 207-208
  housing availability, 222-223
  human capital and, 213-214
  policy questions in, 144-145
  recidivism rate, 211, 212
  recipient termination, 211-215, 212 (figure), 215 (table)
  supplemental assistance programs and, 212-215, 215 (table)
  wealth redistribution and, 178
  welfare-to-work programs, 143-144, 162-163, 196, 204, 205-207
  See also CalWORKs; Inequality

Welfare state: System in which the government provides social services such as income assistance, food, medical care, subsidized housing, and the like to those in need, 203
  maintenance of effort (MOE) funding, 205, 217, 223
  reform of, 204-205
  social programs in, 204
West, C., 49, 52
West Hollywood, 96, 98
Wilson, P., 86, 146, 236, 359
Women:
  AIDS infection in, 88, 89, 119-120, 121 (figure), 125, 126 (figure), 132, 133
  feminization of poverty, 154-156, 155 (figure), 157 (figure)
  homelessness and, 7, 8, 28
  incarceration of, 74
  nonmarital childbearing, 139, 141, 149, 151-152, 152 (figure), 158-159, 160 (figure)
Workforce:
  female workers, 184-185, 187, 196
  foreign-trained professionals, 5, 284, 287
  racism and, 65, 66, 67
  structural transformation in, 10, 12, 25-26, 47, 48, 70

See also Employment; High-tech industries; Human capital; Labor market
Works Progress Administration (WPA): Agency developed to provide jobs during the Great Depression, 141, 171, 221
World Pride march, 108

Youth Corrections facilities. *See* Juvenile justice system
Youth gangs. *See* Urban gangs

Zoot Suit riots: In 1933, riots began in Los Angeles between U. S. servicemen and young Chicanos. The riots stemmed from media response to the killing of a young Chicano in a gang-related incident. The media warned of the dangers of Mexican zoot suit-wearing gangs. In this atmosphere, a group of U. S. servicemen on leave in the city engaged in brutal and dangerous riots that many in the Chicano population describe as race riots, 10, 45, 70, 331

ABOUT THE EDITORS

Charles F. Hohm is Professor of Sociology and Associate Dean of the Division of Undergraduate Studies at San Diego State University. He received his B.A., M.A., and Ph.D. in sociology from the University of Southern California, is the founding president of the California Sociological Association, and is currently vice president of the Pacific Sociological Association. His most recent books include *Population: Opposing Viewpoints* (2000), coedited with Shoon Lio and Lori Jones, and *Global Social Problems* (1996), with James A. Glynn and Elbert W. Stewart. He has published numerous articles on demography, environmental sociology, housing, and child maltreatment in journals such as *Demography, Sociological Perspectives, Sociology and Social Research, Social Science Journal, Journal of Interpersonal Violence, Housing and Society, International Journal of Sociology of the Family,* and *Growth and Change*. He is the past editor of *Sociological Perspectives* and the 1998 winner of the Pacific Sociological Association's Distinguished Contributions to Teaching Award. His current research interests are social demography and the adjustment of refugees.

James A. Glynn is Professor of Sociology for the State Center Community College District. He has taught sociology for more than 30 years, and he currently holds classes at the Madera Center in California's great Central Valley. A contributing author to the first edition of *California's Social Problems* (1997), he also collaborated with Charles F. Hohm and Elbert W. Stewart in writing *Global Social Problems* (1996). Previously, he and Stewart produced four editions of *Introduction to Sociology* (1971, 1975, 1979, and 1984), a text that was translated into Spanish and Portuguese. In 1990, he worked with Professor Hohm and others in organizing the California Sociological Association; he served as its third president and has been executive director for the past 10 years. From 1996 to 2000, he served on the Editorial Board of *Sociological Perspectives,* the official journal of the Pacific Sociological Association, an organization from which he received the Distinguished Contributions to Teaching Award in 1997. For the past 2 years, he has written a weekly column for the *Madera Tribune* in which he frequently addresses issues of sociological importance. Now involved in local community affairs, he serves on the Board of Directors of the Madera County Arts Council and the Citizens Commission to Plan and Implement the Government Center.

ABOUT THE CONTRIBUTORS

Adalberto Aguirre, Jr., is Professor of Sociology at the University of California at Riverside. He received his B.A. in sociology from the University of California at Santa Cruz and his Ph.D. in sociology and linguistics from Stanford University. His research interests are equity issues for women and minority faculty in higher education, social class as a diversifying feature in the workplace, immigrant populations in California, and the sociolinguistic features of the Chicano speech community. He has published several books and many articles.

Robert Enoch Buck is an Associate Professor of Sociology at San Diego State University. His research interests are poverty and inequality and the social history of industrialization. He serves variously as a board member, consultant, urban strategist, research director, and program designer for a number of organizations and governmental agencies working in the area of poverty and welfare reform, both in Southern California and nationwide.

Ronald W. Fagan is a Professor of Sociology at Seaver College, Pepperdine University, Malibu, California. He received his Ph.D. from Washington State University in sociology, and he has an M.A. in marriage and family counseling from Pepperdine University. He has worked with the homeless for more than 25 years, including helping to develop and run a substance abuse treatment program in Seattle, Washington, and serving as a consultant and program evaluator for many nonprofit organizations that work with the homeless. His articles on substance use and the homeless have been widely published.

Robin Franck is a native Californian who currently teaches at Southwestern College in San Diego, where she is Behavioral Science Department chair. She attended San Diego State University and University of California, San Diego. She is active in the Pacific Sociological Association and was president of the California Sociological Association several years ago.

Anne Hendershott is Professor of Sociology at the University of San Diego. She received her Ph.D. in sociology from Kent State University and has published several articles in the areas of family research, evaluation, and organizations. Her first book, *Moving for Work,* focused on the difficulties of relocation for families. Her most recent book, *The Reluctant Caregivers,* explores the stress of caring for elderly family members. Currently, she is completing a manuscript on redefining deviance.

Kevin D. Kelley, an M.A. candidate with the Department of Sociology at San Diego State University, is a researcher with the Henry M. Jackson Foundation for the Advance of Military Medicine. He is currently working with the Naval Health Research Center San Diego.

Phylis Cancilla Martinelli is Professor and Chair of the Anthropology\Sociology Department at St. Mary's College of California. She received her Ph.D. in sociology from Arizona State University. She helped establish the California Sociological Association and served as its fifth president. She has also served in the Pacific Sociological Association. Her areas of research are racial and ethnic relations, and she has published books and articles on these topics. She is doing research reassessing the process of racialization in the Southwest mining industry in the early 1900s.

Peter M. Nardi is Professor of Sociology at Pitzer College of the Claremont Colleges. He is the author of *Gay Men's Friendships: Invincible Communities* (1999); the coeditor of *Social Perspectives in Lesbian & Gay Studies: A Reader* (1998); *In Changing Times: Gay Men & Lesbians Encounter HIV/AIDS* (1997), and *Growing Up Before Stonewall: Life Stories of Some Gay Men* (1994); and editor of *Men's Friendships* (1992) and *Gay Masculinities* (2000). He is the editor of *Sociological Perspectives,* the journal of the Pacific Sociological Association.

Harvey E. Rich is Professor of Sociology at California State University, Northridge, where he was department chair for 7 years. He received his Ph.D. in sociology from Purdue University and was advanced to candidacy toward the

D. Env. in the Environmental Science and Engineering Program at UCLA. He coauthored a series of monographs for the California Air Resources Board and the U.S. Environmental Protection Agency and authored *Student Attitudes and Academic Environments: A Study of California Higher Education* (with Pamela M. Jolicoeur, 1981). His articles have been published in a variety of academic journals. He is a member of the Governing Council of the Pacific Sociological Association and Southern Vice President of the California Sociological Association.

Lorena T. Valenzuela is the daughter of immigrant Mexican parents who stressed the importance of an education. As an undergraduate in liberal studies at San Diego State University, she served as vice president for external affairs, Associated Students, from 1994 to 1995 and was extremely active in voter registration drives. She has appeared in many public forums concerned with higher education, including meetings of the American Sociological Association and the California Sociological Association. She is currently teaching in Los Angeles while doing graduate work at California State University, Dominguez Hills.

Loretta Winters is Associate Professor of Sociology and Coordinator of the American Indian Studies Program at California State University, Northridge. She received her Ph.D. in sociology from the University of California, Riverside. Her research interests include biracial identity, teen pregnancy, and natural disasters. She is working on a book entitled *New Faces in a Changing America: Multiracial Identity in the 21st Century*. Recent publications with Herman DeBose include chapters in Eleazu Obinna's 2nd edition of *Human Geography of African America* and *Foundations of African American Education*.

Robin Wolf received her Ph.D. in sociology from the University of California at Berkeley and is the author of *Marriages and Families in a Diverse Society* (1996). She has taught classes in marriage and family, critical thinking about social and cultural issues, social psychology, and social research at Diablo Valley College over the last 29 years. She is involved in the effort to increase the recognition of undergraduate teaching as an important endeavor and represents community colleges on the council of the Undergraduate Education of the American Sociological Association.

James L. Wood is Professor of Sociology at San Diego State University, where he was department chair for 9 years. He received his Ph.D. in sociology from the University of California, Berkeley and is the author of many journal articles and several books, including *Social Movements: Development, Participation, and Dynamics* (with M. Jackson, 1982), which was nominated for the American Sociological Association's Award for a Distinguished Contribution to Scholarship and the C. Wright Mills Award. Since 1992, he actively supported higher education against serious attacks, testifying before several government committees.